AF704386

Reasons ...

- Nobel Laureate Immigrants
- World Capitalism & cultural/technological [illegible]
- specialized Grad ed — started in 1800s
- Intact industrial and cultural base at conclusion of WW II
- Internet & [illegible]
- Hegemony of US
- Technoscience world leader
- British laissez faire attitude vs US R&D support

And Britain's scientific importance (cradle of industrial rev & empirical science)

(stats from my book

The Dominance of English as a Language of Science

Effects on Other Languages and Language Communities

edited by
Ulrich Ammon

Mouton de Gruyter
Berlin · New York 2001

Mouton de Gruyter (formerly Mouton, The Hague)
is a Division of Walter de Gruyter & Co. KG, Berlin

⊗ Printed on acid-free paper which falls within the guidelines of the ANSI to ensure permanence and durability.

Library of Congress Cataloging-in-Publication Data

The dominance of English as a language of science / edited by Ulrich Ammon.
p. cm. – (Contributions to the sociology of language ; 84)
Includes bibliographical references and index.
ISBN 3 11 016647 X (cloth : alk. paper)
1. English language – Technical English. 2. Scientific literature – Authorship. 3. Technical writing – Social aspects. 4. Communication, International. 5. Communication in science. 6. Science – Language. I. Ammon, Ulrich. II. Series.
PE1475 .D66 2001
428–dc21
2001030388

Die Deutsche Bibliothek – CIP-Einheitsaufnahme

The dominance of English as a language of science : effects on other languages and language communities / ed. by Ulrich Ammon. – Berlin ; New York : Mouton de Gruyter, 2001
(Contributions to the sociology of language ; 84)
ISBN 3-11-016647-X

Printing: WB-Druck GmbH & Co., Rieden/Allgäu.
Binding: Lüderitz & Bauer-GmbH, Berlin.
Printed in Germany.

Editor's Preface

That English is today's dominant language of science is stating what would be called a *Binsenweisheit* in German, a trivially obvious insight. Science has in fact been pointed out repeatedly as one of the main fields contributing to the spread of English as a global language (Crystal 1997: 80f., 107–109; Graddol 1997: 8f.). Many a triviality, however, reveals less generally agreed-upon, or even hitherto unknown aspects upon closer inspection. Thus in the present case, it may not even be clear what we mean by „dominant language". Do we simply have in mind *prevalence,* i. e. the language being used more frequently than others, or do we imply – in the literal sense of the word – dominance of some persons over others by means of the language in question? It seems that both meanings make sense in the present context and can be explored as to their reality.

The degree of prevalence of English in science communication has been investigated in numerous studies of which Tsunoda's (1983) is one of those frequently quoted (for an overview see Ammon 1998). Other widely recognized studies of the topic have been done by Baldauf and Jernudd (e. g. 1983) who both contributed to the present volume with the former presenting new comprehensive data and the latter, with co-authors Wu and Chan, specific data on Hong Kong. McConnell gives here an overview of the prevalence of English over French for East Asia illustrating the situation by maps. In addition, nearly all the articles in the present volume contain new data on the extensive use of English in science communication, at least for the country under scrutiny, as for example Gunnarson for Sweden or Kryochkova for Russia. Nevertheless we are still far from a comprehensive picture. Though any such picture will of course remain ephemeral, its details may at any point of time be crucial for more refined questioning.

Reasonably comprehensive, even historically comparative, data are available for the share of languages in printed science publications, but they are – due to available bibliographical data bases – less comprehensive for the social than the natural sciences and least for the humanities. De Swaan's article on the social sciences or Siguan's that extends into the humanities, in the present volume, are problem-oriented rather than concerned with representative figures; they touch, among other things, upon linguistic relativity asking, how completely terminology in the social sciences and the humanities can be translated from another language into English, or whether the wealth of knowledge available in different languages can be stored and expressed in a single one.

For kinds of communication other than printed publications, like oral presentations at conferences, informal encounters, written and oral correspondence, or science teaching, the available data for the share of languages, or for language choice in the case of international contacts, are still rather sporadic. Nevertheless, the contributions to this volume do fill many a gap. Some demonstrate the ascent of English in recent years by comparing data collected at different points of time, as do Medgyes and László for Hungary. Others show the degree of penetration in different countries and language communities like Murray and Dingwall in their comparison of Sweden and Switzerland.

Most studies go beyond mere description in that they attempt to explain, how prevalence, or dominance, of English in science has come about. It seems safe to predict that such attempts at explanation will never be complete but will still always remain rewarding for deepening our insight into the development and history of the world language system. Kaplan takes on the task of comprehensive explanation in a convincing manner, but other articles contribute likewise, though from different angles, as does Martel's by revealing forces that counteract the prevalence of English in science communication which otherwise would be even more overwhelming. This contribution also deals with questions of language politics and language planning within the framework of states and nations, with reference to the case of Quebec.

A question which peaks through in many articles is the evaluation of the prevalence of just one natural, historical language over all the others. De Swaan puts forth an explicitly positive viewpoint by stressing the advantages, especially for scientific communication, of a world lingua franca in comparison to a linguistic Babel, without being unaware of problematical implications at the same time. Other articles show stronger tendencies towards a critical view. An example is Truchot's contribution on the public debate in France on the trend towards Anglification of science communication and on measures of the state to guarantee the maintenance of the country's own language in functions complementary with English.

It seems to me that negative valuations of the present world language situation are often not frankly expressed in publications, in order to avoid conflicts or criticism of being resentful or simply for reasons of politeness. The absense of explicit value statements is therefore, in my opinion, a questionable indicator for real judgment. Fishman in a way uses such statements in his analysis of the contributions to the volume *Post-Imperial English* by taking authors' expressed „own opinion" on „the absence or presence of English linguistic imperialism" as an indicator for the reality of linguistic imperialism (Fishman 1996: 634–636). This is a useful first approach, but there remains the possibility that such value-loaded views are not readily expressed. Therefore we need more valid, and of course representative, resentment studies with respect to the present prevalence of English.

Many of those who resent the present situation will at the same time admit that the English language community has only achieved what others would have liked to but couldn't. This seems to be a valid assumption in face of practically all the larger language communities' endeavors at spreading their own language or promoting its status and function in the world, at least in recent times. It is mainly after failing themselves that the non-English language communities or countries now start to compare the present situation with conceivably fairer solutions. Prototypical for such a solution would be having Esperanto, or some other generally non-native tongue, as the world language. Though such wishes appear somewhat phony if being put forth after the game has been lost, they may still comprise truthful aspects of the present situation.

A number of contributions to the present volume point out real advantages of the English-speaking world, or its scientists, and disadvantages of the other language communities and their scientists. Here the question of dominance in the literal sense arises, namely dominance of the native speakers of the world lingua franca by means of their language over the non-native speakers, let alone the non-speakers.

It seems almost self-evident that the native speakers of the prevelant scientific language have less difficulty using it passively (in reading, oral understanding) and actively (in writing or speaking) than do non-native or foreign-language speakers and, therefore, have advantages over the latter in communicative situations which require the use of English. It is easier for them to produce utterances and texts in line with the existing, native-speaker norms. Higher investment in language learning and additional costs of producing linguistically adequate texts are additional problems with which the non-native speakers have to struggle. These difficulties extend beyond the individual scientist and scholar to publishing companies or even all firms for which science and scientific communication are economically essential in countries where English is not the native or at least a widely-used official language.

In order to raise awareness of these problems I have postulated, in my contribution to this volume as well as elsewhere (Ammon 2000), the

„non-native speakers' right to linguistic peculiarities." It may appear a rather hopeless postulate considering the well-founded linguistic veneration of the native speaker, but I believe it deserves close examination, also re the possibilities of a political campaign to gather support similar to that for female linguistic rights. The feminist campaign too was far from being taken seriously at the beginning but has certainly had considerable success meanwhile. I am aware that the postulate of equity for non-native speakers of English, to put it another way, faces far more formidable obstacles than did, or does, linguistic gender neutrality. It needs, first of all, adequate specification before it can be taken seriously.

The possibility of disadvantages of scientists who are native speakers of English should not be entirely forgotten. They become noticeable in Baldauf's report on Australia, where scientists typically seem to be less and less inclined to study foreign languages and, as a consequence, are unable to take notice of publications in languages other than English. There is, however, also the tendency from the side of the English-speaking scientists to stress this point in order to calm down complaints about their otherwise enormous privileges.

The English-speaking scientists' advantages extend way beyond what has been mentioned above. They enjoy, for one thing, their prestigious language's halo effect that texts tend to be valued more highly if written in English (cf. Ammon 1998: 194–197; Nylenna/ Riis/ Karlsson 1994). Also, the English-speaking countries, or their scientists and scholars, are credited with inventions and innovations which in reality were made elsewhere but have not become known for language reasons. Willemyns' contribution to this volume demonstrates this with examples from sociolinguistics.

The spread of English, in science and other fields, limits the use of other languages even within their home countries. A domain in question is teaching at tertiary level for which the penetration of English is reported in a number of articles in this volume, like in Spolsky's and Shohamy's on Israel, Smolicz', Nical's and Secombe's on the Philippines, de Cillia's and Schweiger's on Austria, or Dürmüllers on Switzerland. Though perhaps still limited to only a few institutions or certain occasions, like in the case of Israel to the Medical School at Tel Aviv University or to new staff members not yet familiar with Hebrew, respectively, the language is there and often noticably on the advance. Often it makes inroads via graduate and post-graduate studies as in the case of Switzerland. Writing theses in English is widely tolerated, if not the rule, in non-English-speaking countries, especially in the natural but also the social sciences. Many countries have introduced English-language teaching in order to attract more foreign students, who are not always willing to learn the native language. This is even true of countries with quite large and widely studied languages (cf. Inoue's contribution on Japan or my own on Germany).

The preponderance of English in science and other domains does affect the other languages even internally. Jacobson reports on frequent code-switching and code-mixing in scientific communication in Malaysia. In most cases, modernization of terminology occurs mainly by way of loans from the English rather than using indigenous linguistic recources, as Inoue shows for Japan. Will such terminology further distance the lay population from the realm of science? This would hardly be compatible with democratic ideals. In addition, loans from English are – facilitated by the language's prevalence in science – ubiquitous in everyday domains practically all around the globe. Oliveira and Pagano illustrate the situation for Brazil. Contrary to wide-spread opinions, these loans not only affect the receiving languages' lexicon but also their deeper structures. Haarmann and Holman demonstrate this for Finnish. Is there a slow assimilation process of all languages in the direction of English?

The prevalence and dominance of English in science is a global fact, but it varies of course in kind and degree, as well as to its effects, between language communities and countries. Such distinctions are suggested in a rough manner by the way the contributions to this volume have been assembled in chapters. The first group of articles deal with general problems or give overviews of various countries. It is followed by a chapter on countries with a history of English-language dominance. Australia where English is the sole official language has been included since the language has been imported and the population is linguistically diversified even today. The third section comprises articles on countries which have always, or at least in recent times, used languages for science different from, or in addition to, their own indigenous tongue. For these countries adjustment to the recent prevalence of English only means shifting from one foreign language to another if at all. The fourth, final chapter pictures countries and language communities which have until recently had their own international language of science. For them, one would think, it must be hardest to adjust to today's Anglification of scientific communication. This becomes noticeable in some of the articles, especially those on France and Germany. Delimitation between these types of countries and language communities is of course not clear-cut and at times arbitrary.

May the present volume inspire further theoretical and empirical research into language choice for scientific communication and into the consequences of the spread of English as the world lingua franca of science. Insights gained by such studies might help control and alleviate ensuing problems.

I do not want to end without expressing thanks to all contributors to this volume as well as the editor of the series, Joshua Fishman, and the publisher. Finally, I want to thank Gabriele Scheewe, one of the secretaries of Germanistik at the Gerhard-Mercator-Universität Duisburg, for invaluable computer and paper work, and Peter Joy, an Australian Ph.D. student at the same University, for linguistic corrections in line with my suggestion not to eliminate all the non-native-speaker traces.

Ulrich Ammon Duisburg, February 2001

References

Ammon, U.
1998 *Ist Deutsch noch internationale Wissenschaftssprache? Englisch auch für die Lehre an den deutschsprachigen Hochschulen.* Berlin/ New York: de Gruyter.
2000 Towards more Fairness in International English: Linguistic Rights of Non-Native Speakers? In Phillipson, R. (ed.) *Rights to Language: Equity, Power, and Education.* Mahwah, NJ: Lawrence Erlbaum, 111–116.

Baldauf, R./Jernudd, B.
1983 Language of Publication as a Variable in Scientific Communication. *Australian Review of Applied Linguistics* 6 (1): 97–108.

Crystal, D.
1997 *English as a Global Language.* Cambridge: Cambridge University Press.

Fishman, J.
1996 Summary and Interpretation: Post-Imperial English 1940–1990. In Fishman, J. A./Conrad, A. W./Rubal-Lopez, A. (eds.) *Post-Imperial English. Status Change in Former British and Americal Colonies, 1940–1990.* Berlin/ New York: Mouton de Gruyter, 623–641.

Graddol, D.
1997 *The Future of English. A Guide to Forecasting the Popularity of the English Language in the 21st Century.* London: The British Council.

Nylenna, M./Riis, P./Karllson, Y.
1994 Multiple Blinded Reviews of the Same Manuscript. Effects of Referee Characteristics and Publication Language. *The Journal of American Medical Association* 272 (2): 149–151.

Tsunoda, M.
1983 Les langues internationales dans les publications scientifiques et techniques. *Sophia Linguistica:* 144–155.

Table of Contents

I. Overall Perspectives and General Models

English – the Accidental Language of Science?[1]

Robert B. Kaplan

And who in time knows wither we may vent
The treasure of our tongue, to what strange shores
This gain of our best glory shall be sent.
To enrich unknowing nations without store?
Which worlds in the yet unformed Occident
May come refined with the accents that are ours.

1. ELT and the U. S. balance of trade

In 1996, the United States agricultural sector was a key contributor to international trade; it contributed something on the order of US$ 60 billion in exports (Klintberg 1997). In the 1998 budget year, the federal government spent about US$ 1 billion on international education. In the same year, nearly half a million international students studying largely in the tertiary sector in the United States contributed something like US$ 7.5 billion to the U. S. economy and supported something like 100,000 jobs (Levinson and McCarthy 1998). In other words, international students account for about one eleventh the value of agricultural exports. That's a lot of money. When taken together with overseas student numbers in the United Kingdom, Australia, Canada, New Zealand, and other

English-speaking countries, international education and education through English has become a massive business.

However, this trade in English education is only half the story. No calculation has ever been attempted to estimate the additional contribution to the United States balance of trade that derives from teaching English abroad. Such is the demand for English tuition that there are literally thousands of United States citizens -mostly young- teaching English to speakers of most of the world's other languages. Some of these teachers are college-age youngsters, equipped with back packs and a yen for foreign travel, who constitute a cadre of itinerant teachers in virtually every large city around the world. As an untrained teaching force, they really shouldn't be out there – in educational terms – but they have the virtues of being native-speakers of English and of being available on site, thus inexpensive. They serve to answer part of the world-wide demand for English; for example, in such places as Eastern Europe (see, e. g., Medgyes and Miklosy in press) and Asia. Their incompetence is supplemented and modified by the Peace Corps and by a very large number of teachers in programs developed and maintained by the United States Agency for International Development (AID) and the United States Information Agency (USIA, in other countries USIS), by a smaller consort of Fulbright scholars, and by the hundreds of additional teachers in programs mounted overseas by U. S. academic institutions and other organizations.

As with the market in overseas students, the English-language teaching market is not the exclusive preserve of the United States. All the other members of the English-speaking world – Australia, Canada, New Zealand, and the United Kingdom – are out there too. All of these English speaking nations have competed through their agencies for international development[2] by investing to varying degrees in development projects in less developed nations around the world; such development initiatives almost always carry a component in teaching English as a foreign language – much of it targeted at science and technology (see, e. g., Ablin 1991; Crooks and Crewes 1995; Kaplan in Press 1997; Kenny and Savage 1997). Finally, the tertiary academic institutions in the English-speaking nations are also recruiting international students. Not only are academic institutions directly recruiting students to their campuses in the English-speaking world, but they are also mounting programs delivered abroad in conjunction with academic institutions and other agencies in the third world.

This cohort of teachers, of course, requires materials, and the publishing industry produces a plethora of dictionaries, grammars, spellers, course books, readers, audio tapes, computer assisted language learning programs, and a multitude of other resources to meet these needs.[3] Often, these resources require electronic equipment, such as tape recorders, 35 mm. cameras, slide projectors, copying machines, video cameras, videotape players, overhead projectors, CD-ROM players, entire language laboratories; in some instances, even entire computer laboratories equipped to access e-mail and the world wide web are required. In sum, all of this activity generates money, and the teaching of English around the world has become big business. Except for the obvious fact that all this activity generates a lot of money, it would seem important to ask why it exists.

The quotation with which I began this chapter could be argued to constitute the underlying rationale for the teaching of English worldwide. Although this thought was expressed by Samuel Daniel in his poem, *Musophitis,* in 1599, not much has changed in the thinking of English speakers over the past four hundred years.

2. An historical note

2.1. Long history

Before I undertake to address the question why all this language teaching activity exists, let me first set the scene. Over the past four and a half million or so years, the human species has undergone a long series of biological modifications, among them the modifications that made speech possible. But since the emergence of language, the species has undergone three major post-biological modifications: 1) the invention of writing, 2) the invention of printing, and 3) the invention of electronic word processing and the World-Wide Web.

The distinction between biological and post-biological modifications is critical; biological changes constitute a part of the human genetic baggage, while post-biological changes seem not to include any genetic change. All human offspring within the normative ranges have the capacity to speak and to understand speech. But it is not the case that all human children are born with a biologically conditioned predisposition to acquire the skills of reading and writing; on the contrary, reading and writing must be learned anew in each new generation.[4]

From the emergence of language to the emergence of writing, human populations had access to information primarily as it was held in human memory. There is evidence that pre-literate people had the capacity to retain huge quantities of information in memory; nevertheless, memory is an unsatisfactory, unstable, and inefficient mechanism for the transmission of information. Retrieval from memory depends at least on the presence of the owner of memory, on the condition of the owner of the memory, on the audience for whom retrieval was undertaken, on the form in which the information was stored, and on the circumstances in which retrieval occurred.

Once it became possible to write things down – the first post-biological modification – the nature of information must have changed, since it became possible to retrieve information across time and space. Whereas memory probably necessitated the use of memory-enhancing stylistic devices, written text has a different stylistic and rhetorical structure essentially free of memory-enhancing structures. It was no longer necessary for the owner of information to be present, and the condition of the owner of memory and of the form of the stored text became essentially irrelevant. Text could be more widely distributed – granted that the process of manual copying was slow and subject to error, and that the production of large numbers of copies was unlikely. Indeed, the small production resulted in texts that were perceived as works of art and that were quite expensive, limiting acquisition to the elite.

The invention of printing constituted the second post-biological change. Printing permitted much more rapid production of texts and, gradually, over the next several hundred years, reduced the cost of owning texts. As early as the mid-17th century, Samuel Pepys regularly reported on his acquisition of published books and on the various structural modifications he made in his library. The structure *library* gradually became a major repository of texts, making material available not only to scholars, but also to ordinary persons.[5] Gradual improvements in print technology increased the speed of production and the number of copies available.

Electronic word-processing constitutes the third post-biological change. Electronic document production and distribution increased speed of production, served to change the role of the middle man in text distribution (the library), and increased exponentially the amount of material available not merely to the scholar but to anyone with the technical facilities to access the World-Wide Web.

Each of these post-biological changes decreased the effort and cost required to produce, store, and distribute information, and each has, in its turn, caused an information explosion. At the same time, increased availability of printed information is reciprocal with increased desire for literacy; when there is little or nothing to read, literacy is a superfluous skill. Each leap in the availability of information complicates the verification of information, making the veracity of information harder and harder to determine. Each leap in the availability of information seems to be associated with the dawn of a fundamentally new form of human society; it appears that the invention of language and its accompanying genetic changes mark the beginning of what we designate as 'human.' But the post-biological changes are, respectively, associated with the dawn of civilization, with the beginning of modern civilization, and with a new orientation not yet possible to describe or define (or perhaps even imagine). Each has been accompanied by an invention that caused an information explosion. In sum, a limit on the production of information impeded progress in the time preceding each information explosion. As Robertson (1998: 9) suggests, civilization *is* information, and civilizations are perhaps limited more by lack of information than by lack of physical resources. Information limitations are probably quantitative as well as qualitative; limitations on information restrict the number of things a society knows how to do. Unfortunately these phenomena have not been well studied, in part at least because the notion of information as quantifiable is very new – first articulated by Claude Shannon in the 1940s. The French historian Henri Berr, in 1934, writing well ahead of his time and before the invention of the computer, suggested that these post-biological inventions had epochal significance.

Following the invention of the printing press, books became suddenly available in quantities beyond the conception of earlier societies.[6] This proliferation of books carried ideas to a wider audience than ever before. This explosive spread of information (some of it misinformation) offers an explanation for the stunning achievements of the Renaissance.[7] The development of 'scientific method' in the 16th and 17th centuries is really a response to the need to verify information reported elsewhere and to the need to find patterns in large quantities of information.[8] The existence and distribution of the *Philosophical Transactions* of the Royal Society of London helps to explain the explosion in scientific activity in the 18th century. It is important to note that the *Transactions* were published and distributed largely (but certainly not exclusively) *in English*.[9]

The Royal Society was formed in November 1660 as a 'society of gentlemen' committed to the exchange of experimental knowledge under a Baconian[10] plan articulated by Robert Hooke in a letter also written in 1660. The reporting format of experimentation was essentially defined by

Robert Boyle; the venture was to a significant extent a reaction against scholasticism and marked the birth of empirical research. Boyle set the tone for scholarly writing; he may be credited with the introduction of hedging in science writing. In his *Proemial Essay,* he wrote:

> [I]n almost every one of the following [experimental] essays I (...) speak so doubtingly, and use so often, *perhaps, it seems, it is not improbable,* and such other expressions as argue a diffidence of the truth of the opinions I incline to, and that I should be so shy of laying down principles, and sometimes of so much as venturing explications (cited in Atkinson 1999: 103).

But it is not only for the invention of the research article that Boyle is remembered; he and his colleagues in the Royal Society created the overriding code of the gentleman in science – a code which has persisted into the 20th century. Thus, in part, the Royal Society was a 'gentleman's club,' a place where one met and socialized with one's 'own kind.' But quite beyond that, being a British gentleman meant being a financially independent individual

> (...) who cultivated various socially approved pastimes as ways of occupying and improving himself, and sometimes his society, as well, in the eyes of his fellow men and [the] Creator. The study of nature and technology was one such approved pastime (...) In sum, the Royal Society was a society of gentlemen in the fullest sense – run by gentlemen, for genteel purposes, via genteel standards of conduct and communication, as part and parcel of a genteel form of life (Atkinson 1999:17).

It was this orientation which was spread around Europe through the publication of the *Transactions*, made possible by the royal charter of the society, which privileged the group to the patronage of the king, granted permission to publish without government censorship, and assured the right to correspond freely with the citizens of other countries. After all, this was the first "(...) *public institution for the pursuit of scientific research* (...)" (Atkinson 1999:16; italics in the original). As much as facilitating the spread of English, these events facilitated the diffusion of empirical research and the rhetoric in which such research would be reported (see also Prelly 1989).

The electronic storage and retrieval simply increases exponentially not only the availability of information but the mode of its expression. More importantly, it vastly increases the capability to *use* this information. The amounts of information necessary to maintain communication networks, multinational corporations, and international transportation systems exceed the information production and distribution capacities of earlier societies. In short, societies are basically limited by the amount of information they can produce, store, and distribute. This is not an argument for technological determinism based on an assumption that certain levels of information produce certain kinds of societies; on the contrary, the argument is that the *lack of information inhibits the development of certain kinds of societies*.

Science and technology lie at the heart of this discussion, as increased information accessibility encouraged the development of science and technology, and the resulting science and technology reciprocally increased the amount of information.

2.2. The shorter history

Foreign languages have, of course, been taught for as long as there are any records of human societies; Jean Auel, in her yet to be completed four-part series collectively called *Earth's Children*, provides a fanciful notion of multilingualism among the earliest humans. The Greeks taught Greek to the people they conquered, and the Romans taught Latin. During the great expansion of Islam, Arabic was carried to the furthest corners of the known world. In more recent times, Europeans taught French, German, Spanish, Portuguese, and Russian throughout the European world and even further afield in their spheres of colonial activity in Africa, Asia, and South and Central America (see, e. g., Paulston 1998: 2–3). English has now been taught to populations of speakers of other languages for quite a long time-probably since the British Empire was at its greatest expansion. While English (and other languages) have been taught in many places, they have not always been taught from the best possible motivation. It was necessary for the British to teach English throughout their wide-spread empire because they needed people in distant places to speak English so that soldiers could understand their British officers and so that a civil service could be developed to maintain civil order under the leadership of British administrators. Indeed, "[i]t was considered self-evident that the civilizing influence of Britain was a desirable goal, anywhere in the world, and that the English language was an essential means of achieving this end (...)" (Crystal 1997 b: 70).

An interesting point is that, as the British Empire contracted, the teaching of English did not. A set of curious accidents that began with the end of WW I[11] (Ammon 1992) and that were significantly augmented in the middle of this century caused English to thrive. When WW II ended, the United States, generally claimed to be an English-speaking country, was the only major Western power whose educational and scientific infrastructure remained completely intact. The United States participated, with its allies, in dictating the conditions under which the post-war world would be organized. The United Nations, created in the aftermath of the war, chose only five official languages – Chinese, English, French, Russian, and Spanish – basically, the languages of the first four members of the Security Council – the WW II allies.[12] Indeed, the phenomenon had begun to a lesser extent after WW I, but the critical difference in mid-century was the important technological change; i. e., the availability of the computer.

The creation of the United Nations accidentally coincided with the birth of the computer age. The first computer programs were written in English-like languages (e. g., Basic, FORTRAN) and their output was also English, or English-like. Gradually, because so much scientific material had been written in German, the German language had been added to the list as a documentary language. At the same time, the earliest computers could not deal with Chinese characters, and consequently very little was stored in standard written Chinese. By the mid-1970s, the languages of the United Nations were Arabic, Chinese, English, French, Russian, Spanish – and German was also widely used. But the advent of the cold war resulted in heavy political restrictions on the use of Russian – imposed by both sides; i. e., the reluctance of the Soviet Union to share scientific information and the equal reluctance (and inability) of the western states to access material written in Russian.[13]

Because the scientific, technical, and educational structure of the United States remained in tact in the years immediately following WW II, students from the third-world flocked to academic institutions in the United States (see the opening paragraphs of this chapter). United States science and technology flowered in the post war years. It is a 'law' of science that those doing the greatest amount of research both require the greatest quantities of information from the information banks and contribute the greatest quantities of new information to those information banks. Vast numbers of scientists were trained in English, and vast quantities of information were written, abstracted, stored, and disseminated in English.

3. The special status of English

3.1. The special status of English in Europe

More recently, when Britain and Ireland were admitted (1973) into the European Union (EU),[14] English became one of the nine official languages[15] of the EU (Danish, Dutch, English, French, German, Greek, Italian, Portuguese, and Spanish). Against this complex linguistic, cultural, and political background, English has developed a special status within the EU; i. e., English and French are the sole "official" languages of the European Council, while the other languages are designated "working" languages. According to Ammon (1996):

- English has constantly made gains as a language of science over the past fifty years;
- English is the sole working language of the European Science Foundation (which coordinates research projects in EU countries and elsewhere);
- The leading European scientific journals now tend to prefer English as their language of publication; and, in addition,
- English is the most widely taught language in the member countries of the EU;
- There has been a clear shift toward using more English in business-oriented communication among the political bodies of the EU and in the economic domain within EU countries.

3.2. The special status of English in science and technology

Another development which occurred in this same time period was the vast increase in scientific and technical research. While modern science was a child of the first industrial revolution,[16] the heavy dependence on science and technology during the war years resulted in a great growth in scientific activity. The United States, by virtue of the fact that its scientific infrastructure was undamaged by the war, assumed leadership in science and technology. It is an established fact that progress in science depends on the accumulation of a written record of all previous science; that is, science requires great information storage and retrieval systems. The invention of the computer made those information storage and retrieval systems geometrically larger and more accessible. As already noted, it is also a fact that those who do the greatest amount of research *require* the greatest amount of information from those information networks, and they naturally also *contribute* the greatest amount of new information to those networks. Since much of the science and technology research in the 1950s and 1960s was conducted in English, most of the information in the great information storage networks was written in English. The International Federation on Documentation (FID), a world body which keeps track of information distribution, reports that nearly 85% of all the scientific and technological information in the world today is written and/or abstracted in English. (Indeed, FID urges that an article written in another language be accompanied by an abstract in English, German, or Russian.) Scientific and technological journals in countries like Sweden and Hungary publish more material in English than they do in their national languages (Baldauf 1986; Baldauf and Jernudd 1983 a, b, 1986, 1987; Medgyes and Kaplan 1992). Not only is English the undisputed language of science, but because of the importance of the computer in the internationalization of English, the English-speaking nations may hold a virtual cartel on scientific information because the international information systems are organized according to an English-based sociology of knowledge. Even research and development (R&D) functions in non-English states are impacted, since it is necessary to be able to search scientific literature in English and according to its sociology of knowledge. The emergence of a whole new cadre of information managers seems inevitable – both in terms of science – trained balanced bilingual translators at the input end and in terms of comparably trained skilled science readers at the output end. It would seem that R&D functions in third world nations (largely along the North-South axis) cannot survive without such a development. But the emergence of such a new function places another level of intervention between the scientist and the information. The social construction of science information has been extensively researched by discipline and by function, and the implication for language policy and language-in-education policy have been explored (Bazerman 1985 – Physics; Gilbert and Mulkay 1984 – science discourse; Grabe and Kaplan 1986 – language-in-education; Latour and Woolgar 1979 – laboratory science; Maher 1987 – Medicine; Myers 1990 – Biology; Windsor 1990 – Engineering).

Thus the WW II settlements and the birth of the United Nations, the invention of the computer, and the geometric growth of science and technology, all occurring coincidentally at the same time, created the conditions which made English not just an important language but the predominant language of science and technology. At the moment, more people in the world speak English as a first or second language than spoke any other single language in the history of the world (Crystal 1997 a, b) (except written Chinese – which, as I have noted previously, is not [yet] easily computer storable; spoken Chinese is made up of nine mutually unintelligible spoken regional dialects – see Harrell 1993). As a result of all these factors, the teaching of English to speakers of other languages has become a huge industry. However, because of the broad, global distribution of English, and because it has been, and is being, taught in so many places, English is no longer the property of English speakers. Many new varieties of English have developed – for example, Indian English, Nigerian English, Philippine English (ESL varieties), Japanese English, Hong Kong English (EFL varieties). These Englishes are not exactly like British or American English; each one is unique, drawing on the local substrate languages. The growth of other Englishes is assured because in many countries English is frequently taught to children by individuals who are not themselves native speakers of English and who may not have had extensive exposure to native English speakers. In polities like India, Nigeria, Samoa, and Singapore, there are many native speakers of their national varieties – i. e., the local variety of English is their first language. These individuals may also be able to speak an international variety of English (e. g., American or British English) as well.

4. The notion of a standard

This dispersion and diversity makes a mockery of the notion that there is a standard variety (or a number of standard varieties) of English – or of any other language for that matter. A 'standard' language results, generally, from a complex set of historical processes intended precisely to produce standardization; indeed, a 'standard' language may be defined as a set of discursive, cultural, and historical practices – a set of widely accepted communal solutions to discourse problems. Additionally, a 'standard' language is a potent symbol of national unity. If this definition of a 'standard' language may be assumed to be viable, then the 'standard' language is really no one's 'first' language. On the contrary, the 'standard' language must be acquired through individual participation in the norms of usage, and these norms are commonly inculcated through the education sector (with the powerful assistance of canonical literatures and the media – conventional and electronic). But the reality of most linguistic communities is marked by the normative use of a wide range of varieties in day to day communication – i. e., the use of slang, of jargon, of nonstandard forms, of special codes, even of different languages (as in code-switching). Consequently, a 'standard' language constitutes a purely ideological construct. The existence of such a construct creates an impression that linguistic unity exists, when reality reflects linguistic diversity. The notion of the existence and dispersion of a 'standard' variety through a community suggests that linguistic unity is the societal norm; it also suggests a level of socioeconomic and sociopolitical unity which may be contrary to the reality of linguistic diversity. The legal obligation to use a codified standard is likely to cause frustration among minority-language speakers, since the standardized language is for them a non-dominant variety (see the case of Slovakia – Kaplan and Baldauf in press); minority-language speakers probably use a contact variety, likely to be at considerable variance from the 'standard' variety. If this is true within a linguistic community, the variation must be much greater across linguistic communities.[17]

Science is an important candidate for promoting the growth of a standard language because it uses a common set of methods and measurement-standards and is cumulative and self-referential. New research constantly becomes available, but it builds on prior research; thus, there is a need to access both the new and the previous research efficiently if one is to participate in the research activity. Increasingly, as 'special-purposes' language (i. e., in science and technology) has been studied, it has become apparent that the special registers of science and technology are more important than was initially conceived. Translation and the use of technical dictionaries are not sufficient to access science research; in addition, an understanding of discourse styles and rhetorical structures is essential (Burrough-Boenisch 1998; Ventola and Mauranen 1991). A recognition of this issue has led to increasing internationalization and standardization in science writing (Baldauf 1998). Authority and 'gate-keeping' in science lies in the hands of journal editors, referees of papers, and a cadre of self-appointed guardians of 'appropriate' writing. As a consequence, those researchers whose written English-language skills are inadequate find that publication is difficult and indeed may be effectively excluded from participation in the exchange of science information.

As Mühlhäusler suggests (1996: 207–208), language planning efforts, including the world-wide dissemination of English, reflect the cultural view of the West. This view, known as the 'plumbing' or 'conduit' or 'telegraphic' conception of communication, is defined as the translation of messages which exist in the sender's mind into speech signals (coded in linguistic form) which are converted back into the original message by the receiver. Thus, there is a need to identify a single, 'standard' code to assure that this single code is optimally regular, simple, and modern, and to assure that there are optimal channels (postal services, road networks, telegraphs, newspapers, journals, television, etc.) along which the signal can flow. The problem is that this metaphor is not a reliable description of how human beings communicate (see also Wurm, Mühlhäusler and Tryon 1997).

5. On the extinction of languages

While there is no question that a number of new varieties of English have come into existence, a very large number of other – usually smaller – languages are threatened with extinction.[18] Mühlhäusler writes: "Of more than 6,000 languages currently spoken more than 95% are on the endangered list, and the overall rate of language extinction is far greater than that of any biological species" (1996: 206–207; see also Robins and Uhlenbeck 1991). And Crystal (1997 b: 17) reiterates the point:

> No one knows how many languages have died since humans became able to speak, but it must be thousands. In many of these cases, the death has been caused by an ethnic group coming to be assimilated within a more dominant society, and adopting its language. The situation continues today, though the matter is being discussed with increasing urgency because of the unprecedented rate at which indigenous languages are being lost, especially in North America, Brazil, Australia, Indonesia, and parts of Africa. Some estimates suggest that perhaps 80 per cent of the world's 6,000 or so living languages will die out within the next century.

Languages become extinct sometimes because of the decimation of the population of speakers (e. g., the instances of many Native American languages, Australian Aboriginal languages, etc.), sometimes as the result of a period of bilingualism during which a second language is adopted for an increasing number of purposes by a growing number of people (as in the case of Welsh, Irish, and Scottish). (For further discussion, see, e. g., Dorian 1989; Dorian in Bright et al. 1992/3: 135–136; Ingram in Bright et al. 1992/2: 303; Romaine in Bright et al. 1992/4: 21.) Kaplan and Baldauf (1997: 272–273) suggest that languages die for a number of complex reasons:

1. The introduction of a non-indigenous language that, for whatever reasons, takes over some – or all – social functions;
2. The disappearance, for whatever reasons, of the population speaking some particular language;
3. The forceful introduction of a non-indigenous language so that certain functions *must* be conducted in the imposed language.

In sum, other than in the case of the total destruction of a language community, languages die because:

1. Parents are reluctant or unable to pass on a language intergenerationally to their children;
2. The language ceases to serve key communicative functions in the community;
3. The community of speakers is not stable and/or expanding, but rather is unstable and/or contracting.

Where English has been introduced, either as a colonial language or as a commercial language, some or all of these conditions are often met (see, e. g., Phillipson 1992). In science and technology, English has captured the key registers.

It would be unreasonable to assert that the introduction of English is exclusively responsible for wide-spread language death. A great many factors are involved, among them:

1. Population dislocation and redistribution as a result at least of war, revolution, religious persecution, economic development, or urbanization;
2. The spread of world languages other than English – e. g., Arabic, Chinese, French, Japanese, Portuguese, Russian, Spanish (Ammon 1994);
3. The development of supralinguistic functions – e. g., world-wide aviation, tourism, banking, science and technology, etc.[19]

At the same time, it would be equally unreasonable to claim that the huge English-language teaching activities of the English-speaking nations have played no role in language death. The role they have played is, however, not well understood.[20]

6. What global English does

As English is introduced into communities where it has previously had no role (or only a very limited one), and as people perceive English-language ability to provide access to a better standard of living, English is replacing some registers normally reserved to indigenous languages – even some indigenous languages in total. While the register of sports is, perhaps, not particularly significant (though that point is arguable), sports register can serve as an apt illustration. Such phenomena as the introduction of baseball in Japan, of soccer and cricket in much of the former British Commonwealth, or – most recently – of American football in Europe have led to significant language and social change; other phenomena such as CNN news have pervaded the world wherever television is available and have brought with them language and social change. (See Kaplan and Baldauf 1997: 233–235, for a discussion of the penetration of English technical lexicon into Russian.) (Political leaders who want to make a point to a world-wide CNN audience know that it must be done in English.) Still other phenomena such as the multinational corporations (and their accompanying R&D functions – see, e. g., Paradis, Dobrin and Miller 1985), offering good jobs and high wages, have brought language with them and have resulted in significant language and culture change. As noted above, an additional factor in language loss is urbanization (which requires more frequent and more effective communication across a wider range of domains); thus, urbanization is frequently marked by the expanded use of English.

In these circumstances, it is not purely language which penetrates other cultures. Let us consider the case of baseball; it is not merely the game that has penetrated Japan. Rather, the whole panoply of activities connected with the game has also been adopted. The big game, the big star (and the accompanying "star" salary), the live broadcast, even the ubiquitous "beeru" and "hotu dogu," have become part of the Japanese environment. By a similar process, the introduction of any new technology carries with it the language in which the technology was developed; thus, the spread of the technology itself facilitates the spread of English if (as is often the case) the technology arose in an English-speaking polity.

It is unlikely that there is some grand conspiracy among English-speakers to disseminate English world-wide;[21] on the contrary, the spread of English is largely accidental, based in part on the quest for an allegedly better standard of living on the part of receiving populations, and in part on the unconscious press of English on other populations. People talk about the "dominance" of English in certain registers or in certain geographic zones, but the language does not have a will of its own to become dominant, and there is nothing in the natural characteristics of English or of English speakers which would make it inevitable that English should become *the* world language. On the contrary, it is the actions of English-speakers – including journal editors, reviewers and other gate keepers in science and technology – which underlie the spread of English. English-speaking scientists have also contributed to this phenomenon. Again, there is nothing insidious about the actions of English speakers; it is simply a matter of more-or-less benevolent self-interest. After all, English speakers have a distinct advantage in a world that has adopted English as its universal language. At the same time, good scientists who cannot write English to meet the standards of journal editors are deprived of the opportunity to have their views and contributions disseminated through the global information networks; as a consequence, their contributions are not only lost to the scientists themselves, but more seriously are lost to science.[22]

One of the reasons for this advantage is that English is a pluricentric language, and its speakers have never (until very recently[23]) tried to enforce a rigid single standard. Thus, there are American English, British English, Canadian English, Irish English, South African English, and West Indian English – just to name a few varieties. Each creates its own identity and ways of speaking. These are all accepted as English – unlike French, for example, whose speakers try to maintain a single world-wide standard. The fact that English varieties flourish, without being reduced to 'substandard' dialect status, with the only condition on them being that they maintain some level of mutual intelligibility, is one of the underlying keys to the continued success of English as an international language (Baldauf 1998). For example, Iraqi Foreign Minister, Tarak Aziz, may not want to be perceived to be speaking either American or British English, but he does speak English and identifies with a number of other varieties or even with something increasingly recognized as 'international English.'

However, the developments with respect to English are not unidirectional. While the English speaking nations not only push the dissemination of English but also perhaps actively push English in the direction of a uniform 'standard' language for universal communication, the Council of Europe is quietly moving toward multilingualism. As Baetens Beardsmore (1994: 94) suggests:

> A general policy goal [of the 1992 Treaty of Maastricht] is to place the highest priority on educational mobility; the objective is to enhance the level of familiarity of as many European students as possible with other European cultures and languages as an element of quality in education. Language learning remains a top priority, and to this end, member states are encouraged to promote trilingualism; they are advised to make language qualifications desirable for entry into, and compulsory for exit from, higher education; and they are requested to give particular attention to the learning of minority languages.

7. Conclusion

Here one can see the playing out of two conflicting ideologies – on the one hand the acceptance of the fundamental value of multilingualism as an amazing world resource which allows different perspectives and insights and thus encourages reaching a more profound understanding of the nature of the human mind; on the other hand, the perceived fundamental value of a common language as an equally amazing world resource which allows unprecedented possibilities for international co-operation, especially in the solution of scientific and technological problems. Global use of English serves the latter position. But to the extent that global use of English contributes to the death of small languages, the price may be too high; the extinction of small languages is even more catastrophic than the extinction of biological species, precisely because the extinction of languages is an extinction of the means to understand the world and to interact with it. Such extinction narrows the human condition. Some balance between these two views must be struck.

As Koch (1992: 42, cited in Norberg 1994: 156) has said about the destruction of Sorbian:

> I can only imagine the world with my ethnicity in place. Its disappearance signifies loss. Slowly but surely the impoverishment would be perceptible across the country's breadth. Perhaps even continentally and planetarily. One color less. Increase of grayness. One sound less, one language less. Increase of silence.

This is, of course, an emotional rather than a rational view of the problem. While rational approaches involving analysis and description of actual language-loss environments are necessary – indeed, indispensable to an understanding of the problem and of the means of redress – one must not ignore the emotional reactions of the human populations who are being deprived of the right to their language, along with other inalienable rights.

Thus, the spread of English in the registers of science and technology is essentially a coincidence of the confluence of a number of political and economic forces during the last half of the 20th century. But that spread threatens not only the survival of small languages; it also stills the voice of science in languages other than English. While the spread of English – and to a significant extent the widespread use of English in science and technology[24] – has the gravest consequences for the practice of science and technology in other languages – assuring the dependence of less developed nations on the scientific and technological development of a few states (Kaplan 1983), and largely in a single language – in the end, it cannot be said that the ascendancy of English is the outcome of a conspiracy; it is merely the outcome of the coincidence of accidental forces.

8. Notes

1. I am grateful to Professor Richard B. Baldauf, Jr. for his helpful comments and suggestions. I am also grateful to my graduate students at Meikai University, Japan, for their interesting insights on the issues discussed.
2. Britain's Overseas Development Administration [ODA], the British Council, the Australian Overseas Service Bureau [OSB], the Australian Agency for International Development [AusAID], the Canadian International Development [CIDA] and even some agencies of nations where English is not the first language (e. g., the Swedish International Development Cooperation Agency [SIDA]).
3. It is important to note that much of this activity is directed toward science and technology and that the number and diversity of the material results, at least in part, from the absence of a standard variety of English.
4. While reading and writing must be acquired *de novo*, speech must also be acquired, but the acquisition of the capacity to speak is genetically conditioned; it is socialization to the community norms of spoken language that must be taught.
5. The great library at Alexandria about the third or fourth century A. D. probably contained something on the order of 532,800 scrolls. But that library was unique and had a function quite different from later libraries. The state of libraries in China (where printing was invented) has not been well researched, but libraries in China are still difficult to use because classification systems are based on the stroke order in characters.
6. Johann Gutenberg (1398–1468), usually credited with the invention of movable type, worked in Germany (starting printing in 1454 and producing the Vulgate Bible in 1456). William Caxton (1422?–1491) usually recognized as the first English printer, working between 1477 and 1491, issued some eighty separate books from his press at Westminster; a stunning revolution in technology in a period of about 25 years.
7. Renaissance is perhaps a misnomer, since the society emerging from the second post-biological evolutionary change bears little or no resemblance to any preceding classical society.
8. Tycho Brahe, for example, might have been the first person to hold two separate sets of computations based on two different theories; his crucial observations may constitute a reaction to the possession of an unprecedented amount of information.
9. The German-born Henry Oldenburg, whose patron was Robert Boyle, and who was appointed Secretary of the Royal Society in 1662, invented the scientific journal. He had earlier developed a large network of correspondents among those working in the new science; from 1663 on, as scholars throughout Europe became aware of the Royal Society, it was bombarded with letters seeking or offering scientific information. As Secretary of the Society it was Oldenberg's charge to respond, and respond he did; the correspondence, arriving largely in French, German, Italian, and Latin was 'English'd' by Oldenberg and published in the *Transaction*.
10. Francis Bacon was the main conceptual founder of empirical science.
11. Because the delegates of the United States did not speak French, English was introduced into international diplomacy at the Treaty of Versailles (1919).

Furthermore, the English language played a significant role in the League of Nations, whose first Secretary-General was an Englishman.

12. Arabic was added to the list in 1974.
13. It was the Reagan administration in the United States that invented the term 'technology hermorrhage' and that for the first time in recent history began to impose serious restrictions on the free flow of scientific information, not merely in terms of 'national security', but in the context of protecting patents and copyrights.
14. So called since 1993, formerly the European Community [EC – 1967–1993], and before that the European Economic Community [EEC – 1957–1967].
15. The number was increased to 11 in 1995.
16. "Big science" had begun to develop early in the 19th century. The 1875 volume of the *Philosophical Transactions* reports on "a series of experiments carried on by a Committee appointed by the Secretary of State for War ..." (Noble and Abel: 49–50).
17. In a social gathering in Tokyo in 1998, I met a professional economist who had for many years worked for the Japanese Ministry of Economics and who had, in the course of his duties, negotiated on behalf of Japan in a wide variety of other countries. He reported that such negotiation was commonly carried on in English, even though English was not the native language of either side in the negotiations. He complained that the teaching of 'standard' English to Japanese was ineffective and urged that 'broken' English be taught since that was the variety most commonly used in the spheres of activity in which he had been engaged.
18. Of course, when one speaks of the extinction of languages, it is implicit that the speakers of those languages also disappear. The actual human beings may not be gone, but their ability to communicate in the particular language is gone. A discussion of the extinction of languages should not evoke sympathy for the language but rather for its speakers.
19. Because of the need for standardization in scientific and technology information and because, increasingly, problems are recognized as global in nature rather that national or regional, the use of English as a world-wide vehicle has increased.
20. Mühlhäusler (1998) has suggested that perhaps this phenomenon is related to the fact that previous inter-communication languages (i. e., pidgins, etc.) were designed for limited purposes and thus did not infringe on the domains of the contact languages. English, on the other hand, can serve both inter- and intra-communication purposes.
21. This is not to overlook or underestimate the efforts of the British Council or the United States Information Agency to spread English. These efforts appear to be a 'natural' tendency of development agencies to spread the languages in which they operate, sometimes as a matter of official foreign policy, sometimes for economic and cultural reasons. Organizations like the *Alliance Française*, the Goethe Institute, the *Instituto Cervantes*, the Japan Foundation, the Korea Foundation, to name just a few, are engaged with equal enthusiasm in the spread of their respective languages. None of this activity can be construed as a 'conspiracy'; rather, it is a natural outgrowth of economic and political competition (see, e. g., Kaplan and Baldauf 1997: 4–14). It must be noted that, in a historical context, the efforts of language agencies are very recent – almost entirely within the past fifty years – and essentially superfluous.

22. The academic reward structure is heavily imbued with the notion 'publish or perish'. In brief, the reward structure in the academy presses scientists to publish, and in general they must publish in English. Scientists who cannot place their work in English journals are doubly penalized; not only is their voice stilled, but they are denied the conventional rewards available through the academy.
23. The English-Only movement in the United States is an illustrative case. See Gonzales and Melis 1999 for discussion.
24. It is not only science and technology that is influenced by English. Cohen has the following to say:

    In ethically charged areas of international life touching on issues such as conflict resolution, human rights, religion, ethnic identity, security, and so on, English by itself proves an inadequate and biased guide. Thick knowledge of the local language and culture of one's interlocutors is essential if one is to make sense of their needs and concerns. Only unmediated access to the knowledge encapsulated in native discourse can enable negotiators and analysts to identify the congruencies and dissonances between their own assumptions and those of others. If negotiation is an attempt to create shared meanings and understandings where contradictory readings existed before, then there is really no alternative to the attempt to break out of the habits of thought conditioned by the English Language . . .[W]ithin the international relations field on the whole the blatant fact of linguistic and therefore semantic diversity receives little attention and the dominance of English is presupposed. International negotiation is a prime case of neglect, since it is preeminently an activity which brings together individuals who speak different native languages and depict and evaluate reality in different ways (In preparation).

9. References

Ablin, D.

1991 *Foreign Language Policy in Cambodian Government: Questions of Sovereignty, Manpower Training and Development Assistance.* Paris: UNICEF.

Ammon, U.

1992 The Federal Republic of Germany's Policy of Spreading German. *International Journal of the Sociology of Language* 95: 33–50.

1994 International Languages. In Asher, R. E./Simpson, J. M. (eds.) *The Encyclopedia of Language and Linguistics.* 4 vols. Oxford: Pergamon, 1725–1730.

1996 The European Union (EU – formerly European Community): Status Change of English During the Last Fifty Years. In Fishman, J. A:/Conrad, A. W./Rubal-Lopez, A. (eds.) *Post-Imperial English: Status Change in Former British and American Colonies, 1940–1990.* Berlin/New York: de Gruyter, 241–267.

Atkinson, D.

1999 *Scientific Discourse in Sociohistorical Context: The Philosophical Transactions of the Royal Society of London, 1675–1975.* Mahwah, NJ: Lawrence Erlbaum.

Baetens Beardsmore, H.
1994 Language Policy and Planning in Western European Countries. In Grabe, W. et al. (eds.) *Annual Review of Applied Linguistics, 14.* New York: Cambridge University Press, 93–110.

Baldauf, R. B., Jr.
1986 Linguistic Constraints on Participation in Psychology. *American Psychologist* 41 (2): 220–224.
1998 Standardization, Variation and Authority in English: The Impact on Language Diversity. *TESOL in context* 8 (2): 4–10.

Jernudd, B. H.
1983 a Language Use Patterns in the Fisheries Periodical Literatur *Scientometrics* 6: 245–255.
1983 b Language of Publication as a Variable in Scientific Communications. *Australian Review of Applied Linguistics* 6 (1): 97–108.
1986 Aspects of Language Use in Cross-Cultural Psychology. *Australian Journal of Psychology* 32: 381–392.
1987 Academic Communication in a Foreign Language: The Example of Scandinavian Psychology. *International Review of Applied Linguistics* 10 (1): 98–117.

Bazerman, C.
1985 Physicists Writing Physics: Schema-Laden Purposes and Purpose-Laden Schema. *Written Communication* 2: 3–23.

Berr, H.
1934 Preface to G. Weill *Le Journal.* In the series *L'Evolution de l'Humanitè,* ed. H. Berr.

Bright, W., et al. (eds.)
1992 *International Encyclopedia of Linguistics.* 4 vols. New York: Oxford University Press.

Burrough-Boenisch, J.
1998 Writing Science Like an English Nativespeaker: How Far Can and Should Non-Native Speakers of English Go? *Il Jornades Catalanes sobre Llengües per a Finalitats Espec'fiques.* Canet de Mar: Universitat de Barcelona.

Cohen. R.
In Preparation International Negotiation: A Semantic Analysis. Paper Presented to the Annual Convention of the International Studies Association. Washington, DC, February 16–20, 1999.

Crooks, T./Crewes, G. (eds.)
1995 *Language and Development.* Jakarta: Indonesian Australian Language Foundation.

Crystal, D.
1997 a *The Cambridge Encyclopedia of the English Language.* Cambridge: Cambridge University Press.

Crystal, D.
1997 b *English as a Global Language.* Cambridge: Cambridge University Press.

Daniel, S.
[1599] 1956 *Musophilus.* In Lamson, R./Smith, H. (eds.) *Renaissance England: Poetry and Prose From the Reformation to the Restoration.* New York: W. W. Norton. 632–641; (see esp. p. 640., lines 337–342).

Dorian, N. C.
1989 *Investigating Obsolescence: Studies in Language Contraction and Death.* Cambridge: Cambridge University Press.

Gilbert, G./Mulkay, M.
1984 *Opening Pandora's Box: A Sociological Analysis of Scientists' Discourse.* Cambridge: Cambridge University Press.

Gonzales, R. D./Melis, I. (eds.)
2000 *Language Ideologies: Critical Perspectives on the Official English Movement.* Urbana, IL: National Council of Teachers of English.

Grabe, W./Kaplan, R. B.
1986 Science, Technology, Language and Information: Implications for Language and Language-in-Education Planning. *International Journal of the Sociology of Language* 59: 47–71.

Graddol, D.
1997 *The Future of English.* London: British Council.

Harrel, S.
1993 Linguistics and Hegemony in China. *International Journal of the Sociology of Language* 103: 97–117.

Kaplan, R. B.
1983 Language and Science Policies in New Nations. *Science* 221: 4614.
1993 The Hegemony of English in Science and Technology. *Journal of Multilingual and Multicultural Development* 14 (1/2): 151–172.
1997 Review of: *Language Development: Teachers in a changing world.* Kenny, B./Savage, W. (eds.) 1997. London: Longman. *Asian Journal of English Language Teaching* 7: 121–126.
1998 Review of *Language and development.* Crooks, T./Crewes, G. (eds.) 1995. Jakarta: Indonesian Australian Language Foundation. *English for Special Purposes* 17 (3): 317–320.

Baldauf, R. B., Jr.
1997 *Language Planning From Practice to Theory.* Clevedon, Avon, UK: Multilingual Matters.
2000 Not Only English: "English Only" and the World. In Gonzales, R. A./Melis, I. (eds.) *Language Ideologies: Critical Perspectives on the Official English Movement.* Urbana, IL: NCTE.

Kenny, B./Savage, W. (eds.)
1997 *Language Development: Teachers in a Changing World.* London: Longman.

Klintberg, P. P.
1997 Farm and Farming. *The World Book 1997 Year Book*, 207.

Latour, B./Woolgar, S.
1979 *Laboratory Life: The Social Construction of Scientific Fact.* Beverly Hills, CA: Sage.

Levinson, E./McCarthy, J.
1998 Ferment in Overseas Advising Circles, as Concerns Mount About Weaknesses in the Front Lines of Student Mobility. *International Educator* 8 (1): 18–19.

Lynch, M.
1985 *Art and Artifact in Laboratory Science: A Study of Shop Work and Shop Talk at a Research Laboratory.* London: Routledge and Kegan Paul.

Maher, J.
1987 The Role of English as the International Language of Medicine. *Applied Linguistics* 7: 206–218.

Medgyes, P./Kaplan, R. B.
1992 Discourse in a Foreign Language: The Example of Hungarian Scholars. *International Journal of the Sociology of Language* 98: 67–100.

Miklósy, K.
In press The Language Situation in Hungary. In Kaplan, R. B./Baldauf, R. B., Jr. (eds.) Special Issue. *The Journal of Multilingual and Multicultural Development* 20 (5/6).

Mühlhäusler, P.
1996 Ecological and Non-Ecological Approaches to Language Planning. In Hellinger, M./Ammon, U. (eds.) *Contrastive Sociolinguistics*. Berlin: de Gruyter, 205–212.
1998 Layer Upon Layer of Languages. *Journal of Pidgin and Creole Languages* 13 (1): 151–158.

Myers, G.
1990 *Writing Biology: Texts in the Social Construction of Knowledge*. Madison, WI: University of Wisconsin Press.

Noble, R. A./Abel, F. A.
1875 Researches on Explosives – Fired Gunpowder. In *Philosophical Transactions* 165.

Norberg, M.
1994 The Sorbs between Support and Suppression. *International Journal of the Sociology of Language* 107: 149–158.

Paradis, J./Dobrin, D./Miller, R.
1985 Writing at Exxon ITD: Notes on the Writing Environment of an R&D Organization. In Odell, L./Goswami, D. (eds.) *Writing in Nonacademic Settings*. New York: Guilford, 281–307.

Paulston, C. B.
1998 Linguistic Minorities in Central and Eastern Europe: An Introduction. In Paulston, C. B./Peckham, D. (eds.) *Linguistic Minorities in Central and Eastern Europe*. Clevedon, Avon, UK: Multilingual Matters, 1–17.

Phillipson, R.
1992 *Linguistic Imperialism*. Oxford: Oxford University Press.

Prelly, L. et al.
1989 The Rhetorical Construction of Scientific Ethos. In Simons, H. (ed.) *Rhetoric in the Human Sciences.* Newbury Park, CA: Sage, 48–68.

Robertson, D. S.
1998 *The New Renaissance: Computers and the Next Level of Civilization*. Oxford: Oxford University Press.

Robins, R. H./Uhlenbeck, E. M. (eds.)
1991 *Endangered Languages*. Oxford: Berg.

Ventola, E./Mauranen, A.
1991 Non-Native and Native Revising of Scientific Articles. In Ventola, E. (ed.) *Functional and Systemic Linguistics: Approaches and Uses*. Berlin: de Gruyter, 457–492.

Winsor, D.
1990 Engineering Writing/Writing Engineering. *College Composition and Communication* 34: 313–341.

Wurm, S. A./Mühlhäusler, P./Tryon, D. T.
1996 Atlas of Languages of Intercultural Communication in the Pacific, Asia, and the Americas. 2 vols. Berlin: de Gruyter.

When Does Knowledge Have a National Language? Language Policy-Making for Science and Technology

Angéline Martel[1]

1. Introduction

The spreading of English as *lingua franca* of science and technology on the one hand, and of business and finance on the other hand, is increasingly documented[2]. The phenomenon does not seem to leave individuals indifferent; nor does it leave collective organisations, particularly states, inattentive to what some call the invasion of English. Different attitudes prevail: some may accept this fact as a contemporary imperative, others may seek alternatives.

In this context, collective organisations are increasingly drafting language policies, each according to their own socio-economico-political context, according to their linguistic ideology or to their positioning intentions. In effect, science and technology[3], just like business and finance, just like languages, are contemporary objects of regional, national and international positioning in the pecking order of recognised powers. In this information age, the adage that "Knowledge is power" has never been taken so seriously; by individuals (scientists, citizens, bureaucrats, professors, students, entrepreneurs, managers, etc.) as well as by collective structures: states, business firms, industries[4], research laboratories, language communities, etc.

Within the very words "Knowledge is power" lies, however, a paradigm shift for science: from a pursuit of knowledge for itself (science for science) to a pursuit of politics through sciences (politics with science). And so "Knowledge is power" has evolved into another axiom: "Knowledge is the most powerful engine of production" (Schutz 1980). Although this belief is open to debate and has not proven itself in all contexts (Salomon 1990; Limoges 1990), the presumed causal link between knowledge, productivity (wealth) and power has particularly enticed states to institutionalise scientific and technological development[5]. In addition to the traditional physical powers of states (i. e. territory, military forces, demographic weight), sciences and technologies take states into a new grey matter power edge that is couched today into an ideology of innovation, consumerism and competition.

The research presented in this paper is situated at the crossroad of state policy discourses with a vertical axis on science and technology and a horizon on languages. This focus on states is justified in two ways: states are today the most active structural bodies drafting language policies, world wide; and they still remain the major type of human collective governance structure although clear trends towards global integration and regional collaboration are developing. Similar analyses could, and should, however, be drawn with supranational bodies (United Nations, UNESCO, European Union, etc.) as well as with multinationals businesses and with linguistic communities, in a minority situation, for example.

In this paper, I focus on the conditions of emergence, and on the particular configurations, of state language policies for science and technology. Of interest is an understanding of the conditions when the language of science and technology becomes problematic for a state: for what purposes? in whose interests? A study of the nationalisation of science, technology and language reveals important issues about the spreading or diminishing of cultural/linguistic powers, and about its dialectic with economic powers in our modern world (Montgomery 1996: 45). Such a study could also help gain an understanding of the extent to which states value the interconnection between scientific and technological activities and the importance that they give to the cultures/languages that make them possible.

"When does knowledge have a national language?" was chosen as a title to contrast with a more predictive intent: "When *should* knowledge have a national language". In order to begin answering the first question, I provide a descriptive framework and a case study. In order to answer the second question, one could place himself or herself from a particular language *planning* perspective. In the latter case, he or she could utilise one of the language planning typologies reviewed by Kaplan and Baldauf (1998) or by Cooper (1989). But there is yet no framework to guide the interpretation of the former question in the context of science and technology.

In the first part of this paper, I briefly review the paradigm shift that has occurred in the representations of science and technology as they are outlined in public policies in the Organisation for Co-operation and Economic Development (OCED) countries. In the second part, I present a heuristic framework for understanding the dynamics of language policies by repertoiring the sectors that we need to investigate. In the third part, this framework is used to understand the importance, or its lack thereof, of language policies pertaining to science and technology in a case study: Québec. The following aspects are of interest: the sociopolitical context, the linguistic ideology, the scientific/technological paradigm invoked. The fourth part summarises today's studies on the status of French language in science and technology. Finally, I propose some hypotheses on the dynamics of language planning in science and technology by states.

2. Science and technology: increasingly handmaidens of state and economy

Only a quarter of a century ago, most countries did not envisage that scientific and technological policies could be important while today, policies on this subject have been institutionalised in most Western states (Salomon 1990). Technology is now a particularly important dimension of official state discourse.

But, policies on science and technology are not a recent invention. In the IXth century, the calife Abbâssides intended to give Islam means to transcend its state of militarily conquered civilisation with a scientific culture that would position it with the Greek, Syrian and Indian sciences (Leclec 1990: 7). Another example of the merging of scientific and political interest is the famous Manhattan Project, a U. S. government research effort which involved numerous researchers in different countries and which culminated in the production of the first atomic bomb during the Second World War.

In effect, states have become increasingly involved in scientific and technological development since the Industrial Revolution[6], mainly through the allocation of research funding (Salomon 1990; Kaplan and Baldauf, 1988: 243). But what has primarily changed over the last decades is the emphasis placed on the *utilitarian* function that science and technology are to play in the economic, social and political context from which they emerge. This change of emphasis is evidenced by the key words chosen to name policies: from science, to technology, to innovation, each having its own constellation of meanings, actors and financial resources in the relationship between knowledge and utilitarianism.

2.1. The age of "science policies"

One of the first official and foundational policies on science was Vannevar Bush's *Science: the Endless Frontier* (1945). As scientific councillor to the American president, Theodore Roosevelt, Bush proposed the creation of a national agency to support scientific research, thus bringing to public awareness the implication of the state in science. Scientific research (particularly fundamental research) was then seen as contributing to culture and national well-being, as well as being part of education for scientists and for the general population. Such a policy gave priority to supporting research in universities and research centres and to the training of researchers. The quality and value of scientific results was to continue to be judged by peers. Encouragement was also given for the application of research by industry.

This model of science is linear: from fundamental research to applied research to technological development, production and commercialisation. At that time, science policy came down to "research policy". Key words in this paradigm are *free* and *fundamental*, in a context of economic prosperity, demographic growth and tertiarisation of the economy in the Western world. In this model, scientific activities are mostly conducted in universities. The researcher's most prominent need is to communicate freely with peers at national and international levels.

In this paradigm, investigations on the languages used in science focus on understanding the trends with indicators like:

- number, and percentage of total, of articles published in a particular language in national and international scientific journals, according to disciplines;

- number, and percentage of total, of articles published in a particular language in dissemination journals, be they national or international;
- number, and percentage of the total, of citations for a particular country (as indicator of the recognition of its science);
- number of scientific journals;
- official languages, and the languages of presentations at national and international conferences;
- sources and amounts of research funding;
- language of training in institutions;
- number, an percentage of total, of thesis in a particular language;
- number, and percentage of total, of textbooks.

2.2. The age of science and technology policies

The 1960's see the emergence of a discourse on scientific and technological development in most of the OCED countries (Gingras and Godin 1998; Davis and Duchesne 1986: 129). In a context of increased state involvement, of a realisation that there is no pure science, of economic crises and of a revolution in production and consumer systems introduced by the new technologies, science policies were reoriented toward an inclusion of technologies.

Conscious of the trend towards increasing pressures for a merging between science and technology and their increasing subservience to economic imperatives, the OCED published in 1971 a report entitled *Perspectives on Scientific and Technological Policies.* This report called for a more global approach to science and technology deeming that innovation should not be exclusively economic but also social. It asked states to consider "a new approach to research and development, no longer aimed exclusively at improving existing technologies and producing new ones, but also at perfecting conceptual instruments for the better understanding of man and his society and at delineating the requirements of harmony in social and economic development" (Brooks 1971: 62). This call was not heard: the focus by scientific and technological development on production of consumer products and processes deepened.

2.3. The age of innovation policies

Today, the new breed of scientific and technological policies definitely gives emphasis to innovation. As evidence, Table 1 shows the title of recent policies from seven OCED countries. We can note that:

- innovation plays a large role as the subject of policies;
- science and technology are used towards political ends;
- well-being and competitivity increasingly become covert or overt issues in which science and technology play a role through increased production.

Table 1: Scientific and technological policies: Samples from seven OCED countries (1993–1996)

United States	1993	*Technology for America's Economic Growth: A New Direction to Build Economic Strength*
	1994	*Science in the National Interest*
	1996	*Technology in the National Interest*
Great Britain	1993	*Realising our Potential. A Strategy for Science, Engineering and Technology*
	1995	*Competitiveness. Forging Ahead*
European Union	1996	*Green Book on Innovation*
OCED	1996	*Technology, Productivity and the Creation of Jobs*
Japan	1996	*Science and Technology Basic Plan*
France	1996	*La recherche: une ambition pour la France [Research: an ambition for France]*
Ireland	1996	*White Paper on Science, Technology and Innovation*
Canada	1996	*Sciences and Technologies at the dawn of the 21st Century. Federal Strategy in Science and Technology*

Innovation is defined by OCED as the development/commercialisation of an efficient product as a method of production/distribution with the objective of providing new or better services to consumers. In effect, innovation conveys a dramatic paradigm shift, –at the same time as a continuous development–. First, innovation takes science and technology out of the traditional breeding ground of universities and research centres and places it in businesses and industries because of the emphasis on the production of consumer goods and processes. Secondly, it displaces fundamental research from the initial point in the chain of production to a collective process where all stages of creation and production are intermingled.

This paradigm shift radically alters the dynamics of language use in scientific/technological development. First, researchers can no longer claim allegiance to none other than science; states, suprastates organisations, businesses and industries fund the vast majority of research, world-wide. This has as an effect to add linguistic imperatives from the source of funding, be these imperatives overt, covert, intentional or unintentional. The language of science and technology thus becomes increasingly tied to the language of business/finance and, it is too early to tell, but could well then be that of English as a *lingua franca.*

Secondly, the physical sites of research are increasingly moving, from universities to industries, hospitals, public research centres, small businesses, communities and associations, although this transfer is not occurring as quickly as policies may lead to believe (Godin and Trépanier 1995). At first look, one could hypothesise that moving research into society could militate for the use of a national language. But transfer of knowledge between researchers is increasingly cross-national so that a national language may seem as inadequate in knowledge transfer and, in a context where time is money, short cuts with a *lingua franca* are welcomed. In this context, the location of a business or research centre is less important than its ability to communicate world-wide, increasingly through distance communication technologies.

Thirdly, communications between researchers are no longer horizontal among peers but subjected to the vertical judgement of other spheres in society, particularly from businesses and industries. Evaluation of the quality of research is increasingly couched in terms of its usability on the global market. Thus, the notion of scientific community is moving: from a restricted group to a large diffuse body including entrepreneurs, managers, industrials, bureaucrats, etc. and the jargon of science, where incomprehensibility and non-communicability with the general public was a source of power and a weapon of difference and identity (Montgomery 1996) is increasingly denigrated. These factors could also militate for the use of national languages but again, the generalisation of international transfer of knowledge can be favourable to the use of a *lingua franca.*

Thus, the ivory tower is broken. Large portions of research have been industrialised and domesticated. Science, technology, state positioning and economy are intermingled. A new term, *technoscience*[7] has come to represent the a-social and a-ethical development of knowledge as handmaidens of industrialisation and economy. The products of scientific research are increasingly objects with words (software and technologies). Consequently, investigations on the languages used in science, technology and innovation should now use new indicators, complementary to those used to analyse the paradigm of science, indicator like:

- language of work in laboratories, be they situated in industry or in universities, hospitals, etc.;
- language of software used in industry, business, education;
- language of networking, particularly e-mail;
- the number of patents from a particular country as indicator of creative vitality.

At a policy level, we can ask how attention can be focussed on the language of science and technology? How can the increasing spread of English in science and technology, hand in hand with that of economy and finance, be understood and planned? How can the allegiance of scientists to a particular national language or to a scientific/economic community (or both) be analysed? How can the publication dynamics, which remains as the sediment of scientific and technological activities, be understood? – We now turn to a tool designed to help apprehend this complex and moving field.

3. A heuristic framework for understanding language policy dynamics

Language policy-making is a complex process of situation-assessing, discourse-making, consensus-building, forward-planning. With language planning emerging as an area of applied sociolinguistics, there is an attempt to understand both macro and micro factors affecting overt and covert changes favoured by policies. This section of the paper constitutes an effort to conceptualise the foundations of the dynamics of language planning in science and technology[8].

In this section, I propose a heuristic framework (Figure 1) which helps to draw a picture of the type of elements that enter into the configuration of policies for the language of science and technology. As a framework, it does not have a predictive function. Rather, its aim is similar to that of a geographic map: it could guide the study of particular practices (in time and space). Its intent is also sociocritical to the extent that it encourages the uprooting of the ideologies that ground language planning and implementation. Furthermore, it provides spaces for describing policies within changing notions/functions of science and technology towards innovation; as they were and as they are becoming under increased conditions of globalisation, industrialisation and consumerism.

Language planning takes place on two planes: the subjects and the objects. On the first axis of the heuristic framework, the subjects axis is built on a continuum between collectives and individuals. It shows both

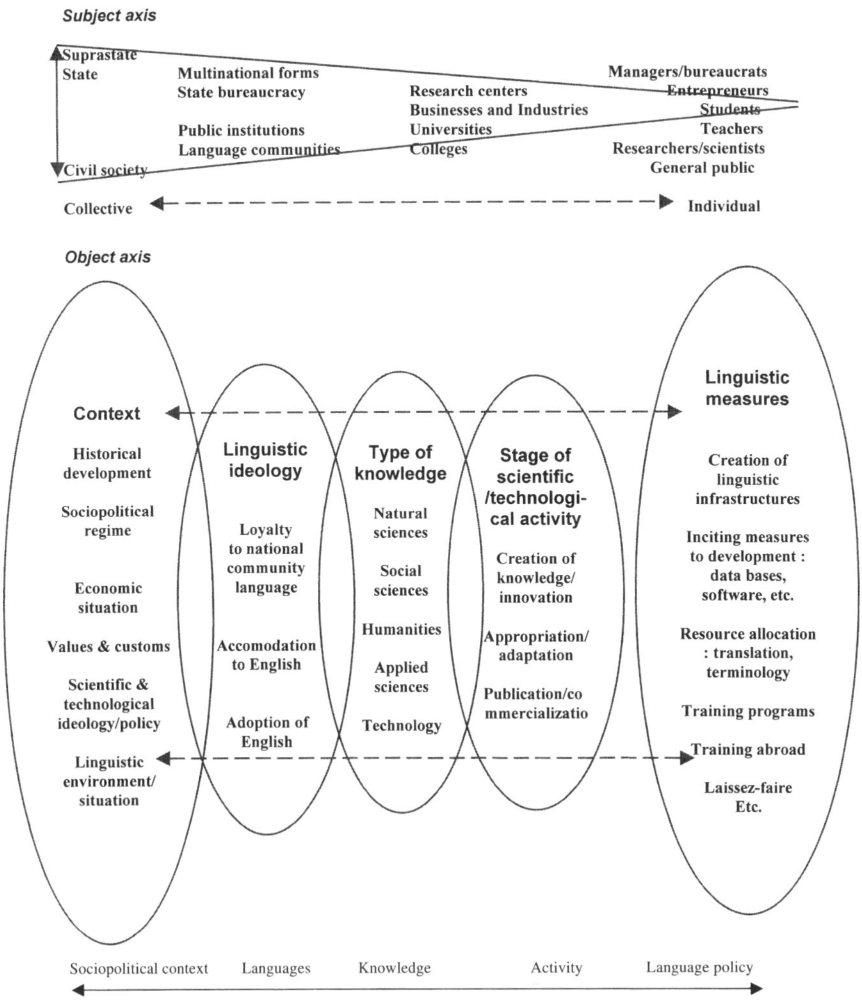

Figure 1: Heuristic framework for understanding the dynamics of language policies in sciences and technologies

the actors (be they bureaucrats, researchers, public institutions, states, etc) and the recipients of language policies, including the general public in a age of increasing discourses on democratisation of knowledge[9] and links science/technology to society[10].

On the object axis are situated four intertwining and inter-influencing spheres, from macro (context) to *meso*, or intermediary, levels of considerations (linguistic ideology, types of knowledge, stages of scientific activity). They all feed into the type of language planning measures, or its lack thereof, that are likely to be announced, published and/or adopted.

The first sphere, the context, provides a broad overview of the environment: the historical development (developed or developing country, industrialised, urbanised, rural, imperialist, colony or ancient colony, etc.), the sociopolitical regime (democracy, dictatorship, etc.), the economic situation, the values and customs, the scientific and technological ideology, the linguistic environment (including its assessment and statistical description).

The two last components of this sphere are of utmost importance. First, the nature of the scientific and technological policies are crucial since the importance accorded to languages will depend largely on the type of scientific and technological culture a state adopts. The most current ideologies for Western states have been described in the previous section: science for science, science and technology, innovation. The link between development and the adopted notion of science/technology is also important. In industrialised states, approximately 10% of the research funds is allocated to fundamental research and it is never larger than 20%. In many developing states, this proportion is between 18% and 32%. In other words, the more spent on fundamental research, the less the research system is integrated in the production system. The states that spend the most on fundamental research are also those whose industrial structures are the least adapted to technological innovation (Salomon 1990: 52).

Likewise, the linguistic environment, in reference to English, will matter in the type of linguistic ideology dominating. Some types of considerations pertaining to the linguistic environment, in reference to English, are presented in Table 1[11].

Table 2: Typology of linguistic environments in reference to English

Highly developed nation/community/agency	Developing nation/community/agency
a) Native of English	a) Native of English
b) Non-native with English as official language	b) Non-native with English as official language
c) Non-native without English as official language i. who had an international language of science ii. who did not have an international language of science	c) Non-native without English as official language

The inclusion of the type of knowledge sphere is an effort to nuance the practices in different disciplines. Typologies of knowledge disciplines differ according to institutions. I have adopted the most commonly used one. But, with a trans-disciplinary trend and the rise of thematic studies, these are likely to be vastly modified in the next decades. The distinction between natural sciences on the one hand and social sciences/humanities on the other is however important. Their distance or proximity with local and general public interests, their ability to reach vast audiences will matter in the type of measures adopted in language planning.

The fourth sphere, stage of scientific/technological activity, allows for considering the number and type of actors involved. The distinctions between creation, appropriation/adaptation and dissemination/use are neither linear nor exclusive. They however can serve to explicit the utilitarian paradigm which focuses on use and dissemination of knowledge.

In the spheres representing linguistic measures, these have been categorised, from the most interventionist ones to the least: from the creation of infrastructures like national research funding agencies, to incitative measures like the participation in data-bases and the use of softwares, to resource allocation (funding for translation) to training programs, national or abroad to laissez-faire.

We now need to test this heuristic framework to see if it can accommodate the description of situations. We have chosen the case of Québec for it is a good example of a modern state that has strong linguistic ideology, attentiveness to language through policy and introduction of the innovation spectrum of science and technological development.

4. A case study: Québec between culture and economy

4.1. Context

Québec, one of Canada's ten provinces, has been a former colony of both, France and Great Britain. This historical context is reflected in its demography with a French majority and English minority population. With migrations, however, its population has gone through important changes between 1951 and 1996 (Table 3). The French-as-a-mother-tongue population has increased by 67% since 1951 to 82.0% of the total population in 1996. The English mother-tongue population has also increased but less rapidly during the same period, +12.2%, and consequently has gone from 13.8% of the total population in 1951 to 9.2% in

1996. The population with a mother tongue other than French or English has greatly increased (+298%), from 3.7% to 8.8% of the total population in 1996.

Table 3: Québec's population, by mother tongue 1951, 1971 et 1991 (percentage of the total population)

	1951		1971		1996		
Mother tongue	Population	%	Population	%	Population	%	Change 51–96
French	3347000	82.5	4867200	80.7	5585600	82.0	+ 66.9
English	558300	13.8	789200	13.1	626200	9.2	+ 12.2
Other	150400	3.7	371300	6.2	598500	8.8	+297.9
Total provincial	4055700	100	6027700	100	6810300	100	+ 67.9

Québec's history accounts for its current nationalising trend. In fact, since the 1950's, Québec has undergone what has been called a *Révolution tranquille*: a mutation and a rapid growth of the state and of its administrative apparatus, a sharing of responsibility and powers with the private sector, a renovation of the educational system, a change in identity from Canadian to *Québécois*.

In 1977, the *Parti Québécois*, a party whose ideology is nationalistic and autonomist/sovereignist came to power. The laws and measures put in place to strengthen Québec have since been constant according to this ideology, despite a ten year interlude with the Liberal party in power (1985–1995).

4.2. Linguistic ideology

Since its Révolution tranquille, Québec has put in place an impressive infrastructure to develop the corpus and the status of the French language. The best known of these measures is the *Charte de la langue française* (*Law 101*) that regulates the language dynamics on its territory and attempts to develop French in a North American environment. It is based on both a developmental and defensive ideology in that it focuses on the principle of the primacy of French but seeks complementarity with the language of the Anglophone community, that of autochtonous populations and of the other cultural communities. To that effect, and with exceptions, it encourages the development and use of French in five fields: legislation and justice, administration, work, commerce and business, and education. It also consolidates a bureaucratic infrastructure dealing exclusively with language: an advisory body, the *Conseil de la langue française* (CLF), an implementation body, the *Office de la langue française* (OLF) and a corpus body, the *Commission de toponymie.*

4.3. Scientific and technological development policies: from science to innovation with a linguistic edge

Historically, the evolution of scientific/technological policies in Québec takes its origins in the demands of universities and continues with a project of technological restructuring of industries in the 1980's. This evolution is accompanied by constant reorganising according to social and cultural challenges (Davis and Duchesne 1986: 125). In effect, Québec's science and technology policies were also cultural and social policies. The objective has been to induce change and make it socially acceptable: towards a social mastery of technological change. This has been done in four stages concurrent with the changes that were happening in the OCED countries at approximately the same time.

As will be shown, there is a convergence between nationalistic ideals, be they implemented through linguistic or scientific and technological policies. They all aim primarily at strengthening the state, the French language, the Québec cultures, its economy, its demographic base, its territory.

4.3.1. Institutionalisation of science: 1965–1975

The first stage of scientific policy development in Québec begins with the annual congress of *l'Association canadienne-française pour l'avancement des sciences* (ACFAS)[12] in 1965 and ends with the advent to power of the *Parti Québécois* in 1976 (Table 4) . Incidentally, in that first year (1965), the ministers of science of the OCED countries meet for the first time in Paris.

At this point in time, Québec's universities take the initiative of the promotion and elaboration of policies favouring research in their walls. Science is still a private affair between universities and industries.

Although it claims inspiration from the UNESCO and OCED, the governmental policy adopted in 1971 (Table 4) is a technocratic justification of one of the themes of the Révolution tranquille: technocratisation,

i. e., catching up and development of the institutional framework of research. A scientific policy is seen as a tool and a motor of progress. Social, economic and political roles are given to scientific development. Interestingly, the policy notes the increase of scientific and technological innovation due to military research; in its opinion, however, to plan and invest in scientific development is to make science subservient to a more social (vs. military) structure, i. e. the state and civil society. The emerging trend toward state-funded industrial research is confirmed by the foundation of the *Centre de recherche industrielle du Québec* [CRIQ, Centre for Québec's industrial research] in 1969.

No mention is made, however, of language at this point in time in the policy but, in 1969, the three French universities (Universities of Laval, Sherbrooke and Montréal) presented a brief to the Federal Senatorial Committee for Scientific Policy arguing that francophone universities were disadvantaged in obtaining research grants[13] because they were not sufficiently developed to compete with anglophone universities. The Federal government allowed extra funds. Québec founded the *Formation de chercheurs et action concertée* [FCAC, Training of researchers and concerted research] which advanced training of researchers and allowed for the preparation of grant submissions to the federal government.

Table 4: First stage of scientific and technological policies in Québec: 1965–1975

Date	Policy	Other action	Language issue
1965		ACFAS congress on *Pour une politique scientique au Québec* [Toward a policy for science in Québec]	
1971	*Les principes de la politique scientifique du Québec* [Principles for Québec's scientific policy]		None overtly

4.3.2. Science for national development and for international recognition: 1976–1982

With the advent to power of the *Parti Québécois* in 1976, initiatives to enlarge research to government and industry subjected scientific and technological development to the sovereignist project. Its 1979 policy proposal (Table 5), *Pour une politique québécoise de la recherche scientifique* [Toward a policy for scientific research in Québec], is a blueprint for the restructuring of the institutions managing national scientific and technological activities. It bases its proposals on the vision of science held by a well-known botanist who, already in 1938, attributed it two roles: (1) international scientific recognition as a means survival for a people and (2) national life developed by science grounding economy.

This 1979 policy proposal makes proposals in four areas: the central co-ordinating role of national instances, the development of new sites for research (public institutes), actions to increase the dissemination of scientific and technological information, and training of research personnel. It proposes, namely, to reorient the institutional framework towards Québec's Research and Development, to multiply student grants and bursaries, to create strategic research themes, to create new research centres. To the scientific culture of experts, it opposes a democratisation of science (scientific culture of citizens), attempting to reconcile the ideal of democracy and technocracy through the proposed creation of museums (arts, ethnology, scientific dissemination), radio programs, school reforms, dissemination journals natural parks historic sites.

The 1979 policy proposal was submitted to a vast public consultation. It resulted in 1980 in a policy entitled *Un projet collectif* [A collective project]. This policy does not differ in its political orientations from the 1979 proposal. Consequently, its four main objectives deal with the democratisation of science, the training of researchers, the co-ordination and planning of functions and responsibilities dealing with research, and the allocation of funding.

A strong nationalistic discourse is well developed throughout the document. The state of Québec is to become the first instance responsible for technological development, as opposed to development through the federal government; Québec proposes to request that all research grants be repatriated under its jurisdiction. And it is only on that issue that preoccupations for French are somewhat present, but they are structural and not linguistic. At this point in time then, language is not problematised as such in science. Instead, language is an issue raised in institutional terms. The policy states that French universities are disadvantaged in the federal repartition of research funding. It denounces, after the fact, the discrepancy in grants allocation by Ottawa in 1968. Language is then a territorial and institutional matter.

Table 5: Second stage of scientific and technological policies in Québec: 1976–1982

Date	Policy	Other action	Language issues
1979	*Pour une politique québécoise de la recherche scientifique*[For Québec's policy on scientific research]	Submitted to a vast consultation	None other than allusion to francophone universities
1980	*Un projet collectif* [A collective project]		Francophone universities as disadvantaged by Federal funding

At the time of drafting the *Chartre* in 1979, the state did not deem it necessary to include science and technology as one of the fields of intervention. It gave, however, its advisory body in matters related to language, the *Conseil de la langue française*, the mandate to monitor the development of the French language and to give advice on changes to be made. Immediately after the adoption of the 1980 policy, in 1981, the *Conseil* began its reflections on the topic of the language of science and technology through an international seminar and with publications called *Avis* (Advice) in order to guide governmental policy. As indicated by Table 6, it clearly raises the issue of the language of science and offers a series of recommendations (Lapointe 1998).

Table 6: Activities of the *Conseil de la langue française* pertaining to the language of sciences and technologies, 1978–1982

Date	Activity
1981	International symposium on the language of science
1981	*La loi 101 et l'avenir de la langue française au Québec* (*Avis*). It reviews the gains made by Law 101 (Charter) and indicates new preoccupations: with the advent of new technologies, it calls for inter-ministerial co-ordination as motor of change, for co-operation throughout *Francophonie* and for funding of scientific journals and software

4.3.3. The technological turning point: 1982–1994

During this third policy period, there is a vibrant policy-drafting activity on the economic, social and political challenges offered by technology

(Table 7). The initiative remains with public administration and with its privileged clients. The accent is on the development of the industrial structure of Québec through technological appropriation and innovation.

Table 7: Third stage of scientific and technological policies in Québec: 1982–1994

Date	Policy	Other activity	Language issues
1982	*Le virage technologique*[The technological turning point]		
1982	*Bâtir l'avenir* [Building the future]		
1983	*Loi pour la promotion du développement scientifique et technologique du Québec* [Law for the promotion of scientific and technological development in Québec]		
1983	*Plan de relance AGIR (Actions gouvernementales pour l'intensification de la relance de l'économie* [Boosting up plan ACTION]		
1988	*La maîtrise de notre avenir technologique* [Mastering our technological future]		
1991			MESS[15] study on the language of scientific publications
1992	*Le développement scientifique au Québec* [Scientific development in Québec]		
1993		*Urgence technologique: pour un Québec audacieux, compétitif et prospère.* [Technological urgency: for an audacious, competitive and prosperous Québec] CST[16] Report	
1994		*Miser sur le savoir: les nouvelles technologies de l'information.* [Building on knowledge: the new information technologies] CST Report	

This period is a turning point in that the applications of science and research are seen as a mean towards another objective: that of making Québec's *industries and businesses* more technologically dynamic. Policies aim at transforming the industrial and business structures. Economic programmes attempt to restructure Québec's industry and support business in times of recession.

The 1983 plan *AGIR* (Action) aims at placing Québec in first place among Canadian provinces in scientific and technological training, at doubling scientific and technological personnel, at increasing collaboration between businesses and universities. Among some of the means proposed are six new applied research centres[14], help to universities whose performances in research with industry and replacement of professors and researchers loaned to industry. In 1989, a fund for technological development is founded.

If the language of science and technology are not a focus of governmental policies, it is, however, an issue raised on numerous occasions by the *Conseil de la langue française.* As indicated by Table 8, the *Conseil* provided quite a number and variety of recommendations pertaining to the language of science. It also included the language of technology as part and parcel of its preoccupations.

Table 8: Activities of the *Conseil de la langue française* pertaining to the language of sciences and technologies, 1982–1994

1986	*L'informatisation et le français au Québec* [Digitalisation and French in Québec]. It analyses the evolution of computers in society and it bases it recommendations on the "normality"of working in French and on the leadership role of the government. It namely recommends to take the means to give access to French software, documentation and data bases, to promote the role of colleges and universities in the formation of computer specialists and technicians, to adopt a policy of multilingual software development
1986	*La place du français dans l'information scientifique et technique* [The place of French in scientific and technological information] (*Avis*). Recommendations on the basis of a shared responsibility between scientists and state, on the respect of the autonomy of the researcher, and on the necessity of inter-French international co-operation
1986	*Avis* on the francisation of businesses. It calls for a francisation of new work technologies and for a computerised access to terminological data bases

Table 8 (Continued)

1988	Position paper on Québec after the Meech Lake Accord. It mentions the rapid changes brought about by technologies and call for a francisation operation
1989	*Le français dans l'information scientifique et technique* [French in scientific and technological information] (*Avis*). Same position as the *Avis* of 1986
1991	*La situation du français dans l'activité scientifique et technique* [The situation of French in scientific and technological activities] (*Avis*). It bases it recommendations on the autonomy and freedom of the scientists and on their responsibility to make the society benefit from their work. Main recommendations: support for quality (primary and synthesis) journals, creation of new journals where Québec researchers excel, proposition by researchers of substantial abstracts (1 page) in French in international journals, creation of scientific data bases in French, development of collaborations among la Francophonie, support for public dissemination of scientific information, francisation of terminology in technical and scientific domains
1991	*Langue, culture et politique culturelle.* [Language, culture and cultural policy] (Position paper). It indicates that a cultural policy cannot be separated from a policy of dissemination of science and technology

4.3.4. Culture and innovation as a tandem: 1994–1998

Today's policy making continues to rests on national positioning and development. An impressive number of activities and policies (Table 9) have been put into place to reflect on the changes to be effected and on the resources to be allocated, particularly on the object of new technologies and on the growth of the Internet.

Of particular interest is a 1998 Report from the *Conseil de la science et de la technologie: Pour une politique québécoise de l'innovation* [For Québec's policy on innovation]. It is heavily influenced by factors external to Québec and attempts to move Québec toward the increasing innovation pole of science and technology. Its five recommendations intend to guide the drafting of a new policy (recommendation #1). The main issue here is: how can a state favour development of activities that are to take place outside of its competence, i. e. in business and industry? The answer that is first given to that question is that the state should insure the development of a global environment favourable to innovation: training of resources, development of a research base, favouring a culture of innovation and creating fiscal and regulatory incitives. Secondly, it should facilitate the development of networks: favouring interactions, circulating resources, facilitating information exchanges and collaborations. For businesses, the state should help them to build their innovative edge. Finally, the state should enter into partnership with the actors of innovation, reinforce the co-ordination between government actors, design evaluation and forecast tools. The Report on innovation encourages the state to play the role of facilitator and not that of main actor. But no innovation policy has as yet been adopted.

The paradigm grounding innovation becomes increasingly operational outside of the state. The main question then becomes: If states must adopt the role of facilitators how can they develop language policies to guide and protect national languages? The question remains whole for Québec. Although this question has not been resolved, Québec has acted on another dimension of innovations: the information highway.

With a 1998 policy on the information highway entitled *Agir autrement* [Acting differently], the government focuses on five priorities:

- to generalise access to the Internet;
- to prepare the new generation to new technologies through education;
- to build French cultural contents into the information highway;
- to accelerate the transition of economy and create jobs;
- to bring the state closer to citizens and businesses.

In this policy, the state, contrarily to the innovation paradigm, sees its role as central: first as a motor and actor in fields that are of its (Canadian provincial) competence (education, health); secondly, as an interstate collaborator to define the juridical framework of the Internet; thirdly, as a supporter of novel ideas and experimentation in the new economy. Forty-nine measures, identifying the ministry or agency responsible, accompany the policy statement.

This policy, evidence of maturity and experience in policy-drafting, marks a turning point – and a continuation, as always. It attempts to reconcile social imperatives with economic benefits. The political positioning of the state is not as important as the development of its own people through the state apparatus.

Table 9: Fourth Stage of scientific and technological policies in Québec, 1995–98

Date	Policy	Other activity	Language issues
1995		*Inforoute Québec: plan d'action pour la mise en œuvre de l'autoroute de l'information* [Québec's information highway : action plan for the implementation of the information highway]. Report of the Comité consultatif sur l'autoroute de l'information	
1995		*Inforoutes, culture et démocratie: enjeux pour le Québec* [Information highway, culture and democracy: challenges for Québec]. Hearing of the Cultural commission of the National Assembly	
1996		*Pour une stratégie de mise en œuvre de l'autoroute de l'information au Québec* [For an implementation strategy for the information highway in Québec]. Report of the Secrétariat de l'autoroute de l'information	
1998		*Pour une politique québécoise de l'innovation* [For Québec's policy on innovation]. (Report from Conseil de la science et de la technologie)	
1998	*Agir autrement. La politique québécoise de l'autoroute de l'information*[Acting differently: Québec's policy on the information highway]		1 out of 5 chapters on culture and language

Today, a sign of the importance that Québec gives to scientific and technological development is the publication of a yearly compendium of scientific and technological activity. This publication includes extensive statistical information on expenses and resources.

Although policies on technology and science had not generally been concerned with language, the 1998 policy on the information highway, *Agir autrement*, is a notable exception perhaps because times have changed and the 1979 *Charter* could not, nearly twenty years ago, foresee the importance of new technologies. In particular, this policy aims to:

- promote tools for French through the development of computers that can process, adapt and understand French;
- emphasise the creation of a critical and attractive mass of content in French;
- digitalise the essential elements of the cultural heritage, according to the needs of the public, the scientific community and the professionals;
- harmonise national laws on intellectual property with international developments in order to respect the rights to intellectual property and the liberty of users to have access to productions and to be able to use them;
- promote the appropriation of NCT by the Québec cultural community;
- commercialise cultural products on the information highway.

Faithful to its orientation and its mandate, the *Conseil de la langue française* provided two important *Avis* during this period (Table 10). Its particular interest lays in the new language resources and tools created by the Internet and digitalisation.

Table 10: Activities of the *Conseil de la langue française* pertaining to the language of sciences and technologies, 1994–1998

1994	*Avis sur les industries de langue dans la société de l'information* [Advice on language industries and the information society] (Avis). It recommends to favour the development of automatic translation, to support emerging entrepreneurs, to develop collaboration with the Francophonie, to consider the importance of the quality of the written language.
1996	*L'arrivée des inforoutes: occasion pour un nouvel essor du français* [The arrival of the information highway: an occasion for a new advances for French] (Avis). It places an emphasis on the production of contents in French for Internet, on generalized access to the Internet, on the development of search engines

4.3.5. Observations

As we can observe from this data, in matters of science and technology, Québec is behaving as a master of its own faith (Davis and Duchesne 1986). Science and technology policies oscillate between economy and culture; they attempt to bring to society the benefits of those activities.

As for the language of science and technology, what can we then observe from the dynamics of language policy-making in Québec? Mainly that :

- over the years, emphasis was placed on the development of the object, science and technology, as means of attaining the status of a sovereign state, mainly by the development of scientific and technological culture, structure and bureaucracy;
- until 1998, no official policy included provisions to protect French as the language of science. This is so perhaps so because issues pertaining to the language of science were dealt with in other manners, for example, important terminological activities provided researchers with terminological tools, support to scientific journals was given through funding agencies, etc.;
- until 1998, the main tool for language development in matters of science and technology remained the *Charte de la langue française;*
- an arsenal of measures for the francisation of business and industry, those very actors who became the main actors of the new paradigm of innovation, may have been seen as sufficient until 1998;
- the advent of the Internet has become the first occasion to introduce language issues in policies on language and technology.

If specific linguistic measures have not been introduced into public policy on science and technology until 1998, one may hypothesise that there were no need to introduce such measures. Let us then turn to the assessment made of this issue by scientific community through some indicators.

5. French as a language of science and technology in Québec

As most issues, the importance, or lack thereof, of the French language in science and technology in Québec is a matter of interpretation, of representation of the reality as expressed by statistics and by personal experience. Thus, whether French is gaining, or is loosing, ground serves to feed linguistic ideologies in this matter and scientists as much as policy-makers place themselves on the terrain of these ideologies. As an example, the most pertinent issue seems to be the number of publications in French, whether in Québec's own scientific journals or in journals published or publishing in French. Scientists usually side with one of these two statements:

- as a scientist, my duty is to make my work known, be it in French, in English or in any other language. In other words, my science has no language;
- as a scientist, I have a duty to my origins in Québec and I must/should/do publish in French.

The question is complex. In Québec, numerous studies[17] have monitored the importance of French in science and technology according to a wide variety of indicators[18]. I have chosen one of these indicators corresponding to the "Publication/commercialisation" phase of the heuristic framework (Figure 1) in order to illustrate the fact that French, as a language of science and technology in Québec, is not flourishing. And the trend towards innovation, as we have seen, may only serve to give a state less and less of an influential role in the matter.

As a preamble, I must, however, indicate that in Québec, even though no official language policy has been established, numerous linguistic measures have been taken in general, and in particular for publications in French: funding to scientific journals, development of resource and funding for the development of specialised terminology and the establishment of an international network[19], funding for scientific dissemination through journal and textbooks, etc.

5.1. *Publications*

Publications serve as indicators of the vitality of scientific development in a state; it is their symbolic value. But they also serve other purposes. For scientists, they are means of communicating research results and of investigating/contributing to knowledge capital; they traditionally have also been an indicator of productivity and of prestige. Thus, before being linguistic objects, publications and patents serve a capitalisation role in a society. This does not simplify the issue since linguistic strength may well be dependent on capitalisation strength. In order to understand this dimension of the issue, I indicated the strength of scientific development in Québec before quoting some of the statistics on the use of French, for all sciences combined first and for natural and social sciences next.

- For all sciences and technologies
  - there are over 100,000 scientific journal in the world and this number is increasing at the rate of 5,000 articles per day adding to the 30 millions existing (Bilan 1996: 171);
  - 3% of scientific production in the world is in French, compared to 89% in English (Bilan 1996: 173);
  - 80% of data-bases are in English and 62% of CDroms are in English (Bilan 1996: 173);
  - 30% of the articles published by researchers in Québec are co-signed, 20% of which are co-signed with international collaborators, 10% with Canadian colleagues. This is three times more frequent than the Canadian national average. This can be seen as a sign of recognition from the world scientific community or as a sign of dependence according to Godin (1996; cf. also Leclerc 1993).
- In natural sciences, engineering and biomedical sciences
  - Québec produces 1% of the world scientific literature, that is more than 5,000 articles per year (Godin and Limoges 1995);
  - in some thematic sectors, scientists from Québec have a greater impact on the scientific community than their counterparts in Canada as measured by the Science Citation Index: in toxicology (15% more often), pharmacology (15% more), neurosciences (12%), mathematics (11%), cardiology (8%), experimental medicine (7%). But in physics and chemistry, they are cited 30% less (Bresson 1996);
  - in natural sciences, Québec has a greater impact factor (dividing the number of citations for a scientist for a university and a country by the world average with an average index of 1), than other Canadian provinces: 2.92 and 2.4 respectively (Godin 1996).
  - publications in French in natural sciences, engineering and biomedical sciences dropped from 13% in 1980 to 4.9% in 1988 to 1.9% in 1993. More articles are published in French in biology (5.1%), mathematics (3.8%), engineering (3.4%) and the earth and space sciences (2.7%) (Godin 1996);
- In social sciences and humanities
  - publications in Québec are ten times more numerous than those in natural sciences (Godin 1996);

in 1993, 16.4% of publications in social and human sciences are in French;
- the disciplines that show the most publications in French are planning and development (41.7%), political sciences (39.1%), public health and social medicine (26.4%) (Godin 1996).

Table 11 summarises the data available for publications in French in Québec and compares with countries where French is an official language. It mainly serves to indicate how French is loosing ground, across disciplines and across countries.

Table 11: Percentages of scientific publications in French: Natural sciences and social sciences: 1993 and 1988 (Source: Godin 1996. The data was compiled from the *Science Citation Index* and from the *Social Science Citation Index*)

		Québec	Canada[20]	Belgium	Switzer-land	France	Total[21]
Natural sciences, engineer-ing and biomedical sciences	1993	1.87% (131)	0.07 (17)	2.25% (150)	1.38% (143)	13.28% (4 629)	1.11% (5 327)
	1988	4.9% (183)					
Social sciences and humanities	1993	16.40% (202)	1.28% (66)	8.14% (39)	4.90% (33)	35.79% (636	
	1988						

5.2. Conclusions

The statistics that I have just quoted indicate clearly that French is not gaining grounds as the language of science and technology in Québec, if we can judge from an indicator like publications. And today, with the paradigm of innovation, businesses and industries, as guided by economics, are having a larger than ever role to play. Things are unlikely to improve for French.

6. Hypotheses on language planning

As we have seen abundantly throughout this text, the language of science and technology is so tied to multiple factors that we cannot separate it from the context in which it is created and disseminated, the type of knowledge developed, the linguistic ideology (Figure 1), etc. How then can an activity like language planning be conducted?

Bourdieu (1945: 45) had a radical position in this matter:

> "Those who want to defend an endangered linguistic capital (...) are condemned to a total fight: one can only save the competence if one saves the market, i. e., the totality of the social and political conditions of production and of consumption." [translation A. M.]

In order to retain/save/develop a national language as language of science and technology, a state would then have to intervene in all of the conditions of creation/innovation, appropriation/adaptation and publication/commercialisation of knowledge and products of science and technology. This would seem to be an immense contract indeed. But one has to ask if such an investment is productive on a number of count : for the development and exchange of knowledge, for the just return on the investment, for the participation of a state in the global advancements, etc.

On the other hand, it cannot be denied that a certain amount of language policy planning should be done by states in order to give their citizens just and equitable conditions of development and to safeguard their part of the global treasures that constitutes their language. But, more fundamentally, linguistic policies in science and technologies should aim at preserving the unique view of the world that each people has developed over the centuries, the cultural dimension of doing science and technology. Otherwise, one view of science and technology, the Western World's, will prevail.

But, as we have seen, even linguistically alert states like Québec may not deem it important to specifically design policies on the language of science and technology even if French is loosing ground in this sector. But of course, Québec has acted on new grounds like computarization.

From these reflections, we can draw some hypotheses using the heuristic framework as an aid to categorise different aspects of the issue. We could say that:

- Science and technology as objects cannot be separated from the larger state context. states will implement policies on new developments to the extent that they serve their positioning. Today, it is less science (and fundamental research) than technology and innovation that attracts their interests.
- states can develop scientific and technological policies on cultural grounds, i. e. the example of Québec and its nationalistic orientation and its 1998 policy on cultural heritage and on the Internet.

- The closer the consumer object that results from science and technology, the more likely states will intervene in matters of language. For example, publication in specialised scientific journals vs. publications in general public magazines (dissemination).
- The closer the field of research or development to the general public, the more likely it is for language to become an issue.
- Conversely, the more the language of science or technology is specialised, the less it is used in daily life, the less the concern for publications in the national language (i. e., the difference between natural and social sciences).
- The more nationalistic the ideology, the more loyal to the community language, the more likely states will intervene with comprehensive measures.
- With the innovation paradigm gaining grounds, it will be increasingly difficult for states to impact of the language of science and technology.

7. Notes

1. Professor of sociolinguistics and languages teaching/learning at Télé-université, 1001 Sherbrooke est, Montréal, Québec, H2L 4L5, telephone 1 514 522 3540, fax 1 514 522 3608. E-mail: amartel@teluq.uquebec.ca
2. For Japan, see Coulmas (1996); for Germany, see Ammon (1996) and Coulmas (1996); for France, see Truchot (1996); for Denmark, see Phillipson and Skutnabb-Kangas (1996); for Spanish American countries, see Marin (1996); for Russian, see Gak (1996).
3. In this paper science is defined as a body of knowledge and experiences with recognised fields and methods. Technology, on the other hand, is the study and production of techniques, tools, and electronic components.
4. We only need to think of the industrial espionage network to be convinced of its importance in business and industry and of the importance that the U. S. National Security Agency has gained today, over the C. I. A.
5. State institutionalisation of science and technology is a new phase following interest in supporting research, first by industry during the Industrial Revolution followed by States shortly thereafter (Kaplan and Baldauf 1988: 243).
6. States were to follow industry in this matter.
7. For an overview of the question, see the thematic issue of *Manière de voir*, 38, march-april 1998 entitled: *Ravages de la technoscience*.
8. This field is grouped under the title of *Specific Purposes Language Planning* by Kaplan and Baldauf (1998).
9. Each State uses the terminology that describes it ideology in matters of democratisation of scientific and technological knowledge. In the United States, the term used is *scientific literacy*, in Great Britain, it is *public understanding of science*, in Canada, it is *public awareness of science*, in France *culture scientifique et technique de masse* and in Québec *culture scientifique publique*.

10. This notion of scientific culture refers at time the dissemination of scientific knowledge (and its hidden agenda), the use of technologies, and the link between technology and social issues (Davis and Duchesne 1986: 124).
11. Inspired by Ammon's proposal for the symposium held during the XIV World Congress of Sociology held in Montréal, July 26th to August 1st 1998.
12. Founded in 1923, l'ACFAS has been a major motor in the development of sciences and of a scientific community in Québec (Gingras 1994). Like many associations that originally served all francophones of Canada, l'ACFAS lived the Canada-Québec identity crisis but remained to serve both ' the Québec scientific community and that of other Canadian provinces.
13. In 1968, they received 11% of the total grants, 8$ million out 72,4$ millions.
14. These are: Centre québécois de la valorisation de la biomasse (CQVB), Centre de recherche informatique de Montréal (CRIM), Centre francophone de recherche en informatisation des organisations (CEFRIO), Institut de technologie du magnésium, Centre de recherche en miroélectionique, Centre de recherche en calcul appliqué, Institut de recherche en pharmacie industrielle (Godin and Trépanier 1995).
15. Ministère de l'Enseignement supérieur et de la science.
16. Conseil de la science et de la technologie.
17. Prominent among these are studies sponsored by the *Conseil de la langue française.* (Godin 1996; Drapeau 1991), others by the *Ministère de l'Enseignement supérieur et de la science* (Gingras and Médaille 1991). Furthermore, in 1996, on the occasion of an evaluation of Law 101, an inter-ministerial committee thought the issue was important enough to published a section in *Le Français langue commune* on the role of French in the globalisation of exchanges and in information.
18. Namely according to those listed in section 2.1.
19. Réseau international de néologie et de terminologie (RINT).
20. Excluding Québec.
21. The other countries included in Godin's survey of publications in French are Germany, United States, Great-Britain and Japan

8. References

Ammon, Ulrich
1996 L'allemand dans les sciences. In Conseil de la langue française du Québec, 209–220.

Ambroise, Antoine/Jacques, Jocelyn
1982 L'appareil administratif. In Bergeron, G. et al. (eds.) *L'État du Québec en devenir.* Québec.

Bourdieu, Pierre
1984 *Ce que parler veut dire. Questions de sociologie.* Paris: Les Éditions de Minuit.

Breeson, Douglas
1996 Le Québec sous la loupe de ISI. *Interface* 19 (1): 45.

Brooks, Harvey (ed.)
1971 *Science, Growth and Society. A New Perspective.* Paris: OECD.

Bush, Vannevar
1945 *Science: the Endless Frontier.* Washington, D. C.: U. S. Government Printing Office.
Conseil de la langue française du Québec (ed.)
1996 *Le français langue scientifique de demain. Actes du Colloque tenu à l'Université du Québec à Montréal,* 19 au 21 mars 1996. Québec.
Conseil de la langue française
1991 *La situation du français dans l'activité scientifique et technique.* Québec.
Conseil de la langue française
1991 *Le français dans l'activité scientifique et technique. Quatre études.* Québec.
Coulmas, Florian
1996 Les facteurs économiques et les langues de la communication scientifique: le japonais et l'allemand. In Conseil de la langue française du Québec, 69–81.
Cooper, Robert
1989 *Language Planning and Social Change.* Cambridge: Cambridge University Press.
Davis, Charles/Duchesne, Raymond
1986 De la culture scientifique à la maîtrise sociale des nouvelles technologies 1960–1985. *Questions de culture* 10: 123–150.
Gak, Vladimir
1996 La langue russe en science. In Conseil de la langue française du Québec, 221–226.
Gingras, Yves
1994 *Pour l'avancement des sciences. Histoire de l'ACFAS 1923–1993.* Montréal: Boréal, 268.
–/Médaille, C.
1991 *La langue de publication des chercheurs québécois en sciences naturelles, génie et sciences biomédicales.* Research done for the Ministère de l'Enseignement supérieur et de la Science. Québec.
–/Godin, Benoît
1998 Le Québec en quête d'une politique de la science. *La Recherche,* Supplément au No 309, 6–10.
Godin, Benoît
1995 *Endangered Species? Une nouvelle estimation de la part du français dans les communications scientifiques.* Study done for the Conseil de la langue française.
1996 a La recherche scientifique et technologique au Québec: du porteur d'eau à l'excellence. *Interface* 19: 1.
1996 b Parle, parle, jase, jase: l'utilisation du français dans les communications scientifiques. In Conseil de la langue française du Québec, 55–68.
–/Landry, R.
1995 *L'avenir de la collaboration scientifique au Québec: une analyse basée sur la convergence d'indicateurs.* Report presented to the FCAR, Québec.
–/Limoges, Camille
1995 *Les revues scientifiques québécoises: une évaluation du programme de soutien aux revues du Fonds FCAR.* Report presented to the Fonds FCAR, Québec.

–/Trépanier, Michel
1995 La politique scientifique et technologique québécoise: la mise en place d'un nouveau système national d'innovation. *Recherches sociographiques.* 36 (3): 445–477.

Gouvernement du Québec
1994 *Charte de la langue française*. Québec.
1996 *Le Français langue commune. Enjeu de la société québécoise.* Québec: Direction des communications et Ministère de la Culture et des Communications.
1997 *Compendium de l'activité scientifique et technologique au Québec*. Québec: Bureau de la Statistique du Québec.

Kaplan, Robert/Baldauf, Richard B., Jr.
1998 *Language Planning. From Practice to Theory.* Clevedon: Multilingual Matters.

Lapointe, Gérard
1998 *Le Conseil de la langue française, 1978–1998.* Québec: Conseil de la langue française.

Leclerc, Michel
1990 Politiques d'innovation et appropriation économique des connaissances. In Leclerc, M. (ed.) *Les enjeux économiques et politiques de l'innovation.* Québec: Les Presses de l'Université du Québec, 1–42.

Leclerc, Michel
1993 La coopération scientifique internationale: l'exemple du Canada. *La Recherche* 252: 350–357.

Limoges, Camille
1990 De la politique des sciences à la politique de l'innovation: l'État incertain. In Leclerc, M. (ed.) *Les enjeux économiques et politiques de l'innovation.* Québec: Les Presses de l'Université du Québec, 61–79.

Marin, Marcial Bonilla
1996 La lengua española en la ciencia. In Conseil de la langue française du Québec, 187–196.

Montgomery, Scott
1996 *The Scientific Voice*. London: Guilford Press.

Phillipson, Robert/Skutnabb-Kangas, Tove
1996 Sociopolitical Factors and Languages of Scientific Communication. In Conseil de la langue française du Québec, 93–110.

Salomon, Jean-Jacques
1990 Les politiques de la science et de la technologie: d'un paradigme à l'autre. In Leclerc, M. (ed.) *Les enjeux économiques et politiques de l'innovation.* Québec: Les Presses de l'Université du Québec, 45–59.

Schutz, T. Willam
1980 The Productivity of Research. The Politics and Economy of Research. *Minerva*, 18: 644–651.

Truchot, Claude
1996 La langue française en science. Un cas de figure: la situation linguistique des sciences en France. In Conseil de la langue française du Québec, 159–168.

English and the Language of Science: On the Unity of Language and the Plurality of Languages

Miquel Siguan

1. The necessity of a common language

In today's global village, information of all kinds is circulating ever faster, and now reaches every corner of the world. Scientific information is no exception. Having a common language to serve as a vehicle for scientific communication and production is therefore a necessity. This is the reason for the extraordinary diffusion of English in all areas of scientific activity; the role of English today is similar to that of Latin in the Middle Ages, a role which lasted until the beginning of the modern era.

The advantages of the diffusion of a common language are evident, but one cannot deny that it has limitations and comes up against resistance. Unlike Latin, English is the first language of the inhabitants of particular countries, and for others it is a second language in which they are unlikely to achieve the same degree of mastery as native speakers. As regards communicative ability, then, those learning English as a second language are placed in a situation of inferiority. Furthermore, English is the language of certain countries and states, and above all of the United States of America, a nation of political and economic power; in many places, then, the need to know and use English is seen as an imperialist imposition. (Truchot 1990) I will not analyze these points here; I will concentrate on a more central issue: to what extent are languages interchangeable as vehicles of scientific production and communication? Equally, to what extent is knowledge – and particularly scientific knowledge – conditioned by the language in which it is formulated?

At the beginning of the twentieth century an intellectual group known as the Vienna Circle formulated an approach to science that invalidates this discussion. Culminating a positivist tradition in the understanding of scientific knowledge they considered that the only statements that can be termed scientific are those whose content can be formally demonstrated, as is the case of mathematics, or defined operatively in experimental form. The various scientific disciplines, then, only qualify as sciences insofar as they meet these conditions. Considered in this way, the meanings of the scientific propositions are wholly independent of the language in which they are expressed, and therefore translatable.

The ideas of the Vienna Circle were taken up by analytical philosophy, so popular in the Anglo-Saxon countries, and have had a profound influence on the thought of many scientists. But the fact is that in many disciplines considered as scientific, neither the concepts used can be defined operatively, nor can their propositions be demonstrated experimentally. Equally, all scientific propositions must be formulated in a particular language, for example English – not "the language of the mind", but a specific language with historical and cultural implications which it is impossible to ignore. So the question is entirely justified.

2. Language and rationality

Probably, the most general and most important thing we can say about language is that it performs a dual function in human existence. On the one hand, language is the instrument of knowledge; all our knowledge has a verbal form and a science can be reduced to a system of propositions. But at the same time language is the means of human communication *par excellence*. The two functions are closely linked, but when we refer to language as an instrument of thought we consider it as something common to all humans, closely linked to rationality. In contrast, when we refer to language as a means of communication, it is immediately clear that only those who speak the same language can communicate with each other. The problem, then, lies in clarifying the relation between the two forms of language, between language in general, the "language of the mind" if we like, and the particular languages in which humans communicate.

In the intellectual tradition of Europe, the beginning of the reflection on language dates back to Ancient Greece, and if one thing characterizes classical Greek thought it is the close relation it establishes between language and rationality. Man is defined as a rational animal, and at the same time an animal that speaks; speech and the ability to reason are thus intimately linked. The laws of logic are at one and the same time the conditions of knowledge and the rules of discourse. Rationalist philosophy has maintained this conception to this day, and it is shared by many linguists, including Chomsky. But this conception has a weak point; all men share the same capacity for speech, but they speak different, mutually incomprehensible languages. How did the Greek thinkers account for this problem?

In fact, they never addressed it. If they had, they would probably have admitted that indeed the barbarians spoke different languages but that these languages were merely convenient means for understanding each other rather than vehicles for expressing developed, rational thought.

Nor did the Romans concern themselves with the languages of the countries they conquered. There was one exception: the language of the Greeks, in which they recognized a superior culture. To be a man of culture in Rome a knowledge of Greek was a prerequisite. But they found a simple solution to the possible conflict of languages and cultures. Just as they assimilated the Greek gods into their own pantheon by giving them Roman names, they accepted Greek culture, assuming that there was one single culture expressed in two languages. It is this conception of the unity of culture – in whatever language it is expressed – that the Middle Ages inherited, albeit in a Christianized form. Renaissance humanism reaffirmed this conception of the unity of culture. If culture was one, whatever language used to express it, then translation is always possible (though potentially difficult) and bilingualism should not present particular problems.

3. Languages and cultures

As the modern era has progressed, the limits of the known world have been pushed back both in space and in time. Europeans discovered that other peoples, in the distant past, or in far-off lands, had constructed refined cultures. This was the background to the emergence of the historicist mentality of the nineteenth century. In his *Discourse on Universal History*, Herder presented the history of humanity as a succession of societies each with its own cultural identity and, among other things, with its own language. Humboldt and Herder both insisted on this link between people, language and culture: a people (*Volk*) is constituted by a set of speakers of the same language with common biological and historical roots and a common spirit (*Geist*) which finds expression in a culture; this culture in turn is manifested in a particular way of conceiving the world and organizing society, in myths, beliefs, customs, and particular artistic creations. Language is thus the first manifestation of the spirit and the culture of a historical community.

The most immediate consequence of this way of understanding language was the emergence of a linguistic nationalism based on the unity of language, culture and nation which called for the political unification of certain countries that spoke the same language and supported nationalist demands inside multilingual states or empires. As a consequence of these demands languages whose use had until then been purely oral were now considered as languages of *culture* – vehicles for literary and even scientific activities.

But what interests us here are not the political consequences of this theory of the singularity of each language but its implications for translation and for cognitive activity itself. If each language expresses a distinct conception of the world, translation will not be merely difficult but actually impossible, and the objectivity of scientific knowledge – which of necessity is expressed verbally and in a particular language – will be challenged.

Before considering this alternative I should stress that what I have called the theory of linguistic singularity is not restricted to the school of German romanticism. It has met with wide acceptance elsewhere. In the twentieth century it was taken up by linguists such as Sapir and by cultural anthropologists such as Whorf (1956) who, studying primitive cultures – Native American tribes – also concluded in favour of a close correlation between culture and language. Even more significantly, in the writings of most researchers in the science of language today statements of this order are frequent and accepted as self-evident. One example should suffice: in his much-read treatise on general linguistics, the well-known linguist Martinet (who died recently) wrote: "each language responds to a particular organization of the elements of experience. Learning another language does not consist only in putting new labels on known objects, but in becoming accustomed to analyzing the object of linguistic communication in a different way" (Martinet 1961).

In an extreme version of this theory, a culture forms a closed system in which each element is only explained in relation to all the others; analogously, the language in which this culture is expressed also constitutes a closed system in which, as in a dictionary, each word is explained by means of other words in the same language. Insofar as this is true, each culture and its corresponding language constitute a closed, incommunicable universe.

4. The limits of the identification between languages and cultures

It is not difficult to show that this unilateral statement is exaggerated. Whorf (1956) states that the nine verb forms in the language of the Hopi Indians correspond to the way in which the tribe thinks and understands reality. In other well-known examples, he shows the close correspondence between the vocabulary of kinship relations in this language and the organization of the family in the tribes that speak it. In the case of the language of the Eskimos, he contrasts the wealth of their vocabulary for expressing shades of white, with the poverty of their vocabulary for other colours. These examples are not designed to show that the Eskimos are blind to other colours because they lack the words to name them, nor that other languages are incapable of expressing the same relations that the Hopi verbs express. They merely show the close relation between the language and the culture of the people that speaks it. Therefore, what we have to decide is whether the cultures are singular and non-transferable or have common elements that allow some form of mutual comprehension.

To begin with a simple observation: the cultures implied in each language do not constitute closed universes. A Frenchman who knows no German travelling in Germany understands nothing of the conversations he hears, or of information presented in written form. But many of the cultural products and much of the way of life of the Germans are familiar to him, and easily comprehensible. He can admire cathedrals or abstract paintings in a museum, he can drive without any difficulty along the motorway, use a public telephone, or go to the supermarket. These elements – and many more – form part of the contemporary culture of both France and Germany, even though they have different denominations in the two countries. In fact, they are manifestations of a common European culture, forged over the centuries. Today it has spread to other parts of the world and constitutes what we know as Western culture.

Along with hermeneuticists like Gadamer, we can accept that all our knowledge is verbal and that the meanings refer to a fabric of significations which constitute our world, but our world which makes sense of our meanings is not limited to the words in the language we use; it includes languages that share a common history. Evidently, the languages of Europe have had, and continue to have, a common history, made up by mutual influences and shared meanings.

This explains why a text on atomic physics can be translated easily from English to Italian, or viceversa, or to be more exact, this explains why a problem in atomic physics can be thought of in the same way in Italian as in English. The same can be said of a legal text, or a text on economics. In Italy, as in other western countries, there is a long tradition of university research and teaching in these areas, using Italian as the vehicle.

But what of other languages such as Tagalog or Inuki, the language of the Eskimos? Is it possible to translate a treatise on atomic physics or legal science into Tagalog or Inuki? In the case of Tagalog the difficulties are clear. It is an indigenous language, spoken traditionally by rural populations; only recently has it been standardized and introduced in teaching. Nonetheless, since the independence of the Philippines, it has become the official language of the country and is now used in government and university teaching. So great efforts are being made to adapt the language to the teaching of atomic physics, law and economics, and to promote local research in these areas. If these efforts are successful, Tagalog will admit scientific translation, and will also serve as an instrument for scientific research. But it is also true that this can only occur within the framework of certain sociocultural changes that somehow integrate Philippine society inside western patterns and dissolve the autochthonous Philippine culture into western culture, or into a synthesis of the two.

The case of Inuki is very different. Like many languages of the world, it is the means of expression of an exclusively oral culture. Oral cultures have extraordinarily rich manifestations of their own but are alien to the concepts on which scientific knowledge is based. This does not mean that the mind of an Eskimo is not able to grasp these concepts; it is easy to imagine an Eskimo child growing up in Ottawa or New York and becoming a leading scientist, and it is possible to imagine Eskimo society undergoing transformations in the future which will bring it closer to western culture; but at the moment it is only a remote possibility. So today it is impossible to translate a manual of atomic physics or a book on political philosophy into Inuki.

5. Language and different intellectual productions

Inside the same cultural environment intellectual products vary widely, due to the role that language plays in them. Even inside what we know as science these differences can be seen.

The axioms and demonstrations that make up Euclid's "Elements" were written twenty-five centuries ago in Greek, and the writers took words and turns of phrase from the language they shared with their fellow citizens. But when they used a word they defined its meaning precisely by referring to certain operations that were not verbal. The set of objects thus defined constituted a closed system, and so did the terminology in which each new term was defined in relation to previously defined terms. It can be said without exaggeration that the "Elements" are not written in Greek but in a special language, the language of geometry. Understood in this way, it is clear that they can be translated into another language by maintaining the same operative definitions.

Since all sciences aspire to be totally rigorous in their vocabulary, geometry could be taken as a representative example. But although neo-Positivism and analytical philosophy have proposed highly rigorous parameters for the qualification of knowledge as scientific (as I mentioned at the beginning), the fact is that sciences differ widely in the form in which they operationalize the definitions of the terms they use.

For a long time, the natural sciences limited themselves to describing as accurately as possible the facts they observed and to label the elements or species they described. This is the case of anatomy and botany. At this level the vocabulary used can be unambiguous and therefore easily translated. In the modern era, the natural sciences have consolidated themselves as sciences by embracing experimentation, since experimentation and the use of mathematical measurement ensure that the concepts applied will also have an unambiguous meaning and can thus be translated from one language to another, even though the level of formalization may not be as high as in mathematical sciences. What is true of pure sciences such as physics is even more true of the applied sciences such as industrial chemistry or computer science, which often define their concepts in terms of their production processes or their uses. In contrast, treatises of theoretical physics do not have an operational definition even when they strive to clarify their basic concepts precisely; they contain implicit cultural references and a particular conception of the world and reality.

The same is true of the social sciences. At one point, attempts were made to establish a radical difference between natural and social sciences: the aim of the "sciences of nature" was held to be to explain the objects they were concerned with, while the aim of the "sciences of the spirit" was to *understand* the objects under study. But most social scientists renounced this distinction some time ago and subscribe to the ultimate unity of science and its methodology. Some, such as linguists and to an extent economists, have devoted their efforts to the formalization of their respective disciplines following the models of formal logic or mathematics. Others such as psychologists and sociologists have insisted on the *empirical* character of their sciences, and have developed a methodology in accordance with this perception of their sphere of activity.

Perhaps sociology presents the best example of this new orientation. As it has progressed in this direction, the dependence of sociology texts on conceptions close to philosophy has decreased. Over the last century, and especially over the last fifty years, as sociology expanded and became specialized in subdivisions, the influence of national schools and traditions in the production of sociologists decreased, making way for a conception of sociological knowledge that is widely accepted and originated in the Anglo-Saxon countries. Much the same can be said of psychology. If fifty years ago there was a wide variety of theorioretical proposals today are limited to the transition from behaviourism to cognitivism; here, too, the predominance of US psychology has been decisive.

In spite of all these changes in psychology, sociology and in the sciences of man and society in general, the concepts used can neither be formalized as in mathematics nor verified by rigorous experimentation as in the natural sciences. To a large extent, they are based on a particular cultural tradition, which may have a broad base but is often expressed in the ambit of a particular language. For example, it is easy to show the influence of the German cultural tradition and the German language in the constitution and development of classical German sociology, from Weber to Habermas. It is even easier to show this dependence in sciences related to law and government, in which the dependence on different cultural traditions is even reflected in terminology which is difficult to transfer from one language to another.

At the opposite extreme of the spectrum to mathematics and experimental science is philosophy. The language in which it is expressed is totally charged with meanings dependent on a particular cultural horizon, dependent in fact on a particular language. Translating a philosophical text by Hegel from German – not into Tagalog or Inuki, but into another western language – is extremely frustrating. How does one translate *Geist*? Hegel's terminology forms a relatively closed system in which each term has a specific meaning depending on the place it occupies in the system (and indeed loses this meaning when considered in isolation); but the terms themselves that Hegel uses are charged with meanings that are products of decades or even centuries of philosophical speculation in German. Although Spanish is a European language, the terms available to a Spanish translator of Hegel have a different intellectual tradition and so the translation will require paraphrase and explanation. Even then, to a reader who has a thorough knowledge of the two languages the translation will not be perfect. In the case of translation into English the difficulties are even greater, because the English philosophical tradition is even further removed from the German tradition than the Spanish is. And although English thinkers (unlike the Germans) have been at pains to use a language close to that used in everyday life, the translation of Locke or Hume nonetheless poses similar problems because the technical terms they use are charged with cultural and personal connotations, and because the language into which the English is being translated also has its own connotations which will influence the meaning of the target text.

6. English and the language of science

So far I have spoken of languages in general and of the relation between scientific knowledge and the languages of the European tradition. But one of these languages has become the language of international communication and the language of scientific production and communication in particular. What are the repercussions for scientific knowledge itself?

At the start of this article I recalled the distinction between language as an instrument of intellectual activity and the various languages which humans use to communicate. The contrast between these two levels has often led to a pursuit of the perfect language that reflects exactly the functioning of pure reason. Ramon Llull in the thirteenth century and Leibniz in the seventeenth are two representatives of this utopia. This utopia can be swiftly dismissed today, but recalling it draws attention to the fact that English, like any other language, is not perfect or strictly rational; its meanings are inscribed in a particular culture. More precisely, it is inscribed in the western cultural tradition, the tradition which saw the emergence of the search for strictly rational, and thus universally valid, knowledge. English is well suited as a vehicle to transmit this knowledge, as are the other languages that share this tradition. The languages joining this tradition today will soon be in a position to do so.

What I have just said accounts for the role that English currently plays in the production and diffusion of scientific thought, and also accounts for the resistance that this situation provokes. Speakers of other European languages which contributed to the construction of this scientific tradition consider that their languages are equally worthy vehicles for scientific communication, and may feel marginalized or disparaged. Speakers of languages that do not form part of this tradition may not only feel belittled but may also believe that the acceptance of English implies the acceptance of a culture that may enter into conflict with their own culture, and which will eventually deform or destroy it.

All this refers to the language of scientific production and dissemination in general but, as we have seen, widely differing forms of knowledge can be termed rational or scientific. The compatibility between scientific reasoning and English or any other western language is evident in the case of the sciences that respond fully to the ideal of the scientific method, such as mathematics or the experimental sciences. But there are numerous areas of knowledge that we term scientific because they aspire to a full rationality but do not share this model; in these areas the concepts they deal with depend to some extent on specific cultural and linguistic traditions.

In all these cases the use of English as the language of scientific production, or the requirement that anything produced in another language must be translatable into English, may put a brake on scientific creation, and may eventually impoverish it. The example of German sociology is a case in point. Abandoning this tradition, or maintaining it solely as far as its concepts can be translated into English, obliges a rethinking of the tradition itself from a different interpretative perspective.

We are back where we started. In our global village we need a common language of communication, and history has bestowed this role on English. But the generalization of the use of a language as a means of scientific communication leads to the generalization of its use as a means of scientific production. This is where the problems arise. English is not a perfect language, an exclusively rational instrument, but a language inscribed within a cultural tradition. It is true that there are scientific areas in which the aim of objectivity means that the language in which they are expressed and the cultural tradition in which they emerge are of no consequence, but there are others in which this is not the case; they retain a certain dependence on their cultural or linguistic tradition. In these areas, the subordination of other languages to English may signify limitation and impoverishment. Evidently globalization has many advantages, but it comes at a price; to what extent are we willing to pay it?

7. References

Chomsky, Noam
1957 *Syntactic Structures.* The Hague: Mouton.

Fodor, J.
1975 *The Language of Thought.* New York: T. Crowell.

Henley, Peter
1972 *Language, Thought and Culture.* An Arbor: The America of Chicago Press.

Humboldt, Wilhelm von
1827 *Über die Verschiedenheiten des menschlichen Sprachbaues und ihren Einfluss auf die geistige Entwicklung des Menschengeschlechts.*

Martinet, André
1961 *Elements de linguistique generale.* Paris: Colin.

Steiner, Georges
1975 *After Babel. Aspects of Language and Translation.* Oxford University Press.

Truchot, Claude
1990 *L'anglais dans le monde contemporain.* Paris: Le Robert.

Whorf, Benjamin L.
1956 *Languages, Thought and Reality.* Cambridge, MA: MIT Press.

English in the Social Sciences

Abram de Swaan

1. Introduction

Let us first praise English, not for its intrinsic virtues, but for the simple fact that it has become the true world language of science, technology, media and business. There now is a language that allows to reach each and everyone who counts in the field of the natural sciences, advanced technology, mass entertainment and corporate affairs. Of course, it would have been nice if this global language had been Latin, or Chinese, or French, or for that matter Dutch. But it is not. It happens to be English.

None of this has anything to do with the immanent qualities of English. Linguists – for lack of any adequate criterion – consider all languages equally worthy by default.

Some sociolinguists believe that the present hegemony of English is the result of a determined conspiracy by the British Council or a host of US foundations and agencies to goad people especially in the Third World into learning English, even if it might go squarely against their wishes or their interests.[1] Of course, English language publishers, language institutes and other peddlers of English have made a strenuous effort to sell their wares.[2] So have their peers who offer French courses and books in the former French possessions[3] or German publishers and language institutes that operated in East and Central European markets (Chaudenson 1991; Djité 1990; État de la francophonie (…) 1989; Ammon 1998).

Students take English courses and study English textbooks because they (and their parents) believe it will profit them, in an intellectual sense, but also, and even more so, in the job market. In the overwhelming majority of cases, they are right. In general, fluency in the standard version of the first foreign language is a key to the higher echelons of employment in any country. Nowadays, only computer courses may be an equally profitable learning investment. So, in most countries in the periphery, the semi-periphery and the core of the World system, parents insist that their children learn English, and on this one issue the children agree with their parents. Nevertheless, people in the (semi-) periphery, even when they have mastered English or another world language, still participate in transnational communication on unequal terms: they are not nearly as proficient as most native speakers of English or the other 'supercentral' languages. Moreover, usually they have been socialized in a variety and an accent that betray their marginal status. In their own society, on the other hand, educated, literate and competent in the formal, written standard of the dominant foreign language as they are, they count among the privileged few in comparison with the vast majority who lack these qualifications.[4] Their linguistic competence serves to underline, to demarcate the prevailing class distinctions and helps to perpetuate them.

In the more developed countries of the world, linguistic competence equally serves as a class marker, as Basil Bernstein (1975) and Pierre Bourdieu (1982; 1984) among many others have argued. Fluency in one or more foreign languages adds to the speakers' linguistic and cultural capital and hence increases their prestige. But in those countries, elementary schooling is well-nigh universal, and as a result, so is literacy. In principle, this allows anyone who so desires to learn their own or a foreign language in the standard version. On the contrary, the illiterates in the (semi-)peripheral societies, numerous as they are, find themselves effectively excluded from that option. As a result, the educated elites in these (semi-)peripheral societies enjoy a group monopoly on the standard language, allowing them to reserve the better jobs for their own kind. It also means that the standard language in these countries can not be considered as a collective good with respect to the population in its entirety, since the necessary condition of non-excludability does not apply in these cases (de Swaan 1998 a).

However, if people have access to education and have the opportunity to acquire literacy, they are in a position to master the prevailing language for transnational communication, and it usually is in their interest to do so. Yet, the aggregate individual choices of so many foreign language students may in the long run erode the vitality of the indigenous language and devaluate the collective cultural capital embedded in that shared language. After all, by the time almost everyone uses the foreign language alongside with the indigenous language, the latter no longer affords communication opportunities which the former can not also provide. People may begin to neglect their original language, even fail to pass it on to their children. As a result, the texts that were recorded or memorized in the indigenous language become unintelligible and inaccessible. The aggregate cultural capital embedded in the abandoned language is lost for future generations (de Swaan 1998 b). Even if the indigenous language is not discarded completely, it may still be spurned in settings of 'high' or 'modern' communication and no longer considered worthwhile to develop for such usage. Apparently, the sizable advantages of English as a world language go together with major disadvantages for the indigenous language as a means of wider and advanced communication.

2. English as the language of the sciences

When it comes to English as a language of science, its advantages for global communication are even more striking than in other domains. Practically the entire scholarly community in the natural sciences reads English, and the vast majority publishes articles in that language. This phenomenon, however, is completely independent of the intrinsic qualities of English, and results from the economic, political and also scientific predominance of the British empire during the nineteenth and the first half of the twentieth century, and since the First World War increasingly the USA. Surprisingly, the technological and scientific vocabulary of English (and the other 'modern' European languages) consists in large part of loan-words from Latin and Greek, with a perverse predilection for bastard formations from both at once: *sociology*, *television* or *automobile*. It is a sign of the popularity of personal computers and the internet that the accompanying neologisms are taken from everyday English (e. g. *file*, *net*, *save*, *site*). In order to demystify these new information technologies it might be useful to adopt equally mundane, indigenous equivalents in other languages.

But, or so the argument runs, if foreign terms are introduced wholesale, people may learn the words without ever understanding their – deeper – meaning. In fact, even in countries with a high level of education, the vast majority has only a practical grasp of these terms, and no more than the faintest idea of whatever else notions such as 'gravity' or 'significance' may denote. Seeking indigenous alternatives for such concepts in no way improves understanding.[5] Nor does the adoption of large quantities of loan-words harm the language in any way.[6] In other words, the vocabulary may change, but the morphology – grammar, syntax and pronunciation – remain mostly unaffected.[7] In everyday speech, the strong jaws of morphology chew the loan-words with surprising ease into well-formed sentences.[8]

Hence, nations with few resources would do better not to waste their time on this aspect of corpus planning or on the development of indigenous terminology, and simply adopt the international terminology constructed from the remnants of long dead languages.[9] The one deficit of the less equipped languages that is mentioned most often – their lack of a science and technology vocabulary of their own – may not be a handicap at all.

In teaching the natural sciences, English as the language of instruction carries with it the risk of widening the cultural rift between the rational world of scientific method and technological procedure on the one side and on the other side everyday life with its common sense notions as well as its magical ideas. A divide that may be even wider in developing countries, but certainly has not disappeared in the West. Scientists and engineers may find it harder to explain concepts and provide instructions to collaborators who have not learned English, thus perpetuating their exclusion from the realm of science and technology, relegating them to practical, routine tasks. This risk is much less in those countries where well-nigh everybody has learned some English in school and the language no longer holds much magic for most. But in many developing countries, fluency in formal English (or French for that matter) is attainable only through extended education and provides exclusive access to the higher echelons of the labor market to a tiny elite, while the vast majority remains excluded.

In such countries, a 'second class English' has emerged and spread: a pidgin (or creole), based more often than not, on an English vocabulary, and serving to link speakers with different mother tongues. But these pidgins, too, function within a linguistic class system and constitute a 'sociolect' that is distinctly 'lower' than standard English (and the same goes for French). Nevertheless, these languages may be quite suitable for teaching the sciences, using loan-words from English (that is neologisms from Latin and Greek) when technical terms are required. Teaching 'high knowledge' in low languages would help to erode class barriers and that is why it will not happen soon.

English has come to serve a number of linking functions, in fact almost all of them and almost everywhere. This is not the result of a grand scheme, although some seem to think so. Nor did it take much persuasion, let alone coercion, to bring people all over the world to learn and use English. The advantages are too obvious. If English is indeed forced upon the world, it is absorbed more avidly than it could ever be imposed.[10]

3. English as the language of the social sciences

Things, however, are much more problematical when it comes to English as the language of the social sciences and the humanities. First of all, these disciplines are much more strongly bound to language. The exactitude that prevails in the natural sciences through the use of formal and quantitative terms and the availability of precise measurements must be achieved in the social sciences and the humanities through meticulous precision in the use of the natural language. In these fields, technical terms are often very close to terms in everyday usage (e. g. role, class, civilization) and it is the continual rubbing together of these different spheres of meaning that conveys to social science writing at once its ambiguity and its richness of meaning.

To give an example: In the early eighties, I set out to write a book on the comparative historical sociology of poor relief, education and health care. In Dutch its working title was, *verstatelijking van de verzorging*, which can be rendered quite adequately in German as *Verstaatlichung der Fürsorge*. Even in French an equivalent title was available: *l'Étatisation du soin*. The latter term, *soin*, is already somewhat precarious in this context, but *étatisation* and even *désétatisation* belong to the accepted vocabulary of political discourse. However, when it comes to translating the title into English, neither *verstatelijking* nor *verzorging* can be rendered satisfactorily. There is no such word as *statalization* or *statification* in English (although the analogous formation *stratification* is accepted sociological usage). The Oxford English Dictionary (Compact Edition) does indeed mention a nineteenth century use of the term *statal* in the U. S. for 'pertaining to the states' as opposed to *national*, but it is obsolete. At present, after a millennium of English, a billion speakers apparently can do perfectly well without a coinage from the root *state* and without an expression for the penetration of society by the state apparatus.

As a matter of fact, German and Dutch, and the Romance languages also, can more easily convey an increase, a process: e. g. *Verhöflichung der Krieger*, as in Norbert Elias's original German version of *Über den*

Prozess der Zivilisation, is awkwardly translated by *courtisation* (Elias 1983: 214).

This is not to say that the much criticized unhistorical and static nature of much of contemporary American and English social science is due to this scarcity of process words, after all American historians excel in their trade, but nevertheless, English lends itself more readily to a 'hodie-centric' perspective on society.

In the natural sciences, most of what can be said in English can also be phrased in mathematics and in formal schemes. But what the social sciences have to say about the social world can only rarely be rendered in mathematical symbols or in diagrams. As a matter of fact, even the translation from one natural language to another presents many difficulties in the social sciences. These translation problems arise at different levels. To begin with, the language of informants and respondents must be transposed into the linguistic register of scholarly articles and professorial courses. Next, the text often is to be translated into an international language of academic discourse, usually English. This requires that the terms for the traditions, mentalities, institutions of a particular society under study be rendered by some English equivalent. How to translate *vrijgemaakt Gereformeerd* (a 'liberated' Reformed church), or what to make of *verzuiling* (which finally became famous enough to be internationalized as *pillarization* in the writings of Lijphart(1968)? How to render *afwerkplek* (a sheltered area, provided by the autorities, where prostitutes provide their services)?

Indeed, much is lost in translation. But without translation nothing is gained at all. One must not be too sentimental about linguistic obstacles. It is a sobering and productive exercise to explain the peculiarities of one society in terms of another. All comparative social science presupposes the intelligibility of patterns from one culture to scholars living in another. Without comparative social science, and thus without social science translation, regional and national exceptionalism will take the place of social theory.

The social sciences study human beings in the social arrangements they constitute with one another. Since their interaction proceeds mainly through linguistic exchange, the vast majority of social science observations concerns human utterances, mostly spoken, sometimes written. This immediately confronts social scientists with the problems of understanding and interpreting statements, of transposing spoken language into its nearest, written equivalent, and of condensing and converting the material into the editorial format and the linguistic register appropriate for professional, academic publications. Social scientists therefore must deal from the outset with language more intensely and more intimately than natural scientists, whose objects of inquiry do not speak.

When moreover, as happens more and more in countries where English is not the first language, the findings are next published in English, an additional layer of linguistic transformation is introduced, with all the problems that go with it.

The social sciences, the human sciences in general, are so closely and intensely tied to language because human beings are their subjects. But for the social sciences (and quite often the humanities, too) language problems manifest themselves in still another respect, as a central preoccupation. In contrast with most natural sciences, the social sciences are not experimental. Only small-scale and short-lasting human interactions can be reproduced in the laboratory, larger and more protracted events are very difficult to stage under controlled conditions. Since the social sciences can not be experimental, they must be comparative and historical. And here, problems of translation manifest themselves once again and to their full extent in the heart of the social science enterprise. In order to elucidate the origins and the working of social phenomena, their development over time must be traced and the differences and similarities with comparable events elsewhere must be assessed. Comparison (both over time and across distances) implies translation, if not between distinct languages, then at the very least between registers, dialects and varieties. Thus not only in collecting and processing the empirical material, but also in situating it in the wider spatial and temporal context, problems of language are central to the task of the social sciences.

This analysis has its consequences for the teaching of the social sciences and even more so for the professional training of social scientists. In the first place, it implies that students must be sensitized to problems of interpretation and translation, as essential constituents of the crafts of their trade: observation and comparison. Secondly, it implies that students can not effectively be taught the social sciences in one language only, whether it be their mother tongue or a foreign, world language, e. g. English. If they are to become adequate observers and interpreters of human interaction in its context, they must be intimately familiar with the language used in each particular setting. This may be some variety of their own mother tongue or an entirely different language (as in much anthropological research). If, moreover, they are to be competent analysts and theorists, they must be capable of situating their own empirical material in the wider temporal and spatial framework of relevant social science findings and that task demands a broad grasp of a language in which all or almost all of these findings are published, i. e. English.

Thus the quasi totality of social science knowledge is contained in the English-language corpus of professional literature. A considerable part is also embedded in French, or German, Spanish or Italian texts, but the literature in these other languages is not nearly as complete as that in English. Hence, English has become the medium of choice for all transcultural and transtemporal comparison and classification in the social sciences. This facilitates comparative analyses that in principle can cover the entirety of human experience and history. It provides a universal corpus and standard of comparison. English as the truly global language presents the social sciences with the opportunity of developing and testing universal models and theories of human societies.

At the same time, this unique and novel possibility of universality entails a risk: in stead of yielding the greatest possible variety of observed human interaction, as recorded in the global language of social scientists all over the world, the English language literature may single out and impose the experience of the English speaking societies, of the United States in the first place, as the standard of human interactions and the model of social institutions: the American experience presented as universal human destiny.

This is a very real risk. Thus, almost without exception, economists nowadays publish in English. But this predominance of English is coupled with a predominance of American economic models and standards across the globe that does no justice to the variety of economies in the world and has led in many instances to policy recommendations that were *malapropos* at the very least, if not downright pernicious in some cases. Similar assessments can be made for other social science disciplines, albeit not with equally dramatic consequences. In other words, English may now be the universal medium of social science, it certainly is not a neutral medium – on the contrary, it favors American ideas, and American authors.

The hegemony of English in the social sciences (and in other fields) has yet another consequence, that is mostly ignored. Academics are required to publish regularly in 'international' and 'refereed' journals. In actual fact, these are almost without exception American and British periodicals. As a consequence, American and British editors and referees judge contributions from scholars all over the world and in so doing – without ever having intended to – exert a major impact on the selection and promotion of academics in other countries who depend on these publications for their career advancement.

Should the scholarly community therefore abandon English and renounce the use of a single, global lingua franca? The point is moot, since there is no global social science community that could adopt and enforce such a decision. Individual scholars would see a personal advantage in aiming for the widest possible audience and might well continue to publish in English, just as they do today. Clearly, the advantages of English as a world language of social science are much too precious, both for individual scholars and for the social science community in its entirety, to relinquish. There now is a coherent worldwide forum where the validity of social science claims to knowledge can be judged. The question is how the bias of this medium in favor of Anglo-Saxon content can be mitigated.

'Il faut désangliciser l'anglais' Pierre Bourdieu once exclaimed:[11] English should be released from the English. For Salman Rushdie, speaking from the Indian experience, this is already an accomplished fact: "the English language ceased to be the sole possession of the English some time ago." (Crystal 1997: 130) English should be taught in a critical vein, emphasizing that the adoption of English need not at all entail the acceptance of American or British conceptions and practices. On the contrary: "the spread of English, if dealt with critically, may offer chances for cultural renewal and exchange around the world." (Pennycook 1994: 325) And indeed, English has proven itself time and again to be a most suitable medium for the critique of Anglo-American institutions and ideas.

But in order to better dissociate the English medium from its bias towards American content, more than mere good intentions are required. International journals should appoint editors and referees that are accomplished but not native speakers of English, so that they may bring their cultural background to bear in their selection and evaluation of contributions. As a matter of fact, the European Union, which tacitly has adopted English as its lingua franca, should set up language academies where all European languages are studied and English is taught by non-native speakers. Gradually, an educated, cosmopolitan variety of English is emerging, sometimes called 'Mid-Atlantic'. One might envisage a kindred variety, Euro-English, supported by the European Union and as autonomous in its development as contemporary Indian English is (Kachru 1992: 355–8). De-anglicizing English of course does not mean taking English away from the English, rather it should mean sharing English between native and non-native speakers on more equal terms.

English is a most suitable and by now indispensable vehicular language for the social sciences and advanced students must learn to use it as the global medium of their discipline. The worldwide availability of English provides the social science community with a single, universal forum. This is a great and precious prize indeed, and it must be used with much circumspection.

4. Notes

1. This view is usually associated with Phillipson (1990). However, the definitions of his main concepts are more circumspect. He defines "English linguistic imperialism" as: "the dominance of English is asserted and maintained by the establishment and continuous reconstitution of *structural and cultural inequalities between English and other languages.*" (p. 41, italics in the original). This, the author continues, is a special case of 'LINGUICISM, which is defined as "ideologies, structures and practices which are used to legitimate, effectuate and reproduce an unequal division of power and resources (both material and immaterial) between groups which are defined on the basis of language".' (*Ibidem*). The problem with these statements is not the contention that between English speakers and other language groups unequal relations of power and prestige prevail, but the notion that this is a consequence of linguistic practices and relations *per se*, and that it is the outcome of concerted and purposive action. In fact, it is a concomitant of unequal political and economic relations. It may be somewhat intensified at most by organized interventions. Although the definition quoted strongly suggest that the outcome is undesirable, in fact, its consequences are manifold, mixed and diverse for different categories of people. Finally, the approach ignores the demand side of the language market: the masses of avid students of English throughout the world.
2. Cf. Pennycook (1994: 155), for a survey of private expenditure on English language learning: 6.25 billion pound in 1988, half of which was spent in Britain and North America, the other half in Europe, Australasia, and the Far East.
3. Chaudenson 1991; Djité 1990, État de la Francophonie dans le monde 1989.
4. Pierre Manessy (1984: 201–2) quotes a *cri de coeur* from J.-P. Makouta Mboukou, in his case with respect to French in former colonial societies: 'L'unique solution humaine est de "vulgariser", de "populariser", c'est à dire de répandre l'art de parler, de lire et d'écrire le français, c'est par-ce-que celui-ci n'est encore l'apanage que de quelques bienheureux. Mais le jour où la population sachant lire et écrire le français passera de 10% B 90%, le courant sera inversé. Et qui sait si nous n'assisterons pas à une métamorphose désagréable 'd'où les coqs majestueux d'aujourd'hui sortiront roitelets malingres et transis.'
5. It so happens that the Netherlands is one of the rare countries where a scientific and philosophical vocabulary was introduced – in the sixteenth century! – constructed entirely from indigenous elements: e. g. *wijsbegeerte* for 'philosophy' (*filosofie*) or *aardrijkskunde* for 'geography' (*geografie*). The Dutch terms are still in use, more frequently in the lower rungs of the school system, whereas the 'classi-

cal' constructs are more current in higher education. Even though the Dutch terms have been completely incorporated into common usage, on the face of it there appears to be no evidence whatsoever that the understanding of Dutch pupils and students is any more profound than it would have been without the Dutch indigenisms.

6. This is a favorite fear of language lovers, who tend to be stronger on vigilance than on understanding. Cf. the discussion between Étiemble and Hagège (cf. Flaitz 1988: 106–107; Étiemble 1973; Hagège 1986).
7. Compare, however, Gopal (1966: 202): 'In commending the use of international terminology, the Language Commission has observed: "In the field of higher sciences, research workers of different countries work as a close fraternity in continuous communion with each other. It is an obvious advantage that in such fields they should be speaking in an identical language; and, in any case since the terms used in these fields do not affect the general language issue, it appears to be manifestly desirable to adopt the terms current in international terminology". English is commonly understood in most countries today, and has facilitated the spread of international terminology.'
8. A telling example: The Dutch sentence *Ik heb de file geseefd* [I saved the file] contains two loan-words from English: *file* and *save*. The verb is conjugated (and pronounced) analogously to the aurally most similar Dutch verb *zeven* [to sieve], with the past participle *gezeefd*. Moreover, the syntax is Dutch, putting the object before the verb, rather than after it, as in English, and so is the choice of tense, the perfect tense rather than the English imperfect (one of the finer points of English grammar that Dutch students find very difficult, but apparently cope with without fault in practice. As to orthography, sentences like these are very rarely written down, they are accepted only in spoken language.
9. Sridhar (1987: 307) quotes Raghuvira, a defender of the Hindi purity movement: "Shall we be anglicized, shall we be turned into Greeks and Latins and shall we then alone pick up the few crumbs thrown to us as refuse (...) by the West? We shall have again our own words... When this is done, Indians will be free of the thraldom of the European languages (...)."But, Sridhar comments: Some of the newly coined terms are ridiculed in the non-Hindi areas, and instead of strengthening Hindi and making it more acceptable, have provided more grist to the opposition."
10. For recent figures of English learning, see Crystal (1998: 53–63): a rough but reasoned estimate yields about 1.3 billion more or less fluent speakers of English. The great majority are non-native speakers.
11. Cf. the debate *Quelle langue(s) pour l'Europe* [which language(s) for Europe] with Pierre Bourdieu, Marc Fumarolli, Claude Hagège, Immanuel Wallerstein and Abram de Swaan, Paris, 18 June 1998.

5. References

Ammon, Ulrich
1998 *Ist Deutsch noch internationale Wissenschaftssprache? Englisch auch für die Lehre an den deutschsprachigen Hochschulen*. Berlin/New York: de Gruyter.

Bernstein, Basil
1975 *Class and Pedagogies: Invisible and Visible*. Paris: OECD.
Bourdieu, Pierre
1982 *Ce que parler veut dire*. Paris: Fayard.
1984 Capital et marché linguistiques. *Linguistische Berichte* 90: 3–24.
Chaudenson, Robert
1991 *La francophonie: représentations, réaltités, perspectives*. Paris: Didier.
Crystal, David
1997 *English as a Global Language*. Cambridge: Cambridge University Press.
de Swaan, Abram
1998 a A Political Sociology of the World Language System (1): The Dynamics of Language Spread. *Language Problems and Language Planning* 22 (1): 63–75.
1998 b A Political Sociology of the World Language System (2): The Unequal Exchange of Texts. *Language Problems and Language Planning* 22 (2): 109–128.
Djité, Paulin G.
1990 The Place of African Languages in the Revival of the Francophonie Movement. *International Journal of the Sociology of Language* 86: 87–102.
Elias, Norbert
[1983] 2000 *The Civilizing Process*. Oxford: Blackwell.
Étiemble, R.
1973 *Parlez-vous franglais?* Paris: Gallimard.
Flaitz, Jeffra
1988 *The Ideology of English. French Perceptions of English as a World Language* (Contributions to the Sociology of Language 49). Berlin/New York: Mouton de Gruyter.
Gopal, Ram
1966 *Linguistic Affairs of India*. Bombay: Asia Publishing House.
Hagège, Claude
1987 *Le Français et les siècles*. Paris: Odile Jacob.
Haut Conseil de la Francophonie (ed.)
1989 *État de la Francophonie dans le monde*. Paris: Documentation Française.
Kachru, Braj (ed.)
[1982] 1992*The Other Tongue. English Across Cultures*. Urbana/Chicago/London: University of Urbana Press.
Lijphart, Arend
1968 *The Politics of Accommodation: Pluralism and Democracy in the Netherlands*. Berkeley, CA etc.: University of California Press.
Manessy, J.
1984 *Le français en Afrique Noire, tel qu'on le parle, tel qu'on le dit*. Paris: l'Harmattan-IDERIC.
Pennycook, Alastair
1994 *The Cultural Politics of English as an International Language*. London/New York: Longman.
Phillipson, Robert
1990 *English Language Teaching and Imperialism*. (Diss. University of Amsterdam). Trønninge: Transcultura.

Sridhar, S. N.
1987 Language Variation, Attitudes, and Rivalry: The Spread of Hindi in India. In Lowenberg, P. H. (ed.) *Language Spread and Language Policy: Issues, Implications and Case Studies*. Washington, D. C.: Georgetown University Press, 300–319.

The Dominance of English at European Universities: Switzerland and Sweden Compared

Heather Murray and Silvia Dingwall

1. Introduction

European universities, like universities in many other parts of the world, are having to adapt to recent trends towards internationalisation.[1] Developments in technology have facilitated communication and mobility among students and researchers and have also led to virtual mobility in the form of distance education. These changes have obliged universities in different countries to compare and, where necessary, to begin to adjust their degree and accreditation systems to allow students to move from one university to another more easily[2]. The European Credit Transfer System, although still far from being standardised, is one such adjustment.

Scientific research has also followed this globalising trend, with more and more large-scale projects demanding international cooperation and funding. Competition for research funding has also become more international, as has competition for space in prestigious journals. For today's scientists, publishing their research findings in international journals means writing articles in English since most such journals, including those published in non-English-speaking countries, now favour or insist on contributions in English.

Increasing communication, cooperation and competition have all affected language practices at universities. In a growing number of countries, courses at both graduate and undergraduate level are being held in English, and/or require reading and writing skills in academic English. However, trends towards using more English are by no means uniform. In Europe, they have tended to be more far-reaching in "small-language" countries such as Denmark, the Netherlands and Finland, and less extensive in countries with large populations. Yet even in a country like Germany, English as an "additional language of instruction" along with German at undergraduate level has recently been introduced in special programmes at a number of universities (Ammon 1998). Such changes may attract international students and improve a country's performance in the race for scientific funding and recognition, but their effects on local students and local languages must also be considered. We have attempted in this paper to shed light on some of the issues these new language practices raise, by comparing the impact of English as a language of science in two countries, Switzerland and Sweden.

2. Sweden and Switzerland compared

Most countries in Europe have their own long and distinct university traditions and cultures. Some have confronted the trends outlined above earlier than others, but all are having to deal with them in some way. We have singled out two countries for special attention: Switzerland, because we know it best, and Sweden, because it offers interesting similarities and contrasts. The two are often confused by people who are not familiar with either country. Both are relatively small and, demographically speaking, the two countries have much in common (see Table 1), particularly in terms of population size, life expectancy, and standard of living. A relatively high proportion of both Swedes and Swiss have tertiary qualifications and a large proportion of the workforce in both countries is employed in the service sector. One difference is that there are proportionally many more foreigners in Switzerland (nearly 20%) than there are in Sweden (around 6%), a pattern which has parallels at the university level, for both students and staff.

Table 1: Switzerland and Sweden compared statistically

	Switzerland	Sweden
Population (approx.)	7.1 million	8.8 million
GDP per inhabitant (US$)	25,130	20,344
Employees in the service sector	69.8%	71.2%
People with tertiary education	22%	27%
Resident foreigners	19.3%	5.9%
Relative Citation Index	1.37	1.24

Most importantly for this study, both Sweden and Switzerland have a very high standing in the international league table for research, at least as far as the Science Citation Index can be used as a basis for judging the quality of research (May 1997). When the number of citations of scientific papers is related to the population of a country, Switzerland had, in 1996, the highest number of citations-per-capita with 79% more than the US, followed by Sweden with 25% more. It would appear, then, that in neither country does the dominance of English in the international scientific community seem to hamper the ability of Swiss and Swedish academics to produce outstanding research. For this reason, we think it could be enlightening to look more closely at how Swiss and Swedish researchers are linguistically prepared for their scientific careers.

2.1. *Languages in Sweden and Switzerland*

Linguistically the two countries are very different: Sweden has one official language whereas Switzerland has four and is often cited as an example of a country where multilingualism works. Three of Switzerland's four national languages, namely German, French and Italian, are official languages in other countries and widely used as second languages in Europe, Africa and the Americas (see Table 2). Swedish, on the other hand, is spoken as a first language (L_1) by only about 10 million people and is, according to Crystal (1997: 37), on the decline as a first language in Finland. Table 2 includes figures for Danish and Norwegian in brackets since speakers of these languages can usually read Swedish without much difficulty. For the group of languages shown, English leads as a world language with well over a thousand million L_1 and L_2 (second and foreign language) speakers worldwide, followed by French and German with close to 350 million L_1 and L_2 speakers, and Italian with around 100 million L_1 and L_2 speakers. The number of people (around 20 million) who understand Swedish is, by contrast, small.

Table 2: English and the official languages of Switzerland and Sweden worldwide (based on Crystal 1997: 443–451 and Coray 1999)

Language	Where spoken	No. of L_1 speakers	No. of L_2 speakers
Swedish	Sweden, Finland, USA	10 million	–
Danish	Denmark, Germany, USA	5.3 million	–
Norwegian	Norway, USA	5 million	–
English	USA, UK, Canada, Ireland, Australia, New Zealand, South Africa	400 million	900 million
French	France, Canada, Belgium, Switzerland, Luxembourg, Monaco,	109 million	250 million
German	Germany, Austria, Switzerland, Liechenstein, E. Europe	123 million	200 million
Italian	Italy, San Marino, Switzerland, Sardinia, Slovenia, the Americas	40 million	65 million
Romansh	Switzerland, Italy	40,000	–

Swedish as a language is numerically not very powerful, but it has long had the support of the Swedish Academy (founded in 1786), which is similar to the Académie Française. It is the only official language in Sweden, but since 1975 the Swedish government has encouraged bilingual school programmes for the immigrant population and the small number of Lapp (or Sami) speakers.

The populations in Switzerland's three main language areas are proportionally unequal[3], but politically the Swiss try to ensure a consensual balance of power between the different language groups. The fourth language, Romansh, is recognised as a national language, but its use is very restricted. Official federal documents are written in German, French and Italian. Goods are sold with labels in all three languages. At intranational gatherings, the language practice according to the Swiss ideal of multilingualism would be for everyone to speak the official Swiss language they feel most comfortable using, on the assumption that the speakers of another Swiss language would at least understand it. Research by Lüdi (1992: 46), however, suggests that, for individuals, Swiss multilingualism may be something of a myth, but it is a myth that has a great deal of symbolic importance for Swiss identity (see, e. g., Coray 1999; Dingwall and Murray 1999). Switzerland's linguistic landscape has also been affected by immigration, so that the reality today is that Switzerland is a plurilingual society (Lüdi et al. 1997).

What is the place of English in these two countries? McArthur (1996: 13), in a paper on English in the world and in Europe, classes both Sweden and Switzerland, but not, say, Finland, Liechtenstein or Luxembourg, as countries where English is "virtually a second language". While the grounds for viewing English as a virtual second language in Sweden are considerable (for example, the length of time it is taught at school, and its presence in the media), a similar claim for Switzerland is likely to be met with resistance, at least among official Swiss circles. Until 1998, the first foreign language taught at schools in all the Swiss cantons was one of the official Swiss languages. But the spread of English across the globe has not left Switzerland untouched, and demand for English has increased enormously, particularly in adult education, but also in state schools. In some cantons, notably Canton Zurich, an earlier introduction of English in primary schools is being piloted, with French coming in later on in the curriculum. In summary, although at the risk of some oversimplification, English has had an official place and has been much more widely used in everyday life in Sweden for much longer than it has in Switzerland. While in Sweden the spread of English may be at the expense of what Preisler (1999: 253), writing about Denmark, calls "functional illiteracy" for the minority of Swedes with little knowledge of English, in Switzerland it strikes at the heart of Swiss national identity. Language and educational policy-makers in Switzerland have to balance the widely felt need for the Swiss to learn and use more English against the need to preserve the unity of the country and not to exacerbate political and linguistic tensions between the different language communities.

2.2. Higher education in Sweden and Switzerland

The Swiss political system is highly decentralised, with a three-tier system of federal, cantonal (regional) and local government. At present the 26 different cantons have 26 different education systems, with even some matters of school policy being decided at the level of the local town or village. Until 1996 there were ten, largely autonomous, universities located in the German- and French-speaking parts of Switzerland: two federal universities (EPF Lausanne and ETH Zurich), and seven large and one small cantonal universities. In 1996 a small university was founded in the Italian-speaking part, and in 1998 the greenlight was given to upgrade seven "technical colleges" and give them university status (similar to the creation of "new universities" from former polytechnics elsewhere in Europe in the 1980s). The universities are small to medium-sized (Zurich with 20,000 students is by far the largest). All students with a *Matura* (university-track high school diploma comparable to the International Baccalaureate) are entitled to study at university, but they have to pay fees and grants are limited. At present the Swiss university system is going through a period of reform, and efforts are being made to ensure, e. g., that student mobility is possible, which involves more coordination between the universities with respect to goals, course content, qualification procedures and so on.

Higher education in Sweden is highly centralised (Kälvermark and Lindström 1996: 7–9), although the university reforms that began in 1993 have promoted greater autonomy among Sweden's seven universities and four specialised institutions of higher education and research. In addition there are 16 university colleges that are state run. Swedish universities tend to be larger than Swiss ones, with the largest (Lund) having 29,000 students. Universities impose their own admission requirements, but university education is free.

Table 3: Comparison of university student and staff numbers[4]

	Swiss universities	Swedish universities
Total full-time students (1996/97)	93,400	176,600
Total teaching staff	11,300	21,100
Total undergraduate students	76,400	160,000
Total postgraduate students	17,000	16,600

As Table 3 shows, Swedish universities have a much higher number of full-time students than Swiss universities, but the student-staff ratio is about the same: just over eight to one. Nearly all the difference is accounted for by undergraduate numbers, which are twice as large in Sweden. In contrast, the number of postgraduate students is about the same for both countries, and suggests a high rate of selectivity between undergraduate and graduate levels.

The reforms that started in 1993 in Sweden have given the individual universities more powers. The Swiss system is adapting to the pressures of globalisation in a way that is difficult for the Swiss government to regulate (see Kleiber 1998), but viewed from the outside it looks as if the higher education systems in the two countries are moving closer together. Nevertheless, the systems still remain differently structured, with that in Sweden being much more centralised.

Given this contrast between the university systems in the two countries, it is somewhat surprising that, at the level of research funding, the pattern is reversed: in Switzerland there is a central funding body (the Swiss National Science Foundation or SNSF), which is still the main source of funds for Swiss research and which thus has considerable weight in influencing what research is conducted at Swiss universities and research institutes. It coordinates research through funding "Swiss Priority Programmes" and will, in 2000, select about ten so-called "Competence Centres" for special support. Unlike Sweden, Switzerland is not (yet) a member of the European Union (EU), but it can participate in EU research and development programmes under certain conditions. Swiss research has also benefited from non-European schemes such as the Human Frontier Science Program (launched by Japan in 1989 to foster international interdisciplinary cooperation on neurobiological research). This high rate of international cooperation is also documented by publications: in 1995, over 40% of all Swiss research publications in the fields of physics, biomedicine, engineering, earth science, mathematics and multidisciplinary research were produced in collaboration with researchers in other countries (Bundesamt für Statistik 1999:412).

The SNSF has played an important role in ensuring that Swiss universities have an international outlook by funding study abroad for its postdoctoral (and sometimes doctoral) students. The number of postgraduates receiving SNSF support to study abroad nearly doubled between 1975 and 1995 (Murray and Dingwall 1997: 58). Study abroad is part of the "Swiss research culture", as one British professor of life sciences at Zurich university put it:

> "[By going abroad], Swiss scientists learn how things are done elsewhere, usually in the USA or UK. Because solid funding allows Swiss scientists to chose which

labs to go to, they pick the best ones. (Most top labs accept postdocs with [their] own funding.) Their English is good or very good and they have to have no fears of collaborating with the international scientific community."

Sweden does not have a single national research funding body similar to SNSF, but, rather, several national research councils. There are also various semi-private funding bodies which play an important role in different fields. Many researchers in the humanities, for example, turn to the *Riksbankens Jubiläumsfond.* Swedish universities appear to have more permanent resources for research than do Swiss ones, and these are drawn on in financing postgraduate study and research. Special government grants are available to promote research links of various kinds and since Sweden joined the European Union it has had easier access to EU sources of funding. Perhaps surprisingly, Sweden's "university teachers traditionally have never been very mobile" (Källvermark and Lindström 1996: 53), but the government is trying to overcome this inertia by promoting a more international outlook at Swedish universities and funding "teacher exchange" programmes. With regard to university staff mobility, the Swiss appear to be more international in outlook than the Swedes.

3. Undergraduate language use at Swiss and Swedish universities

Our comparison of the role of English at Swiss and Swedish universities is based on studies of parts of the two university systems. In the case of Switzerland, it is based on our ongoing survey of Swiss researchers, national statistics on dissertations and research proposals (Murray and Dingwall 1997; Dingwall and Murray 1999), and on a survey of language practices at the University of Berne (Murray forthcoming). In the case of Sweden, it is based on a survey by Gunnarsson and Öhmann (1997) of language practices at Uppsala University (see also Gunnarsson, this volume).

It is important to note that the term undergraduate is used differently in Sweden and Switzerland. Swedish university students generally spend three years as undergraduates before getting their degrees, after which some go on to graduate or research level studies. By contrast, it normally takes about five years to obtain a Swiss degree (a *Lizentiat/Licence/* Scientific Diploma), which includes writing a Master's or Diploma thesis. At this point about 18% start research towards a doctorate. In order to compare undergraduates in the two countries with more precision, we specify whether we are referring to Swiss university courses at the level of "basic studies" (usually the first two years) or "main studies" (the final years).

If we look at the languages of instruction used in courses for undergraduates[5], we find that English is used much more in Sweden than in Switzerland, particularly if we just consider "basic studies". Gunnarsson and Öhmann (1997) report that the percentage of English used in lectures and classes at Uppsala ranges from a high of 13% in pharmacology and the natural sciences to a low of 3–4% in the humanities, social sciences and medicine. At the University of Zurich it appears that no basic studies level courses were offered in English, and a mere 1% of the nearly 2000 courses given at all levels in the winter semester of 99/00 were English-medium, with the highest proportion in biology (10%). Only 4 out of the 485 courses listed by the medical faculty were in English[6]. At Berne about 0.5% of courses offered for undergraduates were in English. In most cases they turned out to be taught by English-speaking professors, who have recently been encouraged by colleagues to give lectures in English instead of in German.

The relatively low number of courses held in English at Swiss universities has been partly a matter of policy. The official teaching language is that of the region where the university is located. English-speaking professors are generally expected to be able to teach in the local language after two years. However, as we noted earlier, regulations and their implementation vary from university to university, and even from faculty to faculty at the same university. For example, federal law in 1991 allowed German, French and Italian as the official teaching languages at the two Swiss Federal Institutes of Technology, but in 1997 the one in Zurich adopted a language policy that actively encourages the use of English in more advanced courses.[7]

Even though very few undergraduate courses are held wholly in English at Swiss universities, the language makes itself felt through required reading and terminology. The acceptance of English terminology is evident in the ease with which it slips into the titles of courses, with a resulting language mix, e. g. "*Coping* bei Kindern" or "*Cleaner Production*: Ökodesign und Nachhaltigkeit". Moreover it seems to be not uncommon for lecturers to use transparencies written in English, but talk about them in, say, German. As far as the use of English course literature in Sweden is concerned, a general introduction to higher education states that:

> "Being a country with a rather small population, Sweden is heavily dependent on other countries when it comes to scientific literature. A consequence of this is the extensive use of books and other teaching materials in English or other major

languages in undergraduate education. All those who enter higher education must have a good command of English since a large percentage of textbooks will normally be in that language." (Källvemark and Lindström 1996:52)

We have yet to carry out systematic research in this area, but we do have data on course literature in English from Berne and the Uppsala Study, which is summarised in Table 4.

Table 4: Percentages of undergraduate course literature in English

	Uppsala undergraduate studies	Berne basic studies	Berne main studies
Natural sciences	87%	20%	90%
Medicine	48%	0%	50%
Humanities and social sciences	47%	10–60%	30–85%

At Uppsala the proportion of course literature in English was found to range from a high of 87% in the natural sciences to below 50% in other areas. The Berne survey found major differences between the basic studies and main studies levels, and these are therefore reported separately. In their first two years, students at the university of Berne are often given the option of reading a textbook in the original English or in the German translation, which means that if they want to, they can avoid doing much reading in English. As their studies become more specialised, however, they are required to read an increasing amount of research literature, which is usually written in English. The Berne survey also found major differences within faculties, especially within the humanities and social sciences, which is why a range of percentages is shown in the table. Departments with a relatively low percentage of course literature in English throughout the undergraduate years were psychology and sociology. On the other hand, the linguistics, cultural anthropology and economics departments revealed percentages closer to the natural sciences. Despite these variations, Uppsala undergraduates do appear to be expected to do more reading in English earlier on in their studies than Swiss undergraduates.

One would expect the figures for undergraduate writing in English to be lower than those for reading, and to a large extent they are. At Uppsala the highest percentages of English undergraduate writing (final essays, seminar papers) were in medicine (57%) and the natural sciences

(48%), with much lower figures for the humanities (14%) and social sciences (5%). The proportion of English papers produced by the Swedish medical students is relatively high, especially in view of the fact that only 48% of their course literature is in English. The Berne figures for undergraduate papers in English are much lower, between 2% and 4%, except in the field of biology, where the figure is about 50%. If we compare figures for Master's theses (*Lizentiatsarbeiten/licentiatavhandlingar*) at both universities, we see that the proportion of English increases: at Uppsala the proportion ranges from 66% English in social sciences and 75% in humanities, to virtually 100% in medicine and the natural sciences; at Berne, the figures range from 5% in English in the humanities and 10% in the social sciences, to 20% in the natural sciences, with the exception of biology, where the proportion is closer to 90%.

This comparison of the role played by English in undergraduate studies at the universities of Berne and Uppsala has thrown up several differences. Compared with their Swiss counterparts, Swedish undergraduates are generally exposed to considerably more spoken English in the form of lectures and classes, are required to read more course literature in English at the beginning of their studies, and produce a much higher proportion of papers, essays and Masters' theses in English.

4. The extent of English domination in postgraduate studies

4.1. Postgraduate courses

It is difficult to use lectures and literature lists to assess English use at the postgraduate level in the same way as the undergraduate level. Graduate programmes in Switzerland tend to be much less structured than in Sweden; in fact, many are totally focused on writing the dissertation and involve no course work at all. Most university departments do, however, organise colloquia or seminars in which graduate students and research staff report on and discuss their research. In the life sciences, many of these graduate colloquia are held in English, even when all participants are native speakers of German or French, purely to provide the opportunity to practise this specialist register of English. In graduate colloquia in the humanities and social sciences this is less the case, but even here the presence of one foreign graduate student or professor who does not understand the local language can occasion a language shift in favour of English ("the default language" as some of our respondents put it). Thus, although it is difficult to find anything like the 70% English instruction for research students in the natural sciences found at Uppsala, figures for the amount of English used in graduate seminars in Switzerland are much higher than those for undergraduate instruction, especially in the natural and life sciences and in economics. It should be noted here that in both Sweden and Switzerland there are special graduate programmes aimed at an international student body, which are held totally or in part in English. In Switzerland such programmes include cell biology (in Basle), public health (in Zurich), international law (in Berne) and international relations (in Geneva). Many such programmes are taught in several languages, e. g., International Business Law (Zurich) and Intellectual Property (ETH Zurich) are taught in German and English, and Work Health in German, French and English. Our impression is that this aspect of Swiss university course provision is expanding extremely rapidly in response to intensified international competition and cooperation. Changes here, however, are being made not at the institutional level but within departments, so they are not easy to monitor. The new universities of applied science may be particularly open to launching innovative courses in English (e. g. the University of Applied Sciences of Northwestern Switzerland is offering an English-language "degree programme in International Management").

4.2. Languages used in doctoral dissertations

As Ammon (1998: 210–11) notes, English is not so much the language of science (*Wissenschaftssprache*) as the language of research (*Forschungssprache*), and this is borne out in the way in which English dominates Swiss and Swedish researchers' writing practices. First of all, if we look at doctoral dissertations, Gunnarsson and Öhmann found that in 1993/94, virtually 100% of the Uppsala theses and doctoral dissertations were written in English in the fields of medicine, pharmacology, and natural sciences, and that in theology and the humanities about 75% were written in English, followed by 66% in the social sciences. As we have shown elsewhere (Murray and Dingwall 1997), the trend towards using English for doctoral dissertations in Switzerland is definite, but the percentages are much lower. Based on the 1991 dissertation figures, we calculated that 11% of all Swiss dissertations in the humanities and social sciences were in English, 31% in the natural sciences (including biology) and about 12% in medicine, dentistry and veterinary medicine. These figures require some explanation, however.

First, there are considerable differences in the proportion of dissertations written in English between the French- and German-speaking parts of Switzerland. For example, in 1991, the highest proportion of natural science dissertations in English produced at a French-speaking university (Neuchâtel) was a mere 19%. If we look at German-speaking Swiss universities for the same year and field, we obtain substantially higher figures, e. g. 47% of the natural science dissertations at Zurich University were written in English.

Further insight into the growing dimensions of English use can be gained by focusing on individual fields of research. At Berne University, departments of both plant and animal biology and microbiology all now require dissertations to be in English, the reason being that a dissertation in these fields usually consists of several articles by the doctoral candidate, all previously published in refereed journals which are, needless to say, in English. Other Berne departments that have recently started to require dissertations in English are computer science, economics, physical geography, and certain areas of chemistry.

One might expect high rates of English for medical dissertations, as was found at Uppsala, but in fact Swiss dissertations in medicine are not all written with a research career in mind and are therefore by no means all in English. Hence the surprisingly low figure of 12% for Switzerland in 1991, although it has no doubt risen since then.

The humanities and social sciences present a very uneven picture of English use for dissertations. We have already noted the new pressure to use English in economics dissertations. English appears to be a rarely used option in all other fields (except linguistics), and is typically only available as an option to psychology students who are conducting research with a foreign collaborator or under a foreign supervisor. Among German-speaking psychologists, opinions on the necessity of publishing in English are divided, partly because of differences in cultural values and publishing customs between German and American researchers. Recently, however, a group of 23 leading German psychologists has come out strongly in favour of publishing in English, even proposing that PhD students should have the option of presenting a doctoral dissertation consisting of several published articles, as is common practice among chemists and biologists, adding that, "in order to improve international visibility and enhance career opportunities, these publication-based dissertations should be written in English" (Gigerenzer et al. 1999: 103 – our translation).

4.3. Other research publications

Beyond the dissertation, English dominates Swiss research to an even greater extent. We have attempted to measure this dominance by means of a questionnaire survey of both pre- and post-doctoral researchers at Swiss universities. In the questionnaires, which were completed by doctoral and post-doctoral researchers at seven Swiss universities, and of which 250 have been tabulated to date, respondents were asked to say what proportion of their research papers produced over the past 5 years were in English. Table 5 gives a summary of their responses and includes figures from the Uppsala study for comparison.

Table 5: Proportion of research papers produced in English

	Swiss doctoral researchers	Swiss post-doctoral researchers	Uppsala researchers
Humanities	29%	35%	49%
Social sciences	22%	27%	53%
Natural sciences	60%	79%	96%
Applied sciences	42%	73%	n.a.
Biomedical sciences	73%	86%	89%
Veterinary medicine	44%	59%	n.a.

Table 5 shows that, not surprisingly, post-doctoral researchers produce more papers in English than do doctoral researchers. Furthermore, we can perceive the by-now familiar pattern of relatively low English use in the humanities and social sciences, with much higher levels of English for publications in the natural and biomedical sciences. Although the Swiss figures are clearly lower than those for Swedish research paper production, except in the field of biomedicine, we can still detect a dominance of English in research in the natural and life sciences, whether theoretical or applied.

In Murray and Dingwall (1997) we described how the proportion of project proposals written in English and selected for funding by the Swiss National Science Foundation (SNSF – the main research funding body in Switzerland) dramatically increased between 1975 and 1995. Thus, only 6% of all successful projects were written up in English in 1975, rising to 27% in 1985, and 63% in 1995. As we predicted, the number of projects in English continued to rise over the next three-year period:

– in the humanities and social sciences, from 7% in 1995 to 9% in 1998.

- in the natural sciences, from 58% in 1995 to 73% in 1998, a considerable increase in just three years.
- in the life sciences, from 91% in 1995 to 95% in 1998.

Fig. 1 shows that the rates of change have been different for the three major groupings of discipline, with English making itself most felt in the so-called life sciences, where nearly all the projects funded had English titles. It is now SNSF policy that the whole of a project proposal should be in English in the life sciences and part of it in the natural sciences so that applications can be evaluated by international experts. In the social sciences and humanities, there are no regulations or requirements about formulating proposals in English. French, German and Italian applications are equally acceptable. In practice about 15% of proposals in the humanities or social sciences are submitted in English. These are mainly from the fields of psychology, economics or political science. It looks as if, in future, larger SNSF research programmes, such as those linked with proposed "Competence Centres", will usually require project descriptions to be written in English.

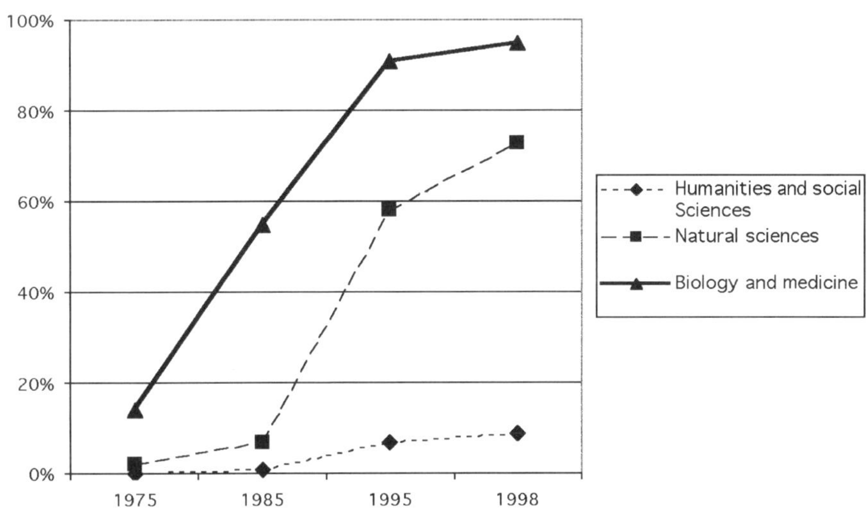

Figure 1: Percentages of successful SNSF proposals written in English

Unfortunately we do not have comparable figures for Swedish research projects – they may be difficult to collate given that research funding is less centralised in that country – but the pattern is likely to be similar, with an even more pronounced favouring of English in some fields.

As mentioned in section 2, analyses of the citation rates of scientific articles indicates that the standard of published research produced by both Swedish and Swiss researchers is high. At Swedish universities students and researchers appear to get much more practice and be provided with more help than their Swiss colleagues with writing and presenting papers in English, although Gunnarsson and Öhmann (1997:76–77) call for more courses and more translation help. We have noted that, among Swiss researchers, many feel that improving their English is the responsibility of the individual, and a widespread view is that this is best done by studying in an English-speaking country.

5. Attitudes to the use of English

As is by now clear, English is the dominant language in research publications worldwide and therefore plays an influential role in researchers' careers. Precisely to what extent scientists' knowledge of English can affect their careers still remains to be demonstrated. Part of our investigation of Swiss researchers focused on their attitudes to English as the dominant language in science. Specifically, we asked:

> "In scientific communication, English is used more than any other language. How do you feel this affects your career? *(Please choose one:)*
> – It is a major handicap. – It is a slight disadvantage.
> – There is no effect. – It is an advantage."

This question was based on a question from Truchot's (1990) investigation of natural and life scientists in the region of Strasbourg, and thus designed to enable us to relate at least part of our findings to those from another European country. The Swiss researchers' responses to this question are shown in Fig. 2.

The responses indicate that relatively few Swiss researchers consider the dominance of English to be a major handicap (6%), although a fairly large proportion (41%) think it is a slight disadvantage. Most of the comments appended to these two choices alluded to English being a foreign language in which it was inevitably more difficult to express oneself than in one's first language. Over a quarter of our respondents (26%) thought that the dominance of English had no effect on their careers, and finally, a slightly larger group (27%) actually felt that the dominance of English

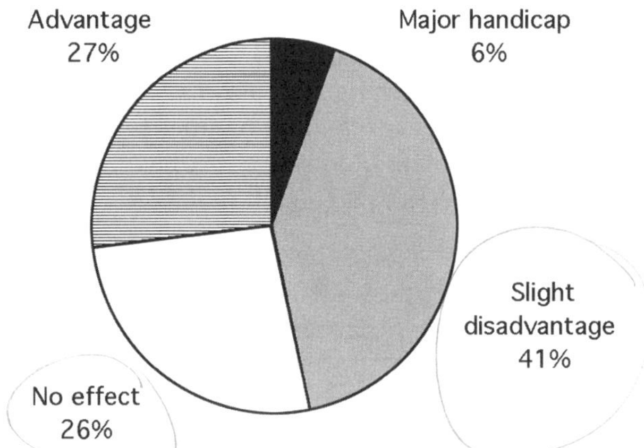

Figure 2: Swiss researchers' assessments of the effects of English on their careers

was an advantage. Explanations for the "advantage" response fell into one of three categories:

(a) that it is an advantage to have a single agreed language of scientific communication;
(b) that English is much easier for speakers of Swiss languages than, say, for speakers of Asian languages;
(c) that English as a Swiss intranational language does not allow German or French speakers an "unfair" native-speaker advantage.

Truchot's survey produced somewhat different responses to the same question: he found that 52% of his Strasbourg scientists viewed the dominance of English as a major handicap, and 8% considered it an *alienation*. In other words, 60% of his respondents, who were all involved in research in the life or natural sciences, felt that the dominance of English was a disadvantage of some kind, whereas only 43% of our life and natural scientists did.

When we sort our Swiss respondents according to field of research we find some interesting differences between researchers in the humanities and social sciences on the one hand, and those in the natural and life sciences on the other, as shown in Fig. 3. The proportion of researchers in the humanities and social sciences who view the dominance of English as a major handicap (17.9%) is about eight times higher than the proportion of respondents answering similarly in the other two major fields. Moreover, the proportion of researchers in the humanities and social sciences who view the dominance of English as an advantage is significantly lower than in the other two fields. Indeed, as several of our respondents have pointed out, since language plays such an important role in most writing about social science and humanities topics, it may well be a disadvantage and involve extra linguistic effort having to formulate complex arguments in these fields in a foreign language.

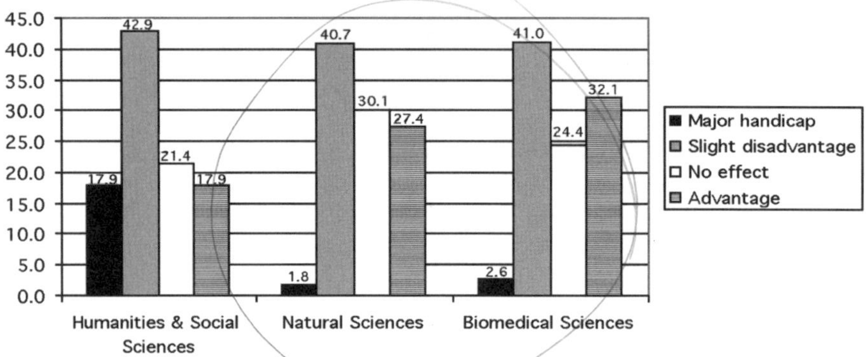

Figure 3: Swiss researchers' assessments of the effects of English on their careers, by field of research

Our results partially recall Ammon's (1990) survey of German researchers in the natural and social sciences. In that study, the scientists were asked (among other things) "Do you feel your ability to communicate is restricted by the dominance of the English language?" Slightly less than half the total sample (44.9%) felt either "very much" or "somewhat" restricted in their scientific communication, while 55.1% experienced no disadvantage whatsoever. When Ammon divided his sample into natural scientists and social scientists, however, he found significant differences between them: a considerable portion of the social scientists (13.3%) felt "very much restricted", while none of the natural scientists did. Surprisingly, at the other end of the response spectrum, 60% of the social scientists felt "not at all" restricted by English while only 53.7% of the natural scientists felt this way.

One explanation for the discrepancy between our results and Truchot's is that our survey sample contains a high proportion of German-speaking Swiss scientists (78%), while Truchot's sample was mainly French-speaking. If we separate our sample into German-speaking and French-speaking respondents, we discover a tendency among the French-speaking ones (of whom 12.9% marked English as a major handicap, 54.8% a slight disadvantage, 12.9% no effect, and 19.4% advantage) which is closer to

Truchot's results. In other words, French-speaking researchers in Switzerland tend to feel greater personal disadvantage from the dominance of English than do German-speaking researchers.

Returning to Gunnarsson and Öhmann's study, and to their report on Swedish attitudes to languages in research, we find that researchers in all fields (with the possible exception of pharmacology) either support the overwhelming dominance of English in their field, or, if English dominance is less than complete, want to aim for an even higher proportion of research papers in English. However, these attitudes to English in research contrast sharply with attitudes to the proportion of English in undergraduate studies. The Uppsala study found that a number of departments questioned the need to expand the use of English in undergraduate studies, although faculties with a very low use of English (theology and law) wanted more. In fact, people generally appeared to feel less of a demand for more English than for more Swedish at this level.

6. Summary of similarities and differences

Before discussing what we think our findings imply about the effects and feared effects of the domination of English at European universities, it could be helpful to summarise the main similarities and differences we have found between Swiss and Swedish universities with special reference to language preferences and practices.

6.1. National language situation

The fact that Swedish is spoken by relatively few people and that German, French and even Italian have much larger numbers of speakers means that the availability of textbooks and other instructional materials will be higher – and that their cost will be lower – in the Swiss national languages than in Swedish. The implication here is that if textbooks are not available or too expensive, English textbooks will be used instead.

On the other hand, it is probable that both Swedish and Swiss children grow up with the idea that they will someday have to communicate effectively in a foreign language: in the Swedish case, because there are so few speakers of Swedish, and in Switzerland because of the multilingual Swiss identity.

6.2. Higher education and research

Swedish universities are highly centralised and transparent compared with Swiss ones. This means that countrywide policies can be developed and implemented relatively quickly. In contrast, the Swiss university system seems lumberingly slow and fragmented, with each faculty acting autonomously. A single Swiss university language policy would seem to lie decades away.

As far as research and grants for study abroad are concerned, however, the situation is somewhat different. Research and grant money are to a large extent dispersed by the Swiss National Science Foundation, a central institution which seriously encourages both Swiss competitiveness and international cooperation, and includes a well-developed programme of sponsoring postgraduate study abroad. Swedish universities and private research foundations also provide substantial support to academics, but this is less coordinated and Swedish researchers are not so internationally mobile as their Swiss counterparts.

6.3. Patterns of linguistic dominance

English dominates foreign language learning for Swedes from the beginning of their schooling. In Switzerland it is usually the second foreign language after a national language, and therefore has not up to now been accorded as much time in the school curriculum as in Sweden. Thus, although both countries are equally affected by globalisation and the concomitant linking of English with career advantage, foreign languages in Sweden are dominated by English more than they are in Switzerland.

English also plays a clearly stronger role in undergraduate studies in Sweden than in Switzerland, being used much more in instruction, in course literature and written production. This pattern continues at graduate level, where the dominance of English in Sweden is even stronger. At postgraduate or senior researcher level there may still be a somewhat greater dominance of English among Swedish than among Swiss academics, but in the natural and life sciences the pattern of English dominance is probably about the same.

6.4. Patterns of change

Differences in the positions of the national languages and in higher education and research probably affect patterns of change. It seems as if

Swedish universities are at least ten years ahead of those in Switzerland as regards accommodation to scientists' language needs. In fact, the spread of English in Swedish academia may have reached saturation point, causing some students and faculty to call for more Swedish in undergraduate education. In Switzerland, the use of English in undergraduate education (except for course literature) is still the exception, but change, led by research pressures, could become more rapid in the next ten years.

6.5. Attitudinal differences

It is difficult to compare Swedish and Swiss attitudes to the dominance of English directly due to a lack of parallel studies. What we can say is that, as a vehicle of scientific communication, English enjoys high acceptance in both countries. It is probable that the majority of Swedish researchers, like their Swiss counterparts, tend to feel that having to use English is at most a slight disadvantage and not a major handicap. In other European countries, scientists might well report more problems with English. Truchot's survey in France discussed in section 5 at least hints at this. In the final section we discuss some issues that our Swiss-Swedish comparison has brought to light, which we thought could be relevant to other countries trying to decide how to deal with English as the dominant language of science.

7. Some issues raised by the dominance of English as a language of science in Europe

European universities face new linguistic situations and difficult linguistic decisions as a result of the growing internationalisation of curricula and research. Often an increase in the use of English for spoken or written communication is involved, which may mean a related decrease in the use of the local language(s). From a sociolinguistic perspective, this may be construed as a move towards diglossia, and give rise to fears that the local language(s) will inevitably be diminished. From the perspective of the individual speaker, on the other hand, it may be regarded as widespread asymmetrical bilingualism, which would call for a re-organisation of studies to help students with the linguistic demands placed on them. Another issue that has implications for language policies at universities is the expectation that universities and researchers should be accountable to the wider society. Turning to diglossia first, we can ask the following questions.

7.1. Need diglossia be a threat?

We have seen that English has become, in many fields, a very important language for writing (and to a lesser extent talking) about research in Sweden and Switzerland, and all the signs point to it becoming more important. Some linguists[8] have gone so far as to label the situation at Scandinavian universities as diglossic and as marginalizing the state language. While it is possible that neglecting Swedish in scientific discourse may mean that scientific registers in Swedish will not develop as fully as English ones, we do not believe this functional distribution of languages need make it more difficult to hold scientific discussions in Swedish. In the German-speaking part of Switzerland we experience diglossia (Swiss German vs. Standard German) on a daily basis, which is why Switzerland is nearly always referred to in discussions of "classical diglossia" in Ferguson's (1959) original sense. The following observations we have made in diglossic Switzerland may reassure those who fear that English will take over as the sole language of scientific communication in countries like Sweden.

Swiss German is flourishing as a spoken language even though it is rarely written and no standardised written form exists. As Sieber and Sitta (1984:15) say (our translation): "Everything that can be said in Standard German, can also be said in Swiss German." In other words the various Swiss-German dialects are in a constant process of development to keep up with the needs of their users, who include scientists – but of course only in the spoken forms.

Students appear to cope well with a "vernacular" language used in informal study contexts (labs and tutorials) and a "vehicular" language used in lectures, seminars and all written work. Some feel awkward using Standard German in informal discussions, and this awkwardness may be aggravated if English becomes the default language to use when there are people present who do not understand Swiss German. Swedish students are not faced with this problem as Swedish is, of course, the language of everyday life in Sweden.

As far as we have been able to assess, Swiss German researchers have no difficulty discussing highly specialised topics in Swiss German. If they borrow High German terms, these will often be made to sound Swiss

German using familiar morphophonemic transformations. English terms[9] are less likely to be subjected to such modification, but in principle English expressions could be adapted to "sound" Swiss or Swedish (although Ljung (1989) found that Swedish speakers differ in their willingness to use Swedish plurals with English loan words).

The fact that Swiss national identity is tied up with the notion of bilingualism may partly explain why we found so little resistance to English as "the language of science" among Swiss researchers. The Swiss tend to be linguistically tolerant and accommodating, and code-switching is more likely to be valued than seen as taboo. Thus, judging from the Swiss experience, a diglossic Swedish scientific community should not lead to the demise of Swedish for academic or scientific purposes, especially as it has, unlike Swiss German, a long written tradition to draw on.

It may well be, however, that written Swedish in formal natural and life science articles will become rare, but this language function is arguably also becoming more restricted and formalised in English as in many fields the tendency is to produce ever shorter articles consisting mainly of tables and figures. How formal written scientific Swedish (like formal written scientific German and other languages with a rich written scientific tradition) will develop in the future is unclear. We would argue that more "popular" scientific articles and textbooks will and should continue to place demands on Swedish (and German, etc.) as a written scientific language in the interests of improving communication between researchers and the general public.

7.2. Is the spread of English in academia a threat to democracy?

There is new pressure on universities to dissociate themselves from their previously prevailing ivory tower image and to become more "open" to the needs and expectations of the wider society (see, e. g. Kleiber 1998). They are also expected to encourage more openness between the disciplines and interdisciplinary and transdisciplinary[10] approaches to research. The spread of English in academic circles in both Sweden and Switzerland seems to go against this trend. Indeed, it is likely even to widen the already sizeable communicative gap between the scientific and non-scientific communities (scientific texts, in particular formal ones, are difficult for non-specialists to understand in any language), and thus lead to further social stratification. Members of the large immigrant communities in both countries, who are already in most cases struggling with several languages, are likely to feel the effects of this gap most strongly.

If universities are serious about encouraging dialogue with non-scientists, linguistic effort will be required to ensure the communication channels are kept open. This means that national languages will continue to need some form of scientific register and should not delegate this function to another language.

7.3. Are English-medium courses during undergraduate studies desirable?

The undergraduate years at university are a time of learning to use the sometimes complex concepts upon which a particular academic discipline is based. The undergraduate period is also a time when students extend the discourse varieties in their linguistic repertoire by learning to understand and use academic discourse in its various forms. However, if undergraduates, who have only had general English in high school, first encounter key disciplinary concepts formulated in scientific or academic English, they will at the very least require more time to understand and, above all, process those concepts. Moreover, there would seem to be a greater risk of imperfect understanding. In our use-of-English survey at the University of Berne, American or Australian textbooks were sometimes prescribed as introductory reading in fields such as microbiology, economics or anthropology. In one case, first semester students, some of whom had had minimal English training before starting university, were given research articles from English-language journals as part of their first compulsory reading packet. We found that the students in these courses who reported struggling with the reading could not tell whether their main difficulty lay in understanding the concepts or the English.

Examples such as these, however isolated, point to the very real possibility that some university staff are or may become insensitive to the language problems involved in requiring beginning students to read academic discourse in a second language. They may think that they are helping students by providing practice in a language that will become crucial in their subsequent studies and career, but they need to be made aware of the cost of this practice and of the fact that beginning students may be overwhelmed when discourse and concepts are both new and, at the same time, more complex than those previously encountered. If linguists and applied linguists do not take responsibility for informing colleagues in other disciplines about the consequences of certain linguistic practices, it seems fairly certain that, in Europe at least, the scope of English use will grow until it also permeates undergraduate studies, which we think would be undesirable for both students and local languages.

Thus the answer to our third question above is: probably not. English medium-courses appear to be neither necessary nor generally desirable in the early years of undergraduate studies. There is no doubt that, if non-English-speaking researchers are to write articles and books, and give lectures and presentations in English, they need ample exposure to these genres first. However, it is doubtful whether a scientist's academic training, which lasts seven to ten years, is improved by being in English from the first semester. Experience suggests, in fact, that intensive exposure, at the postgraduate level, is a more effective means of improving researchers' command of English scientific discourse.

8. Acknowledgements

Our thanks first and foremost to all our informants (researchers, librarians, and administrators) in Switzerland who took the trouble to fill in our questionnaires, answer our sometimes detailed questions and/or keep us informed about developments in their fields. We are also very grateful to Christina Ljungberg-Stücklin for her help with Swedish and other useful input, and to Mimi Bryant, Jeannette Regan and Paul Sunderland for helping us to distribute the questionnaires. Needless to say no one but ourselves can be held responsible for any errors that may have slipped into this manuscript.

9. Notes

1. For example, a recent government bill on research policy in Sweden specifically stated that "the entire education system ... [should] prepare pupils and students for international relations and employment" (Kälvermark and Lindström 1996:14)
2. According to Bennell and Pearce (1998: 21): "The rapid globalisation of the world economy is generating massive demand for internationally acceptable and negotiable academic and vocational qualifications both among individuals and the corporate sector."
3. According to the 1990 census, the proportions are 63.6% German speakers, 19.2% French speakers, 7.6% Italian speakers, and 0.6% Romansh speakers. See also Dürmüller, this volume.
4. Figures based on Arsrapport för universitet och högskolor 1997 and Bundesamt für Statistik 1998.
5. In what follows we exclude reference to courses in the language and literature departments and in language teacher education.
6. Just counting the titles in English would have suggested that more courses were being held in English, but spot checks have shown that a few courses with an English title are actually taught in German.

7. One professor teaching in medicine wrote: "My understanding is as follows, that in teaching students in the first two years of their studies, German is the preferred (read, required) language, and that if one teaches these students, perhaps the 2–3 year 'rule' is in effect. However, in the final three years of a student's studies, it is my understanding that there is no preferred language, i. e. that German and English are equally acceptable, [but] in fact I have been told that it is desirable that the courses in these years be taught in English."
8. For example, Gunnarsson and Öhmann (1997) or the following statement from Phillipson and Skuttknab-Kangas (1996: 446): "Evidence in western and eastern Europe shows that diglossia, with English as the intrusive dominant language, may be imminent. If the state language is construed or presented as unable to function adequately for certain purposes, for instance as the medium for higher education …, [the result will arguably be] the spread of the dominant international language, English, in a diglossic division of labour that marginalizes the state language. There are trends of this sort in Scandinavia … and in former communist states." This usage extends, following Fishman (1972), the term "diglossia" to refer to two distinct languages rather than two dialects.
9. Since scientific terms are being developed by the international, and not just by the Anglo-American scientific community, they do not have to be of English origin, but merely the terms that are used in that field.
10. "Transdisciplinarity" in essence means: "joint problem-solving among science, technology and society."

10. References

Ammon, Ulrich

1990 German or English? The Problems of Language Choice Experienced by German-Speaking Scientists. In Nelde, P. H. (ed.) *Language Conflict and Minorities/Sprachkonflikte und Minderheiten.* Bonn: Dümmler, 33–51.

1998 *Ist Deutsch noch internationale Wissenschaftssprache?* Berlin: de Gruyter.

Årsrappport för universitet och högskolor

1997 Kortversion. Högskoleverkets rapportserie 1998: 25 R.

Bennell, Paul/Pearce, Terry

1998 The Internationalisation of Higher Education: Exporting Education to Developing and Transitional Economies. *IDS Working Paper* 75. Institute of Development Studies.

Beretta, Isabella

1999 Geburtstagsfest auf drei Kontinenten. *Vision* 3/99: 49–51.

Bundesamt für Statistik

1998 *Statistik der Schweiz: Bildung und Wissenschaft.* Bern.

1999 *Taschenstatistik der Schweiz 1999.* Bern.

Coray, Renate

1999 "Sprachliche Minderheit": ein Grundbegriff der schweizerischen Sprachenpolitik. *Bulletin suisse de linguistique appliquée* 69 (1): 179–194.

Crystal, David

1997 *The Cambridge Encyclopaedia of Language.* 2nd ed. Cambridge: Cambridge University Press.

Dingwall, Silvia/Murray, Heather

1999 The Future of English in Switzerland: A Majority/ Minority Problem? *Bulletin suisse de linguistique appliquée* 69 (2): 189–206.

Ferguson, Charles A.

1959 Diglossia *Word* 15: 325–40.

Fishman, Joshua A.

1972 *Sociolinguistics: A Brief Introduction*. Rowley, MA: Newbury House.

Gigerenzer, Gerd et al.

1999 Internationalisierung der psychologischen Forschung in Deutschland, Österreich und der Schweiz: Sieben Empfehlungen. *Psychologische Rundschau* 50 (2):101–113.

Gunnarsson, Britt-Louise/Öhmann, Katarina

1997 *Det internationaliserade universitetet: En studie av bruket av engelska och andra främmande språk vid Uppsala universitet.* Uppsala: Forskningsgruppen för text- och fackspråksstudier.

Kälvemark, Torsten/Lindström, Anne-Marie

1996 *Financing and Effects of Internationalised Teaching and Learning*. Stockholm: National Agency for Higher Education. http://www.hsv.se/english/pdf/HsV199603S.pdf

Kleiber, Charles

1998 Towards a Society Founded on Education, Science and Culture: A Programme for the Future of Swiss Higher Education. Presented on 12 January 1998 by the Swiss State Secretary for Science and Research. Reprod. in *Vision* 1.

Ljung, Magnus

1989 Social Determinants of the Use of English in Sweden. In Fasold, R. W./Schriffrin, D. (eds.) *Language Change and Variation.* Amsterdam/Philadelphia: John Benjamins, 119–132.

Lüdi, Georges

1992 Internal Migrants in a Multilingual Country. *Multilingua* 11 (1): 45–73.

Lüdi, Georges/Werlen, Iwar/Franceschini, Rita et al.

1997 *Die Sprachenlandschaft Schweiz, Auswertung der Eidgenössischen Volkszählung 1990.* Bern: Bundesamt für Statistik.

McArthur, Tom

1996 English in the World and in Europe. In Hartmann, R. (ed.) *The English Language in Europe.* Oxford: Intellect Press.

May, Robert

1997 The Scientific Wealth of Nations. *Science* 275: 793–796.

Murray, Heather

Forthcoming Englisch als Wissenschaftssprache an der Universität Bern: Wieviel Englisch brauchen und wollen Studierende?. In Watts, R. J./Murray, H. (eds.) *Englisch in der Schweiz.* Bern: Akademische Kommission der Universität Bern.

Murray, Heather/Dingwall, Silvia

1997 English for scientific communication at Swiss universities: "God helps those who help themselves". *Babylonia* 4/97: 54–59.

Phillipson, Robert/Skutknabb-Kangas, Tove

1996 English Only Worldwide or Language Ecology? *TESOL Quarterly* 30 (3): 429–452.

Preisler, Bent
1999 Functions and Forms of English in a European EFL Country. In Bex, T./Watts, R. (eds.) *Standard English: The Widening Debate.* London: Routledge.
Sieber, Peter/Sitta, Horst
1984 Schweizerdeutsch zwischen Dialekt und Sprache. *Kwartalnik Neofilologiczny* 31: 3–40.
Truchot, Claude
1990 *L'Anglais dans le Monde Contemporain.* Paris: Le Robert.

The Expansion of English as a Language of Science and Communication: East and Southeast Asia[1]

Grant D. McConnell

with the technical assistance of Suzanne Lapointe, programmer and Stéphane Ogeron, geographer.[2]

1. Research design and results

Although English has made monumental strides throughout the 20th century and particularly in its second half, we can hardly assume at this point in time that English is universally dominant in all countries and regions of the world and in all domains of use. "The trouble with most appropriated models of language change is that they fail to reflect the multidimensional and multifunctional nature of language dynamics" (Mackey 1999). By first examining to what degree and where English stands as a Language of Science (ELS), we can then more easily describe and measure the effects of its so-called *dominance* on other mother tongue communities. In this paper I attempt to describe and measure how English rates as a language of science in East and Southeast Asia. This is a necessary preliminary step to further investigation into the use of other languages in science.

When referring to the dominance of a language, what do we mean? The study of dominance is not just the compilation of a checklist of countries according to their official languages; nor is it just a list of languages according to the number of mother tongue speakers or even total speakers. It further includes gathering data on language *products* (e. g., newspapers, books, hours of radio broadcasts) and on language *functions* (e. g., learning/studying *through* English as a medium; learning/studying

English as a subject matter by educational level); and that, for different socio-political and socio-administrative entities (e. g., countries, provinces, autonomous regions, etc.). Such data can then be arranged by socio-institutional units (*domains*) and vitality measurements can be calculated on the basis of language functions (uses)[3]. If trends are to be discovered, these data also need to be monitored and compared over a period of time. Admittedly, such a research entreprise presupposes a considerable investment in time and in cost. It is no wonder that most macro research efforts usually end up focusing on a very few domains, e. g., one or two institutions, for a sub-section of the population of a country and usually for one period of time. This less than holistic approach forces one to extrapolate, to infer and to generalize. In so doing, we as scientists hope to have enough hard evidence to establish a clearcut description of the phenomenon studied, so as to be able to explain general changes or trends with some confidence.

My approach here is to widen somewhat the scope of the enquiry by examining and comparing selective sets of data, so as to be able to generate hypotheses and also arrive at some conclusions, not only about the concept of English language dominance, but also about the even more dynamic concept of language *spread* for one major region of the world. The world region that I examine here is *East and Southeast Asia*, which covers 18 polities (Map 1: Asia, with East and Southeast Asia countries indicated in shading). For each of these polities we have compiled a demolinguistic data base which includes an estimated 1,500 languages spoken throughout the region, and for a period extending from the present to as far back to the beginning of the century as possible. These figures were then grouped into ten year periods, more or less according to the decadal censuses per country, in order to test for profile patterns for each decade. This was done so as to eventually arrive at a "*most characteristic*" language profile pattern for each polity. Data evaluated were based on the criteria of "*total speakers*" rather than "*mother tongue*" speakers, as the former reflect better the overall communication patterns of a country. This is especially true for English and other lingua francas of the region.

Earlier, two different sets (typologies) of language profiles were generated from this data base. This amounted to the testing of two sets of criteria, with each set representing a specific configuration pattern. These two tripartite typologies were labelled:

1) *Dominant A, Dominant B, Mixed* and
2) *Unipolar, Multipolar and Non-Polar* (McConnell 1998: xiii–xix).

Without having English specifically in mind, the above were prepared earlier in order to generate linguistic composition typologies for the countries of the region. But as English is used for both speaking and writing in highly varying degrees in most of the countries of East and Southeast Asia (but more particularly as a second language and lingua franca), from a demographic point of view, either one of the typologies could help us generate hypotheses, regarding the gains and even the future prospects of English in this region.

For the present paper, the first typology: *Dominant A, Dominant B and Mixed* was retained and the polities of the region were classified and grouped accordingly (Map 2a: East and Southeast Asia: Country Types, shows this *Dominant A, Dominant B and Mixed* typology by country). The criteria (definitions) for this tripartite typology are as follows:

1) *Dominant A* countries:
The numerically strongest language has twice (or more than twice) the number of speakers, as the second largest language, and all other languages represent less than 10% of the total population.

2) *Dominant B countries*:
The numerically strongest language has twice (or more than twice) the number of speakers, as the second largest language, but at least one other language represents more than 10% of the total population.

3) *Mixed countries*:
All countries *not* included in the first two categories. These are mostly countries with *no* single numerically strong language.

Given the above, we can then hypothesize that:

A) *Dominant A type countries should be less open to the dominance of English as a language of science, than Mixed type countries.*

The rational here is that countries with a single, local, dominant (official, national and written) language would require English much less for internal use, than would a country having no single dominant language. Of course, there are always attenuating circumstances to any hypothesis. For example, the above would certainly be less true in Dominant A countries with a colonial past linked to the United Kingdom or the USA (i. e. an Anglo-colonial historical imprint) – (MAP 2 b: East and Southeast Asia: Historical Language Spheres, shows the colonial imprint by country of four major western languages: English, French, Dutch and Portuguese). The results of analyses based on the above typology, will enable us to generate lists of countries in East and Southeast Asia, that would be most or least potentially open to the dominance of English as a Language of Science, depending on their overall linguistic composition in terms of total speakers. The hypothesis here is based uniquely on a macro demolinguistic data set.

The above typological analysis should go a long way to answering the earlier question of "where", that is, "in what countries", English would be a good or a less good potential candidate for becoming the language of science, or explain, at least partly, "why" English is already more or less dominant. As useful as the above demolinguistic analysis by country types may be, other types of data sets need to be examined, to see if they reinforce the tentative assumptions of this hypothesis.

As mentioned earlier, other useful data sets could be closely tied to cultural *products* and *functions* of languages, so as to give a sense of what dominance means concretely in everyday life. This type of information can be found in a second data set constituted earlier on a worldwide basis (McConnell and Roberge 1994). This set deals specifically with functions of languages in the school domain, but more particularly for English and French used as a *teaching medium (studying through the language) and as a subject matter (studying the language)*. This second data set is based on the demographics of the whole school population per country, which in the survey is examined both quantitatively and qualitatively in terms of the use of English or French in the above named functions.

From the above data set, we can then hypothesis that:

B) *Countries with a high percentage of English language usage, in either or both of these two language functions (learning through English or learning English), are more subject to the dominance of English as a language of science.*

In some countries and regions in the world, school data of this type may not always be valid for such analyses, especially when community attitudes to the school language may not be positive due to imposed usage (e. g. Russian in Eastern European countries), i. e., the functional position of the language is upheld by coercive language policies based on socio-political constraints.

The results of the use of English in the school domain (McConnell and Roberge 1994) also need to be examined on a country by country basis, so that they can be compared to the first data set above, which is based on a demolinguistic configuration for the same countries. Where there is concordance between both data sets, i. e. widespread use of English in the schools (Hypothesis B) in those countries also suspected or known to be configured macro-demographically in such a way that promotes English (Hypothesis A), our initial hypothesis would be further reinforced. Because we are specifically enquiring about the dominance of English as a language of science and communication, one would presume that learning/studying through English and learning/studying the English language would both have an important role to play in this type of dominance. Hence, using and learning the language in the schools should be a major source of this scientific thrust.

For a demonstration of the above we present an analysis in table form of the *Dominant A, Dominant B and Mixed* typology, listing the countries not only by these three demolinguistic configurations, but also by their historical language spheres (e. g., Outside the Anglophone sphere; Inside the Anglophone sphere) and by functions (e. g., English as a medium; English as a subject). For a typology of country types according to these criteria see Tables 1 to 6 and Maps 3 a to 6 b.

For each of the criteria of this typology, we hypothesize *that the use of English is more preva*lent: 1) in the countries of the *Mixed* scenario, 2) in the countries historically *inside* the Anglophone sphere and 3) for those studying *through* English (Table 3). The data seem generally to support hypotheses A and B.

In moving from the Mixed to the Dominant A scenario, the percentage drop in English usage is particularly accentuated for the criteria: 1) those countries historically *outside* the Anglophone sphere“ and 2) for those studying through English”. As predicted, the use of English is less on these two variables in the Dominant A type countries (Table 1).

For the group of 10 Dominant A (non-Anglophone sphere) countries (Table 1:1), there is a general absence of studying *through* English, except at the post-secondary level (only 3 countries) and even here, percentages are low (2% to 21%). English is then not generally *internalized* as a medium of study and hence represents less threat in this domain to the dominant languages of these countries. On the other hand, in 7 of the 10 non-Anglophone countries English is widely studied as a subject matter at both the primary, but especially the secondary levels. Only in three countries (Mongolia, North Korea and Cambodia) does English seems to be nowhere present, even as a subject studied. This state of affairs can readily be explained by other attenuating historico-political factors that have impeded the spread of English in these countries.

A breakdown of the countries, across the three configuration types, by colonial affiliation (Outside the Anglophone sphere; Inside the Anglophone sphere) and according to the two functions mentioned (studying *through* English; studying *the* English language), is presented in Tables 1, 2 and 3 by scholastic levels as a percentage of the whole school population. Substantially the same data are repeated in Tables 4, 5 and 6 giving the actual number of pupils. Table 7 is a summary of Tables 4 to 6 and gives total figures for English-using and English-learning pupils by scholastic level for Anglophone sphere and non-Anglophone sphere countries.

English is studied as a language at the secondary level in 15 of the 18 countries of the region (both inside and outside the Anglophone spheres) with three countries excluded (Cambodia, North Korea and Mongolia). As a language of teaching/medium in the schools at the post-secondary level (both inside and outside the Anglophone spheres), English is less prevalent than as a subject, but it is still used in 9 of the 18 countries (4 of the 9 are outside the Anglophone sphere). In the late 1980's over 15,000,000 students were studying through English on all three levels, mostly in Mixed or Dominant B type Anglophone sphere countries. As for the numbers studying the English language in primary and secondary schools, this amounted to over 100,000,000 in the same time frame, with 65,000,000 in China and Japan alone These numbers may not represent virtual speakers, but they do represent potential readers or users in scientific areas.

The above analysis is indicative rather than conclusive. In order to fully substantiate the claims of the hypotheses, a larger number of countries need to be included in a still broader study, covering all Asia and even other continents, As already pointed out, because of the macro nature of such a study, a consolidated set of international data needs to be built up and carefully codified according to their demolinguistic and geopolitical attributes over time.

Unfortunately, such data sets are not readily available. When it comes to dealing with ethnic communities and language speakers there is always a dearth of hard and comparable data. Also, at the present moment, there are no international bodies, that systematically gather and analyse macro type sociolinguistic data, on dominance, spread and other language trend problems. In the light of the above, individual researchers must do the job to the best of their abilities and capacities. However, that questions on the spread, dominance and survival of languages are being asked in many scientific forums, would suggest that answers to such questions are ever more relevant in the rapidly changing theater of world communications.

2. Tables

Table 1: Dominant A type countries

1) Countries historically **outside** the Anglophone sphere:

Countries	Those learning *through* English* (*Map 3a*)			Those learning English** (*Map 3b*)	
	Primary	Secondary	Post-secondary	Primary	Secondary
Cambodia	-	-	-	-	-
China	-	-	+ (5%)	+ (10%)	+ (75%)
Korea, North	-	-	-	-	-
Korea, South	-	-	?	-	+ (100%)
Japan	-	-	?	-	+ (90%)
Laos	-	-	-	-	+ (17%)
Macau	+ (8%)	+ (12%)	+ (21%)	+ (100%)	+ (100%)
Mongolia	-	-	-	-	-
Thailand	-	-	+ (2%)	+ (40%)	+ (100%)
Vietnam	-	-	-	-	+ (50%)

2) Countries historically **inside** the Anglophone sphere:

Countries	Those learning *through* English* (*Map 3a*)			Those learning English** (*Map 3b*)	
	Primary	Secondary	Post-secondary	Primary	Secondary
Brunei	+ (64%)	+ (92%)	+ (100%)	+ (100%)	+ (100%)
Myanmar	-	-	-	-	+ (100%)

Table 2: Dominant B type countries

1) Countries historically **outside** the Anglophone sphere:

Countries	Those learning *through* English* (*Map 4a*)			Those learning English** (*Map 4b*)	
	Primary	Secondary	Post-secondary	Primary	Secondary
Indonesia	-	-	?	-	+ (100%)

2) Countries historically **inside** the Anglophone sphere:

Countries	Those learning *through* English* (*Map 4a*)			Those learning English** (*Map 4b*)	
	Primary	Secondary	Post-secondary	Primary	Secondary
Hong Kong	-	+ (83%)	+ (67%)	+ (100%)	+ (100%)

Table 3: Mixed type countries

1) Countries historically **outside** the Anglophone sphere:

Countries	Those learning *through* English* (*Map 5a*)			Those learning English** (*Map 5b*)	
	Primary	Secondary	Post-secondary	Primary	Secondary
Taiwan	-	-	+ (50%)	-	+ (100%)

2) Countries historically **inside** the Anglophone sphere:

Countries	Those learning *through* English* (*Map 5a*)			Those learning English** (*Map 5b*)	
	Primary	Secondary	Post-secondary	Primary	Secondary
Malaysia	-	-	+ (17%)	+ (100%)	+ (100%)
Philippines	+ (100%)	+ (100%)	+ (100%)	+ (100%)	+ (100%)
Singapore	+ (70%)	+ (98%)	+ (100%)	+ (100%)	+ (100%)

* Refers to the percentage of the total school population by country, that has English as a medium of instruction.
** Refers to the percentage of the total school population by country, that has English as a subject matter.

Table 4: Dominant A type countries

1) Countries historically **outside** the Anglophone sphere:

Countries	Those learning *through* English† (*Map 3a*)			Those learning English†† (*Map 3b*)	
	Primary	Secondary	Post-secondary	Primary	Secondary
Cambodia	-	-	-	-	-
China	-	-	100,000	14,000,000	40,000,000
Korea, North	-	-	-	-	-
Korea, South	-	-	?	-	5,000,000
Japan	-	-	?	-	10,500,000
Laos	-	-	-	-	19,000
Macau	4,000	2,000	1,500	43,000	14,000
Mongolia	-	-	-	-	-
Thailand	-	-	21,000	3,000,000	2,200,000
Vietnam	-	-	-	-	2,100,000
TOTAL	**4,000**	**2,000**	**122,500**	**17,043,000**	**59,833,000**

2) Countries historically **inside** the Anglophone sphere:

Countries	Those learning *through* English† (*Map 3a*)			Those learning English†† (*Map 3b*)	
	Primary	Secondary	Post-secondary	Primary	Secondary
Brunei	33,000	19,000	750	52,000	21,000
Myanmar	-	-	-	-	1,700,000
TOTAL	**33,000**	**19,000**	**750**	**52,000**	**1,721,000**

TOTAL 1+2	**37,000**	**21,000**	**123,250**	**17,095,000**	**61,554,000**
GRAND TOTAL			**= 181,250**		**= 78,649,000**

Table 5: Dominant B type countries

1) Countries historically **outside** the Anglophone sphere:

Countries	**Those learning *through* English† (*Map 4a*)**			**Those learning English†† (*Map 4b*)**	
	Primary	**Secondary**	**Post-secondary**	**Primary**	**Secondary**
Indonesia	-	-	?	-	9,000,000
TOTAL	-	-	-	-	**9,000,000**

2) Countries historically **inside** the Anglophone sphere:

Countries	**Those learning *through* English† (*Map 4a*)**			**Those learning English†† (*Map 4b*)**	
	Primary	**Secondary**	**Post-secondary**	**Primary**	**Secondary**
Hong Kong	-	372,000	82,000	535,000	450,000
TOTAL	-	**372,000**	**82,000**	**535,000**	**450,000**

TOTAL 1+2	-	**372,000**	**82,000**	**535,000**	**9,450,000**
GRAND TOTAL			**= 454,000**		**= 9,985,000**

Table 6: Mixed type countries

1) Countries historically **outside** the Anglophone sphere:

Countries	Those learning *through* English† (*Map 5a*)			Those learning English†† (*Map 5b*)	
	Primary	Secondary	Post-secondary	Primary	Secondary
Taiwan	-	-	222,000	-	1,800,000
TOTAL	-	-	**222,000**	-	**1,800,000**

2) Countries historically **inside** the Anglophone sphere:

Countries	Those learning *through* English† (*Map 5a*)			Those learning English†† (*Map 5b*)	
	Primary	Secondary	Post-secondary	Primary	Secondary
Malaysia	-	-	17,000	2,500,000	1,400,000
Philippines	9,000,000	3,205,000	1,600,000	9,000,000	3,500,000
Singapore	200,000	224,000	45,000	285,000	300,000
TOTAL	**9,200,000**	**3,429,000**	**1,662,000**	**11,785,000**	**5,200,000**

TOTAL 1+2	**9,200,000**	**3,429,000**	**1,884,000**	**11,785,000**	**7,000,000**
GRAND TOTAL			**= 14,513,000**		**= 18,785,000**

† Refers to the school population by country, that has English as a medium of instruction.
†† Refers to the school population by country, that has English as a subject matter.

Table 7: SUMMARY

1) Countries historically **outside** the Anglophone sphere:

Countries	**Those learning *through* English†** (***Maps 3a, 4a, 5a***)			**Those learning English††** (***Maps 3b, 4b, 5b***)	
	Primary	**Secondary**	**Post-secondary**	**Primary**	**Secondary**
Cambodia	-	-	-	-	-
China	-	-	100,000	14,000,000	40,000,000
Indonesia	-	-	?	-	9,000,000
Korea, North	-	-	-	-	-
Korea, South	-	-	?	-	5,000,000
Japan	-	-	?	-	10,500,000
Laos	-	-	-	-	19,000
Macau	4,000	2,000	1,500	43,000	14,000
Mongolia	-	-	-	-	-
Taiwan	-	-	222,000	-	1,800,000
Thailand	-	-	21,000	3,000,000	2,200,000
Vietnam	-	-	-	-	2,100,000
TOTAL	**4,000**	**2,000**	**344,500**	**17,043,000**	**70,633,000**

2) Countries historically **inside** the Anglophone sphere:

Countries	**Those learning *through* English†** (***Maps 3a, 4a, 5a***)			**Those learning English††** (***Maps 3b, 4b, 5b***)	
	Primary	**Secondary**	**Post-secondary**	**Primary**	**Secondary**
Brunei	33,000	19,000	750	52,000	21,000
Hong Kong	-	372,000	82,000	535,000	450,000
Malaysia	-	-	17,000	2,500,000	1,400,000
Myanmar	-	-	-	-	1,700,000
Philippines	9,000,000	3,205,000	1,600,000	9,000,000	3,500,000
Singapore	200,000	224,000	45,000	285,000	300,000
TOTAL	**9,233,000**	**3,820,000**	**1,744,750**	**12,372,000**	**7,371,000**

TOTAL 1+2	**9,237,000**	**3,822,000**	**2,089,250**	**29,415,000**	**78,004,000**
GRAND TOTAL			**= 15,148,250**		**= 107,419,000**

† Refers to the school population by country, that has English as a medium of instruction.
†† Refers to the school population by country, that has English as a subject matter.

3. Maps

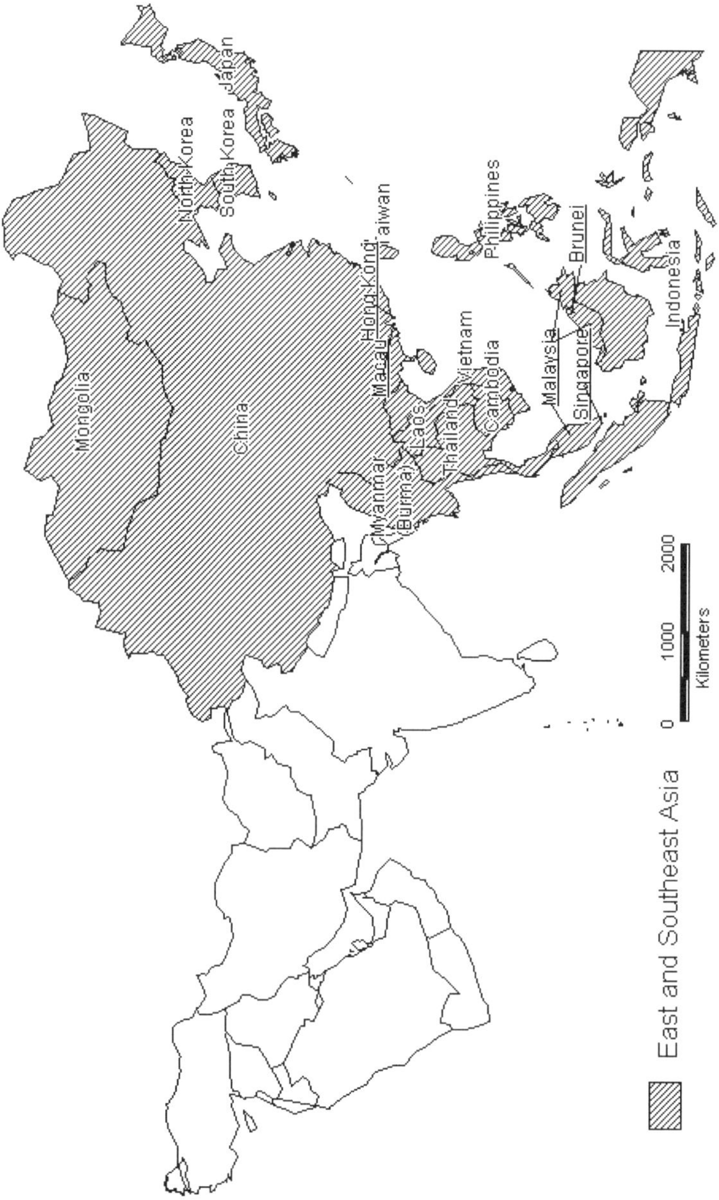

Map 1: Asia

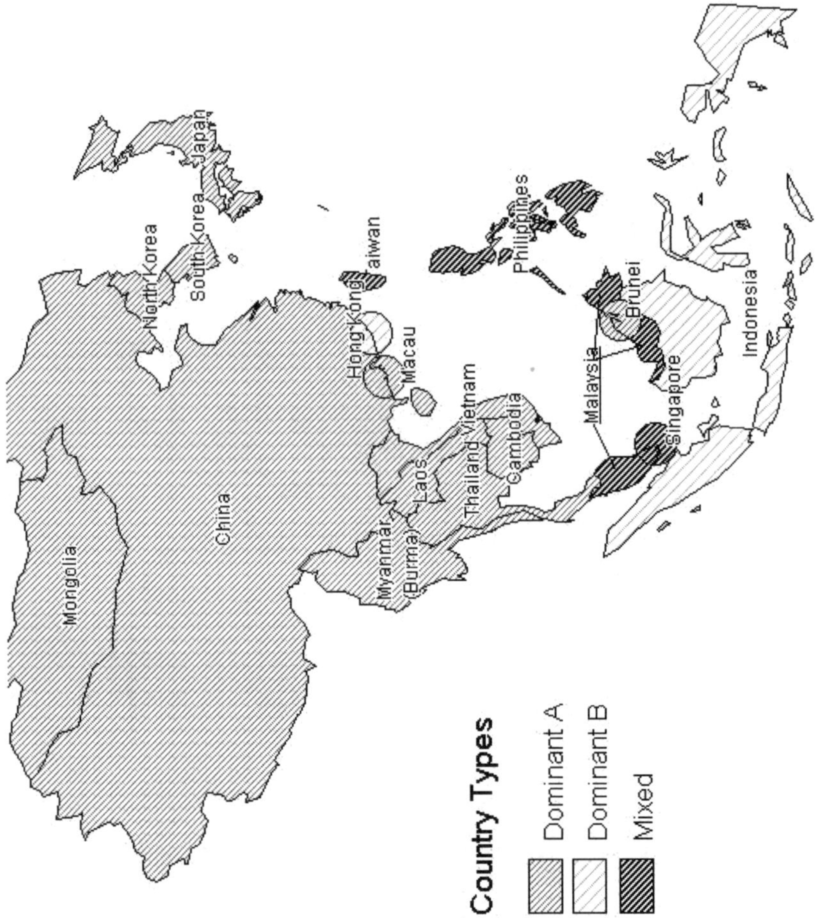

Map 2a: East and Southeast Asia

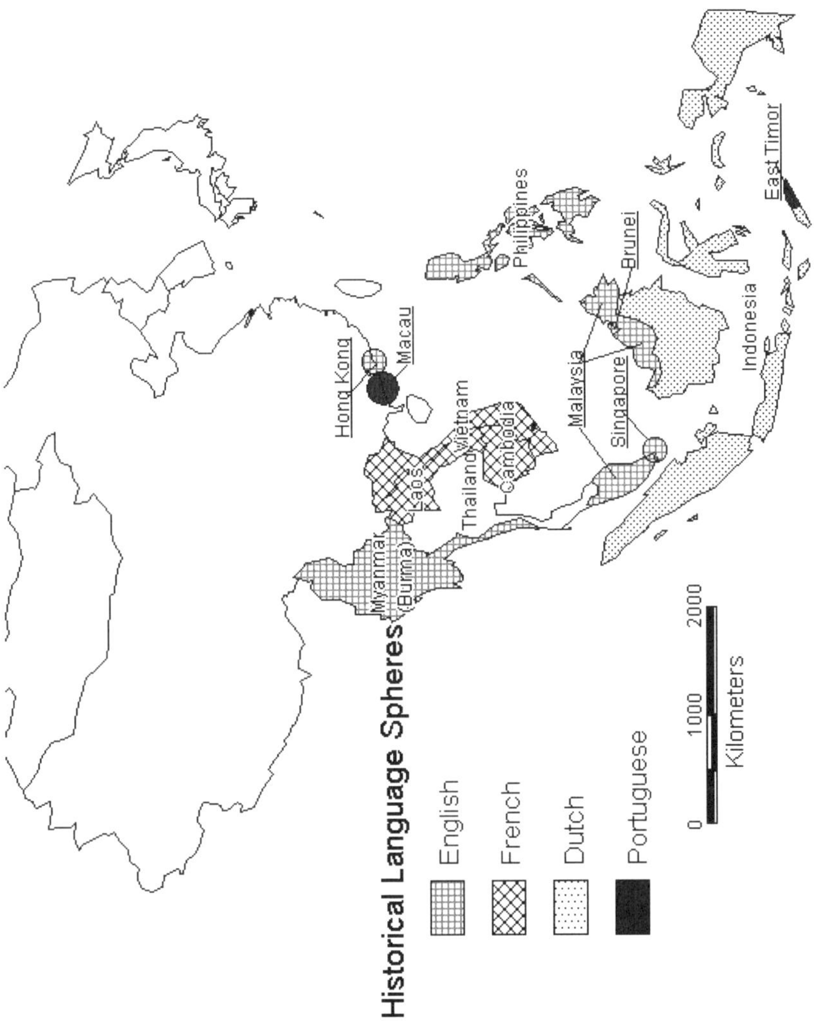

Map 2b: East and Southeast Asia

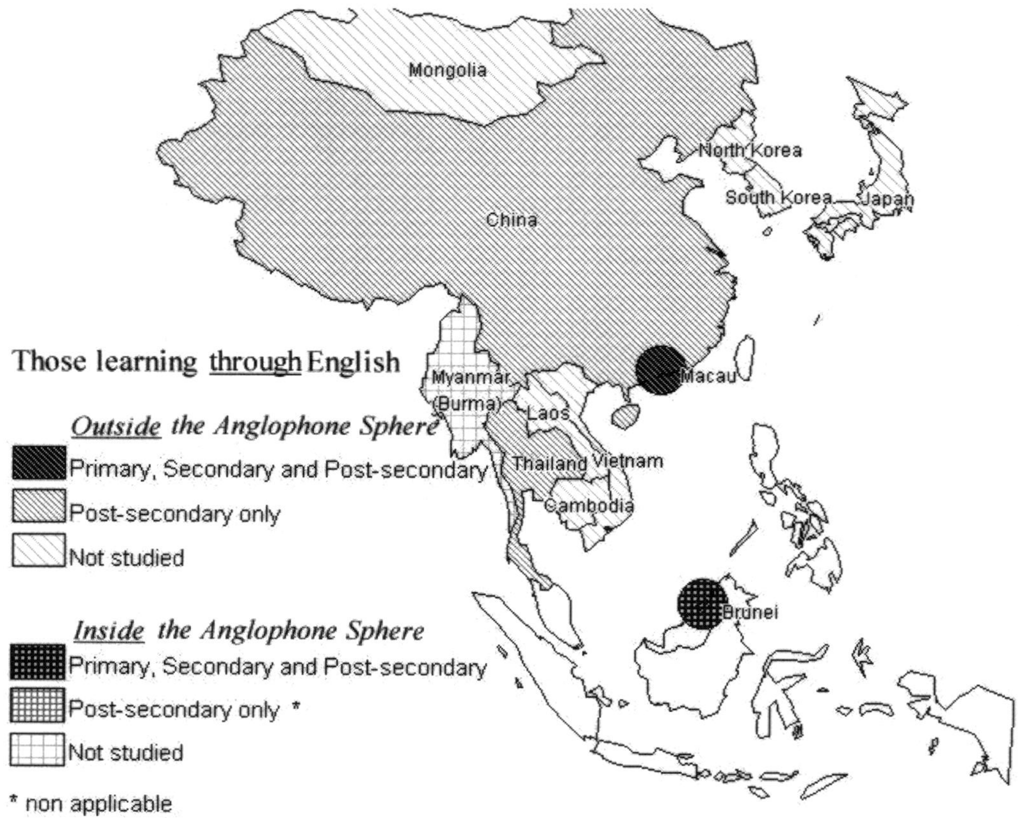

Map 3a: Dominant A type countries

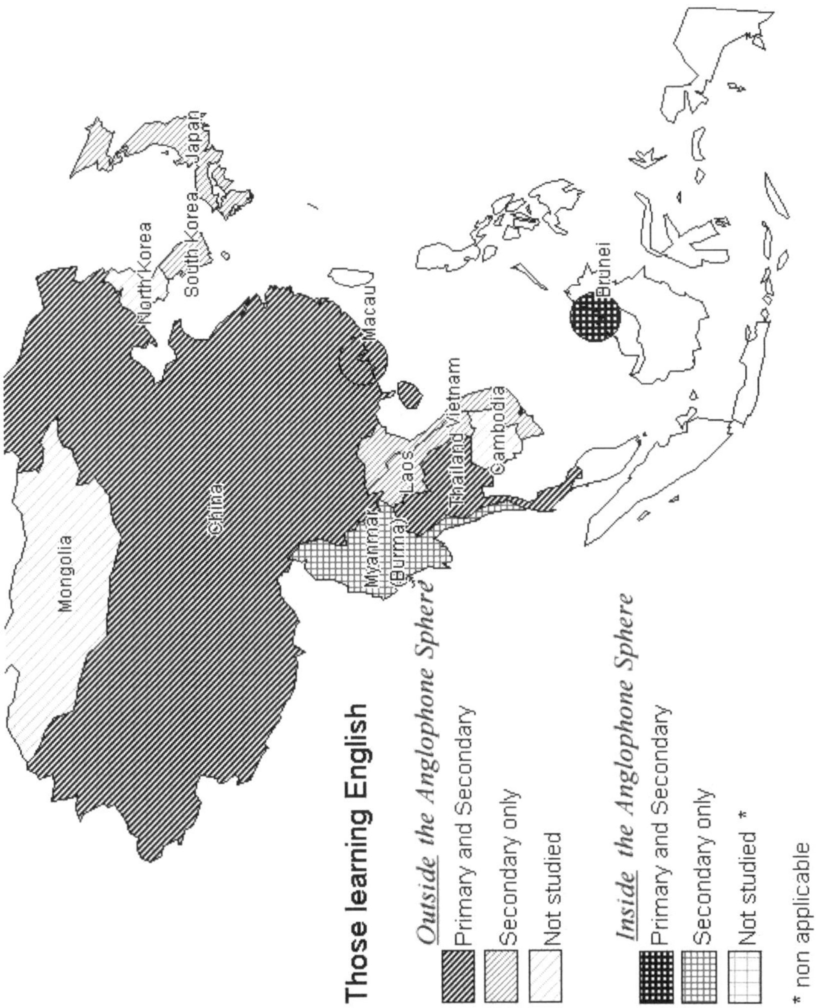

Map 3b: Dominant A type countries

Map 4a: Dominant **B** type countries

Map 4b: Dominant **B** type countries

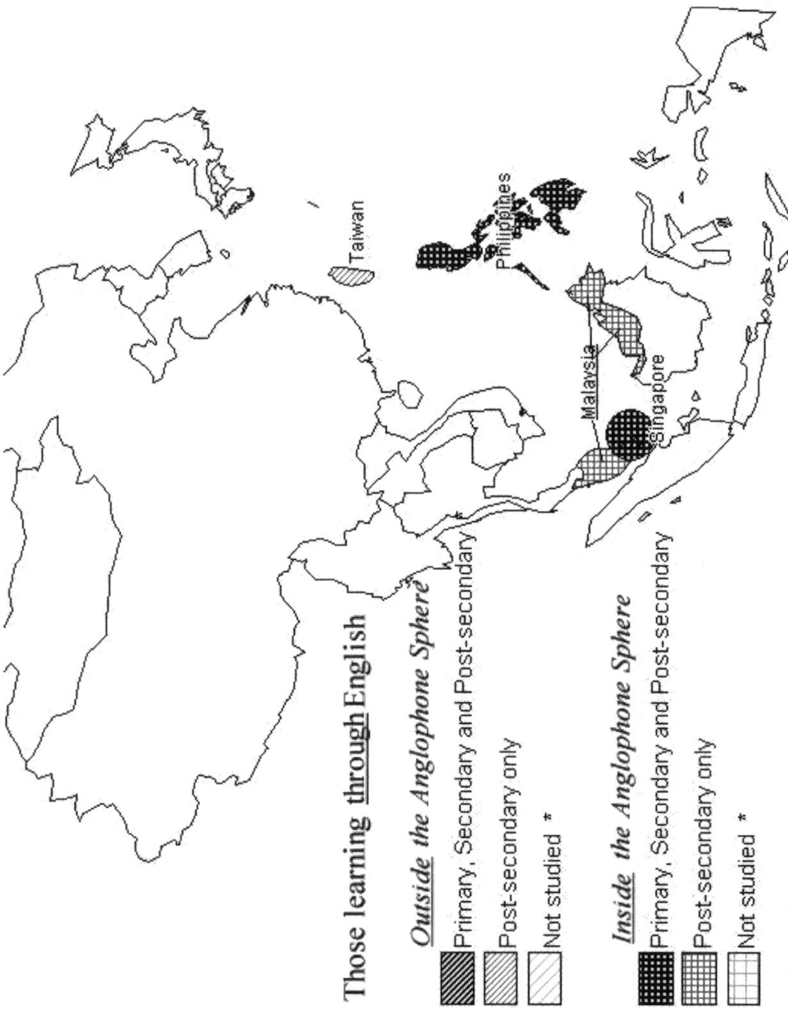

Map 5a: Mixed type countries

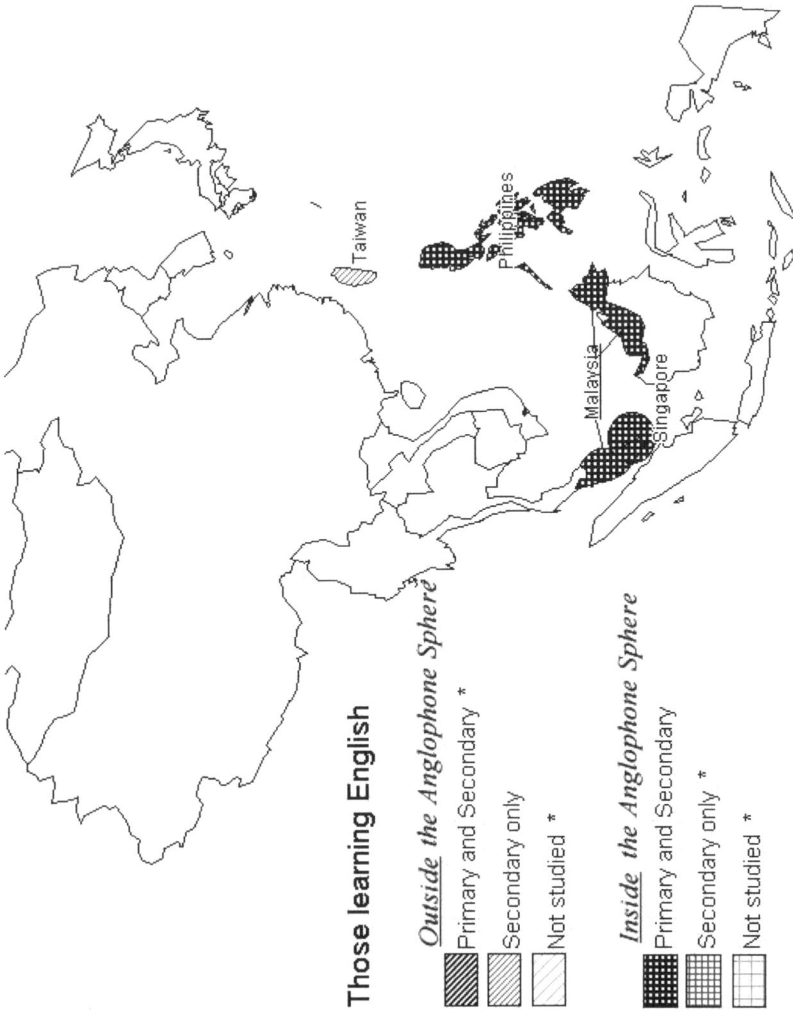

Map 5b: Mixed type countries

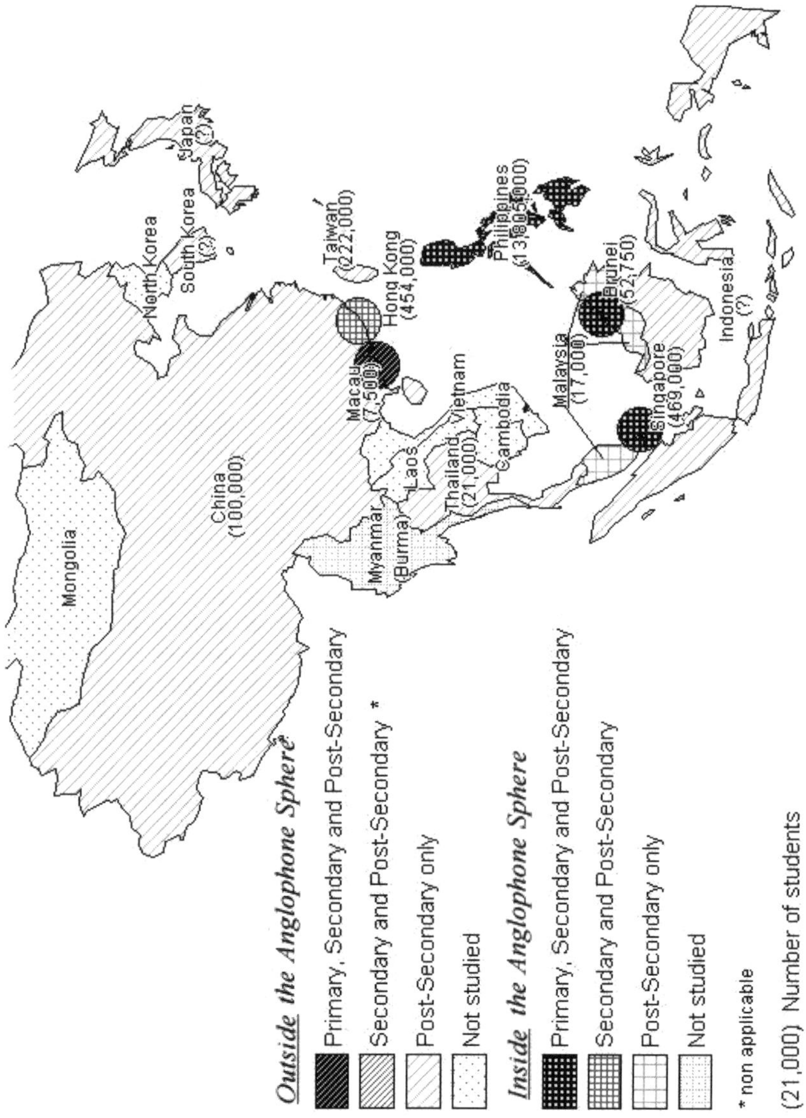

Map 6a: Those learning through English

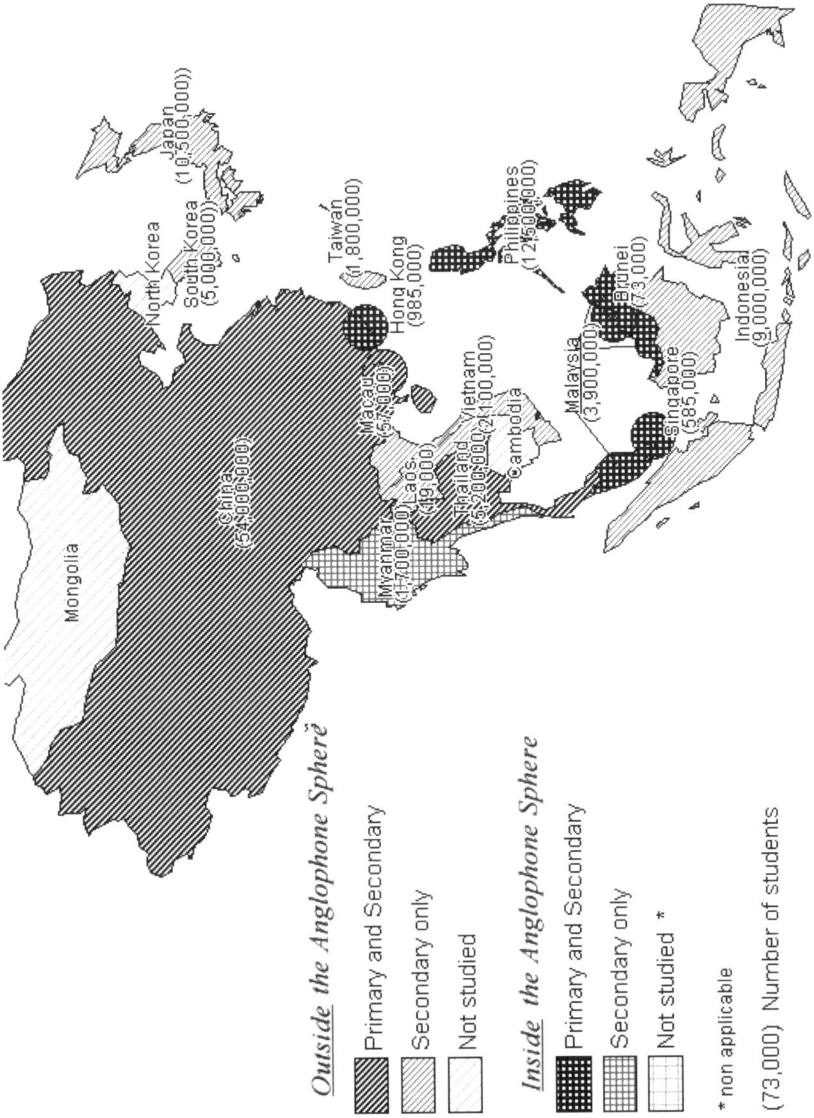

Map 6b: Those learning English

4. Notes

1. This paper was originally presented at the XIVth World Congress of Sociology in Montreal, Canada from July 26th to August 1st 1998 for the Research Committee on Sociolinguistics under the topic: *Effects of the Dominance of English as a Language of Science (ELS) on the Non-English Language Communities.*
2. Maps were produced using the MAPINFO software at the Southeast Asia Unit (Cellule de l'Asie du Sudest), Geography Department, Laval University, Quebec City.
3. Other publications compare vitality rates with numerous other demographic variables and indicate specific variable linkage to these rates. As examples, see Mackey 1976 and McConnell 1997 in the Reference section.
4. In: McConnell and Roberge 1994, the section on Asia covers all countries of the world, whereas in this presentation and analysis, we have focused on the 18 polities of our East and Southeast Asia data set.

5. References

Laforge, Lorne/McConnell, Grant D. (eds.)
1990 *Language Spread and Social Change: Dynamics and Measurement / Diffusion des langues et changement social: Dynamique et mesure.* Sainte-Foy: Presses de l'Université Laval.

Mackey, William F.
1976 Three Concepts for Geolinguistics. In *Sprachen und Staaten, Nationalitäten und Sprachenfragen in weltpolitischer Perspektive.* Teil II. Hamburg: Stiftung Europa-Kolleg, 169–239.

Mackey, William F.
1999 Forecasting the Fate of Languages. Paper Presented at the *Language in the Twenty-First Century* Conference, held on June 6th and 7th at the Whitney Humanities Center and the Center for Documentation on World Language Problems, Yale University.

McConnell, Grant D.
1997 Analyses et comparaisons des situations de contact en Inde. In Labrie, N. (ed.) *Études récentes en linguistique de contact.* Bonn: Dümmler, 294–317.

McConnell, Grant D. (ed.)
1998 *Linguistic Composition of the Nations of the World.* Vol. 6: *East and Southeast Asia.* Quebec City: Laval University Press.

McConnell, Grant D./Gendron, Jean-Denis et al.
1988 *Dimensions et mesure de la vitalité linguistique.* Vol. 1. Quebec City: Université Laval: Centre international de recherche sur le bilinguisme (CIRB).
1993 *Atlas international de la vitalité linguistique / International Atlas of Language Vitality,* Vol. 2: *L'Europe occidentale / Western Europe.* Québec, Université Laval: Centre international de recherche en aménagenment linguistique (CIRAL).

McConnell, Grant D./Roberge, Brigitte (eds.)
1994 *Atlas international de la diffusion de l'anglais et du français: L'enseignement,* Québec/Paris: Presses de l'Université Laval/Klincksieck.

II. Countries with a History of English-Language Dominance

Speaking of Science: The Use by Australian University Science Staff of Language Skills[1]

Richard B. Baldauf Jr.

1. Introduction

Scientific endeavour is an international phenomenon that brings together those working in science through professional organisations, conferences, books, journals, and personal contacts to work on common problems and to share ideas and findings. Science is cumulative and thus, despite some rivalry among individuals, sharing and debating the nature of knowledge is essential to scientific progress. Communication and the languages of that communication are, therefore, critical to the proper functioning of scientific endeavour. Those outside the scientific communication links, for whatever reason, find it difficult to participate in cutting edge science. Languages form one possible barrier to individuals' ability to participate fully in science (Baldauf and Jernudd 1983). As Kaplan and Baldauf (1997: 246 f) have shown, this issue is of such national importance that a small number of nations (e. g. Japan, Saudi Arabia, Israel, Taiwan and Malaysia) have undertaken national strategies to solve the problem language poses for information access.

The languages used in scientific communication can be thought of as forming a linguistic ecology. Mühlhäusler (1996: 206–7) has defined such ecologies in terms of

> [...] structured relationships between [the linguistic] co-inhabitants and between the individual inhabitants and components of the ecological support system. [He goes on to argue that] the key property of any ecology is structured diversity [where] diversity is defined not so much in terms of numbers but in terms of the quality of the meaningful relationships. [...] The aim of ecological language planning is to sustain structural diversity.

The notion of a linguistic ecology suggests that a set of relationships exist between languages that are used to communicate and develop meaningful ways of interacting and which may vary across domains. In the domain of scientific communication, Ammon (1998) has shown the degree to which English has become the dominant international language of science (ELS) and, to a lesser degree, also the humanities. This raises questions about how the linguistic ecology of scientific communication has changed over the last few decades, what effect changes have had on various disciplines and individuals working in those disciplines, and what will the emerging (and stable?) ecology look like?

Jernudd and Baldauf (1987) have examined the specific problem of written scientific communication and have developed a hypothetical model that illustrates some of the ecological variables that influence *language selection* for communication[2]. Briefly, the model (Figure 1) suggests that the sociolinguistic *type* of community[3] will at a minimum influence the *individual* skills that are developed, and that individual differences together with the pervasive characteristics of language use in the sociolinguistic *type* will contribute to determining language selection decisions. *Institutional* constraints and/or *network* contacts will often intervene between *type* and *individual* language selection decisions, as will availability and use of language management *resources* (e. g., library and database services, translation services, language editors, etc.). The various communities' *expectations* to which the individual belongs will also influence the language and text that the individual produces. The factors interact with the *domains* in which communication of scientific information can occur. These factors lead to *language selection*, which in turn have implications for the particular *discourse strategies* that are employed to produce the scientific communication (cf. Swales 1990). Taken together an understanding of the model's components and their interrelationships can help us to better understand the factors which influence language choice and the way in which scientific information is communicated.

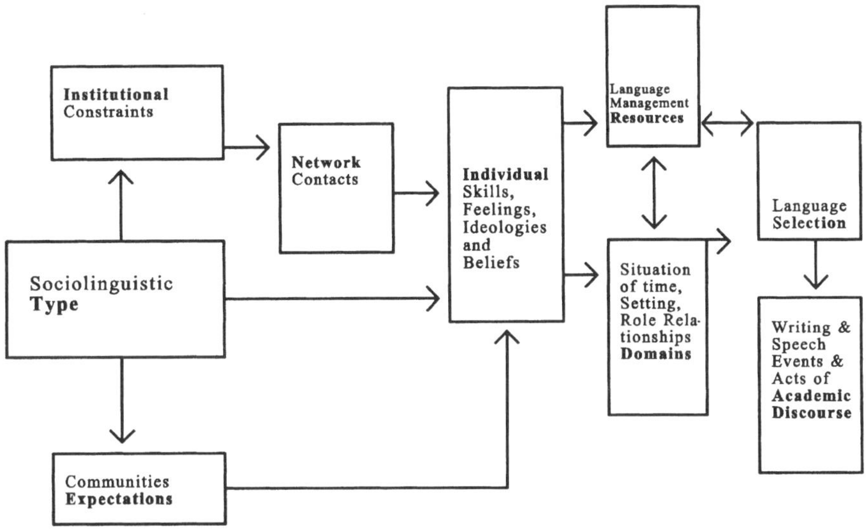

<---Macro Sociolinguistic Level---> <Individual> <MicroSociolinguistic Level>

Figure 1: Model of Language Selection in Scientific Communication (Jernudd and Baldauf 1987: 172)

Thus, it is argued that language forms a potential communication barrier for many scientists which they must overcome if they are to actively participate in their field and interact with their colleagues. In this paper, the role, growth and dominance of English as the language of science (ELS) is briefly reviewed (as both a problem and a solution) before examining how ELS has impacted on some English speaking scientists in Sydney, Australia.

2. English as a language of wider communication

The reasons for the rise of English as a language of wider communication (LWC) for technology and science, industrialisation, international business, transportation, and world-wide communication are not hard to understand, although they are sometimes obscured by accounts which stress social, historical and political factors (Crystal 1997; Wallerstein 1995), rather than those related to economic and scientific motivation (Kaplan 1987; Grabe 1988; Kaplan and Baldauf 1997). The growth of English has not occurred because it is efficient or flexible, but rather as a matter of historical accidents which have lead to the use of the language for these purposes. Although English has been spread as part of the British and American empires, it was the language of the elite and in most countries where it had spread, not the language of the home, of solidarity or of personal identity. Thus, military and political imperialism alone are insufficient bases for the phenomenal spread of English that we have seen since the end of World War II. Furthermore, other imperial languages such as Spanish and Russian have not become *lingua franche* in the same way that English has. What then explains the growth of English?

At the end of the Second World War, the United States was the only major industrialised nation whose industrial and educational infrastructure remained intact. The impetus provided by the war, post-war immigration and the cold war lead to the rapid expansion of higher education, the development of research universities and the development and use of the computer, first as a means for scientific data processing, but increasingly for information processing and retrieval. This co-occurrence of "military development, economic stability, research universities, R and D capacity, and the invention of computer networking" meant that the United States emerged as the greatest contributor to and user of the information pool (Kaplan 1987: 139) and that that contribution was in English. As the major contributor to and user of the system, the United States effectively became the manager of the "information cartel" and provided its organisation. Thus, English became not only the language of science and technology, but the language of information management as well. It is no accident that most computer languages use English terminology and concepts.

This coincidental dominance of English, needs to be seen in the context of the nature of recent scientific research and its relationship to modern economies. Kaplan (1987: 139) and Grabe (1988: 66) describe this relationship in terms of six basic assumptions about science and information that can be summarised as follows:

- Scientific information is cumulative;
- The time from the discovery of new information to its application has decreased markedly;
- The rate of information increase is nearly exponential with the quantity doubling every four years and with the cycle length decreasing;
- A sophisticated management system is needed to handle this information growth;
- Much of the technical information available world-wide is available in English; and

– Countries must have access to this information if they are to be economically competitive.

Based on these historical circumstances and the economic and scientific need to access information, ELS came into being. The need to use and the continued use of English has become a self-fulfilling prophecy. English has become the language of political and social opportunity. It provides important cultural capital, access to information, and it is the language of the multinational political economy[4]. This need for English access has created a thriving English-as-a-second language market worldwide (Kaplan and Baldauf in press). Such has been the demand for English that non-native speakers of English now outnumber those for whom it is a first language. Evidence for the growth of ELS is reflected in the languages used in scientific journals and we now turn to a review of that work.

3. English as the language of science

One measure of the extent to which English has become the dominant language of science can be seen from some surveys conducted on languages of articles contained in major abstracting services. For instance, in an early survey Wood (1967) found that nearly half the literature abstracted in Chemistry, Mathematics and Medicine was in a language other than English, while Biology, Physics and Engineering abstracted about 20–25% in a non-English language (see 1965 data in Table 1). However, in the 1960's the idea that English might become the dominant language of science was still not clearly foreseeable. Wood predicted that "as the underdeveloped areas (three quarters of mankind) catch up technologically..., it seems likely that the proportion of foreign language literature will increase rather than decrease" (1967: 117).

To test this hypothesis, Baldauf and Jernudd (1983) replicated that particular aspect of Wood's study using the articles abstracted in the 1981 issues of five science databases[5] to determine if the changes Wood predicted had occurred. However, as the 1981 data in Table 1 indicates, the proportion of articles in English dramatically increased, rather than decreased, since 1965 while in general the proportion of articles in other languages declined. Specifically, in the two abstract databases which are most equivalent over the sixteen year period, Chemical Abstracts with

CA Search and Mathematical reviews with MATHFILE, the use of English as the language of publication increased from 50.3% to 66.95% and from 54.8% to 69.3% respectively.

A longitudinal study by Tsunoda (1983) that covered publications in mathematics, biology, chemistry, physics and medicine over the course of the century (1880–1980) shows that until after World War I, publication in English was roughly on an equal footing with French and German, but that from 1930 onwards English gained at the expense of the other two languages.

A further replication of Wood's study was undertaken to collect language of publication data from abstracts for 1988 from the same five databases as well as from Sociological Abstracts. Large increases were found in the numbers of articles being abstracted in four of the five databases, reflecting the growth in publication in these fields over the seven-year period. The proportion of articles published in English remained relatively stable, the proportions declined slightly in Biosis and CA Search, but increased in INSPEC, MEDLINE and MATHSCI. The proportion of publications in Russian also remained stable, but from a much smaller base[6]. The proportion of publications published in French and German both declined slightly, while those in Japanese and Chinese rose. Thus, much of the expansion in the science literature continues to come in the English language domain. Eight languages accounted for between 95.2 and 99.7 percent of the publications listed by these abstract services.

Table 1. Language breakdown for abstracted literature indexed in six databases[1] for *1965**, 1981** and 1988.

Language	Chem	Biol	Engin	Med	Math	Socio
1965*						
English	50.3	75.0	82.3	51.2	54.8	
Russian	23.4	10.0	3.9	5.6	21.4	
German	6.4	3.0	8.6	17.2	8.7	
French	7.3	3.0	2.4	8.6	7.8	
Japanese	3.6	1.0	0.1	0.9	0.7	
Chinese	0.5	1.0	0.0	0.4	0.2	
Other	8.5	7.0	2.7	16.1	6.4	
Total N	–	–	–	–	–	

Table 1. Continued

Language	Chem	Biol	Engin	Med	Math	Socio
1981**						**1976**[2]
English	66.9	85.7	84.6	73.1	69.3	46.3
Russian	12.7	3.9	3.8	5.9	18.1	11.3
German	5.5	2.5	3.9	5.5	3.6	5.7
French	1.9	2.1	2.0	4.0	4.8	14.3
Japanese	9.9	1.9	1.5	3.0	0.3	7.1
Chinese	0.9	0.2	0.6	0.7	0.5	+
Spanish	0.2	0.7	0.3	1.2	0.3	+
Italian	0.2	0.6	0.5	1.6	1.3	+
Other	1.8	2.3	2.8	5.0	1.8	15.3
Total N	252,409	300,024	167,618	258,941	35,876	4,827
1988						
English	63.9	84.7	85.9#	75.6	71.1	82.4
Russian	11.2	3.8	1.9	6.1	18.2	1.1
German	4.5	2.4	3.1	4.1	2.0	4.8
French	1.2	1.9	1.4	3.2	3.4	3.4
Japanese	12.6	2.2	2.6	3.2	0.2	0.1
Chinese	2.5	0.5	1.1	0.8	3.2	0.0
Spanish	0.5	1.0	0.3	1.1	0.6	1.1
Italian	0.2	0.8	0.3	0.9	0.7	1.8
Other	3.0	2.2	3.0#	4.6	0.3	4.8
Total N	465,603	520,000	238,522	183,111	54,876	22,832

* Wood (1967: 119); ** Baldauf and Jernudd (1983: 99); + not given separately, included in other; # Estimate

[1] CA Search (Chemical Abstracts), BIOSIS (Biological Abstracts), INSPEC (Engineering Abstracts), MEDLINE (Index Medicus), MATHFILE (Math Review), Sociological Abstracts.

[2] Thogmartin (1980: 6) 22.8 per cent sample of UNESCO *Bibliography' of Social Sciences.* This is not a comparable source as it not a database and was selected because it was likely to have more non-English citations.

Baldauf and Jernudd (1983) also examined the database MATHFILE by language of publication over a seven year period. While English language articles increased proportionally to the total for each year using 1973 as the base line year, Russian, German and Japanese language articles were relatively stable in terms of absolute numbers of articles published. Extension of this data through 1988 revealed similar trends-growth occurring mainly in English with other languages remaining relatively stable. If this data is representative of other databases[6], it suggests that while the English language was becoming proportionately more dominant in the scientific literature, there continued to be a large and relatively stable non-English language scientific literature.

Of course, Wood's hypothesis may still have been right. The proportion of foreign literature may have increased during this period vis-à-vis English – although no data on this is available, but its representation in the abstracted database literature may have declined. Part of this decline can be attributed to the fact that some non-English language journals became English[8] or bilingual (often effectively becoming English only). Part of the decline is also undoubtedly due to the fact that abstracting services, when faced with an increasing literature and limited resources, have opted to select what they consider to be high impact international journals over journals from the Third World written in "exotic" languages (Garfield 1985: 8–9). University library research collections as well, contribute to how the literature is shaped and used. When faced with selecting from a growing list of 80,000 journals, and with shrinking budgets, libraries are having to become more selective. While journal material is still accessible by other means (e. g. on the internet[9], interlibrary loan), the loss of ability for academics to physically browse the literature directly means that different computer-based skills need to be developed to avoid loosing touch with the literature.

In the context of the previous section where Kaplan and Baldauf (1997) noted that database users control the information cartel, the move toward English language dominance reflects the dominance of English language users' contributions to the system and their need for highly relevant information for themselves. While databases are often treated as if they were representative of the world's scientific literature, they are not, nor with their present limitations are they likely to be. Since the large, well-known databases must be selective for practical reasons, database selectivity is a problem that all scientists must face[10]. The growth in coverage and number of databases over the last fifteen years may in part be a response to the need for access to the wider literature. Yet, most of these databases are in English[11].

4. Language selection within national boundaries

The dimensions of the language problem are not just between English- and non-English language using scientists. Scientists working in a particular language area also tend to cite more articles from their own language than do scientists from another language area. For example Louttit (1957) analysed the citations taken from the 1952 issues of journals in psychology, chemistry and physics. As Table 2 shows English and German-speaking scientists made a high proportion of citations to work in their own language. For the French, the trends were more diverse, but nevertheless the French cited French language sources more frequently in their articles than did German or English speakers.

To take another example, Lange (1985) has examined the language of citation for the theoretical sections of psychology journals in four countries. In the Russian journal, 10 percent of the articles cited were in English, in the two American journals 89.1 and 94.2 percent of the articles cited were in English, while in the two German language journals the percentage of citations in English were 38 (GDR) and 69.4 (FRG).

Table 2: 1952 Language of citation by language of publication (Louttit 1957: 315)

Language of Publication	Language of Citations % English	German	French	Other
PSYCHOLOGY				
English	92.5	5.2	0.7	1.6
German	2.7	91.1	6.2	0.0
French	25.6	7.4	64.0	3.0
CHEMISTRY				
English	79.4	11.9	4.7	4.0
German	22.8	64.0	6.2	7.0
French	36.7	27.9	29.1	6.3
PHYSICS				
English	86.2	5.9	2.0	5.9
German	33.5	58.1	1.4	5.2
French	49.8	15.3	30.7	4.2

In another example with a somewhat different focus, twenty-six matched pairs of English-French articles from the *International Journal of Psychology* were examined. French-language articles contained fewer references (459 to 618) than English language ones. Only 66% of the French language articles' references were in English while 96% of the English language articles' references were in English (Baldauf 1986). If we use external citations as a measure of importance, in the period between 1969 and 1984, 78 external citations were recorded for the French language articles whereas 171 were found in *SSCI* for the English language ones. Eleven of the twenty six French language articles had no external citations whereas only six of their English language counterparts suffered the same fate. Self-citation rates were similar and accounted for an additional 18 and 22 citations in each group respectively (Baldauf and Jernudd 1986). These data strongly suggest that language of publication provides differential access to and use of information. For those French authors there were significant disadvantages in publishing a cross-cultural psychology article in French rather than in English, at least in terms of citations.

Large (1983) cites a number of other instances from German, Spanish and Soviet journals of the language-citation pattern of citation. However, this phenomenon may apply only to scientists working in international languages. As Baldauf and Jernudd (1987) have noted, the citation patterns of many Scandinavian Psychologists are not that different from their English-speaking colleagues.

5. Business and language skills in Australia[12]

When examining a problem in one domain of language use, it can often be useful to examine that problem in other domains as the work in those domains may shed light on community and professional attitudes and trends on language use in society as a whole. Such trends may help to explain the milieu in which attitudes scientists have toward language have developed and may help to explain language use patterns. In Australia, economic rationalism has been the driving force behind many of the social and political changes that have occurred over the last decade, and so an examination of the use of language in business may be helpful in understanding the mix of attitudes toward language and culture held by many Australians.

Australia has changed dramatically in the last fifty years in its economic and social structure. Prior to World War II it was primarily a pastoral country, with very limited industry and an Anglo-Celtic monocultural and monolingual view of itself. World War II forced Australia to reexamine this heritage and the post war migration, mainly from Europe, was intended to develop Australia's industrial base. In the past twenty-five years Asian migration has added to the country's multilingual and multicultural diversity. While Australia's industrial and services sectors of business have expanded since the war, its internal market is small and it needs to move more forcefully into the manufacturing and services sectors and/or export more to a wider range of markets.

In response to this growing ethnic diversity, Australia has developed a series of language policy statements since the early 1980s directed at meeting these national economic and social needs (Kipp, Clyne and Pauwels 1995; Lo Bianco 1990; Ozolins 1993). At the Commonwealth level, the *National Policy on Languages* (Lo Bianco 1987) was developed to help achieve four broad social goals: *equality* within a multicultural society, *economic* development and trade, *enrichment* of social and cultural life and *external* relationships. In 1991 this was followed by the *Australian Language and Literacy Policy*, in 1994 a National Asian Languages and Studies in Australian Schools Program (COAG 1994) and in 1996 the government took the first steps toward developing a national schools literacy policy (Lo Bianco and Freebody 1997). Progressively, each of these language and literacy policy documents has stressed the underlying theme of employment and economic development. State governments (e. g., Queensland) in framing their school language policies have also suggested economic motivations for their actions (Baldauf 1993; Djité 1994) and school systems and universities have been keen to export their educational expertise to countries in the region. Perceived economic benefit (economic rationalism) has been a major driving force behind these government language policies and the accompanying public rhetoric. In this context, increasingly universities are being seen as just another business with the need to be more productive; i. e. to teach more students with fewer staff and resources while at the same time doing more research and fund-raising consultancy work.

Business and industry interest in language, to the extent it has existed, has also been based on economic considerations although Australian companies have generally been inward-looking "passive exporters" (e. g., Stanley *et al.* 1990: 46)[13]. As engagement with language and cultural questions has grown nationally, two perspectives of how to deal with this business-sector language problem have emerged.

5.1. Language, multiculturalism and economic outcomes

From the "multiculturalist perspective" the challenge for business has been to change the culture of Australian industry to make it active and world-oriented with an understanding of the role that language and culture play in export markets. In a market which increasingly emphasises selling products from the secondary and tertiary sectors of the economy, more interpersonal encounters with potential customers are required because individualised products are being marketed to individuals or small groups of customers and not, as with primary produce, in bulk quantities to government agents. These strategies require much more personal interaction with customers and a knowledge of their tastes and preferences. As many of these new customers are assumed to be non-English speakers, these changes have obvious policy implications. A demand should be created for second language and culture education and for the proficient and confident use of language skills since secondary and tertiary economic activities necessarily involve the need for diverse language and cultural competencies.

This scenario broadly describes one view of Australia as a multicultural country (see Clyne 1997). An analysis of the 1991 census data (Kipp, Clyne and Pauwels 1995) indicates that; 14.8 per cent of the population uses a language other than English (LOTE) at home. This figure possibly underestimates this potential language resource but reflects an increase from 12.3 per cent in 1976 and 13.6 per cent in 1986. However, there is often a gap between what many immigrants are able to contribute linguistically and culturally because either their English is not up to standard or the Australian authorities are unable to recognise their professional qualifications because they were obtained in another country and are not deemed to be equivalent. This means that Australia is not able to use to the fullest extent the language and cultural resources these people bring.

Stanton, Aislabie and Lee (1992) have suggested that trade and commerce, tourism and small business are the three ways in which multiculturalism could contribute to economic objectives of the polity. However, they conclude that there is little direct evidence from the literature in any of these areas about the contribution that multiculturalism actually makes to economic performance. Stanton and Lee (1996: 510) go on to point out that Australian access and equity programs have long been part of multicultural policy and that such programs may provide economic outcomes. However, they seriously question any attempt to use "the contribution of cultural diversity to Australian export performance [as] one element of an attempt to establish an economic agenda as a part of Australia's multicultural policy."

5.2. *English as the language of business*

The alternative perspective suggests that, a foreign language is becoming less important for business and that trade and commerce can be developed around English as the language of business.

> It is a deliciously Australian irony and so typical of our planning from the beginnings of white settlement that Australian governments should discover the economic importance of languages at the moment in our history when languages have never been less important on purely economic grounds. Exactly when the pendulum is swinging toward English among the educational and trading elites of Asia, for example, our Governments are concentrating their rhetoric on providing language courses of unspecified content and outcomes. The primary wish of these trading partners in language terms is to acquire English proficiency. (ALLC 1994: 9)

A summary of some twenty studies in this area by the ALLC (1994: 33–50) suggests that while personal skills in a language other than English are useful for social intercourse and in marketing, many business people seem to feel that a command of another language is unnecessary for business purposes, other than for interpreters and translators who would be hired as part of the negotiation process.

For example, in a study by Stanley *et al.* (1990), two thousand companies nominated by Austrade were surveyed by questionnaire, with a 25% response rate, to try to establish what the relationship was between language other than English (LOTE) skills and export success. The survey found that there was generally a low level of awareness of the relationship between LOTE skills and export success, as evidenced by the fact that, while knowledge of LOTE skills ranked as least important in the survey, obstacles related to the lack of LOTE skills (knowledge of foreign markets) were rated as the most serious. Even when companies did recognise that LOTEs were important, they were not sure which might be important for their own purposes. Exporters viewed the following nine languages as being in demand: Mandarin, Japanese, Arabic, Indonesian, Korean, Thai, Spanish, German and French, roughly in that order.

In another report on trade with Latin America, Austrade expressed the view that:

> It is difficult for Australian companies, particularly those which are fairly new to the export business, to get into a market like that. Their attitude is why go through all this hassle with...different languages...when they can go to closer markets such as South East Asia, New Zealand or whatever. “ (SSCFADT 1992: 278–279).

In a third study interviews with 60 of the biggest export companies in New South Wales found that Japan, the United States, New Zealand, Papua New Guinea, the United Kingdom and Ireland were the companies' most important clients in 1988 in terms of income earned. By 1998, they expected Hong Kong, China and Thailand to join that list. This suggests that Australian exporters have a limited vision of what markets there might be for their products (Valverde 1992).

Although these companies represented the "exporters" and therefore were supposedly ones with some expertise in this area, most companies were represented by monolingual/monocultural individuals with very limited notions of geographical and cultural differences (i. e., couldn't necessarily differentiate between countries in South East Asia and those in Oceania). Only two export managers said they had heard anything about the National Association of Australian Translators and Interpreters (NAATI)[14], and interpreters and translators were seldom used. Although there was a lot of interest shown in many of the issues presented in the survey, there was very little evidence of an awareness for the need for language and cultural skills.

Exporters suggested that Japanese, Chinese, French, Korean, Spanish, Indonesian and German were the main languages of the future. The survey also showed that these companies actually had multilingual employees fluent in many of these languages, except Japanese where the upper management was directly hired from Japan. Thus, attitude, rather than resources seemed to be a prime problem for these exporters; There was a belief that English is enough. Even when language skills were used, in only one case were these skills recognised by better pay.

This perspective from business suggests that views about language and culture in the Australian community are divided. Government policy has stressed the importance of languages as a resource, of multiculturalism as the Australian way of life and that Australia is a part of and increasingly a partner in Asia. However, despite the rhetoric there is little direct evidence that multiculturalism has substantial economic benefits. While the multiculturalists have been politically dominant for the last decade, it is clear from election results in the late 1990s (e. g. in Queensland) that some Australians have not been persuaded by this line of thinking and are feeling marginalised by it as it seems to exclude rural, "Anglo" and working class groups. Although Australian multicultural language policy has always been based on an "English plus" model, these views undoubtedly mean that many Australians believe that if it is important, it will be reported or written in English (i. e. the "plus" element of the policy is not essential).

6. Languages at Australian universities

How language is viewed in a community also depends on the availability of language to those who wish to learn it. Experience has shown that most immigrant communities are unable to sustain their languages without further in migration. While the Commonwealth and States all have language policies which support the teaching of languages in schools, this support does not extend directly to universities. Under current policies by 2004, all students in years 3–10 are expected to study a language other than English in school. A number of reports have looked at language teaching in Australian Universities (Leal, Bettoni and Malcolm 1992; Bettoni and Leal 1994; Baldauf, *et al.* 1995; White, Baldauf and Diller 1997; Baldauf and Djité 2000) and all have found support for language teaching to be inadequate. In 1997 the 36 Australian universities offered 50 languages (not including Anglo-Saxon or Old English). The most frequently taught languages were Japanese (35), Chinese (26), Indonesian (23), French (21), Italian (16), German (14), Spanish (13), Classical Greek (11), Modern Greek (10) and Latin (10). Other languages offered at more than 3 institutions include Arabic, Hebrew, Hindi, Korean, Russian, Sanskrit, Thai and Vietnamese. Thirty-one languages were offered at only one or two institutions.

While the relatively large number of languages offered might suggest that language offerings in Australian universities are healthy, and in some instances growing (e. g. Japanese) when compared to previous years, this would be a mistaken assumption. Continuing cuts to university funding by the Commonwealth government has severely affected the Humanities in particular. In an economic rationalist environment, subjects with smaller numbers that are more expensive to teach are targeted for closure. Potentially, this places languages, and especially low candidature languages, in a precarious position.

Another major concern is the substantial staff attrition rate that pervades nearly all language departments in all universities. While specific language offerings may be the same in 1997 as, for example, 1994, often they are being taught with far fewer staff or with fewer contact hours. A related concern is the loss of senior academic language staff as they either retire or are forced to accept redundancy packages from beleaguered universities, and then are either not replaced at all, or replaced by non-tenured staff, often part-time lecturers or casual staff, leading to the loss of a whole generation of language specialists. An Academy of the Humanities study currently in progress estimates that nearly 100 full-time academic positions in languages have disappeared between 1996 and 1999.

Underlying this rather grim environment is a concern by many of the low esteem in which languages are held by university colleagues (often senior administrators); the lack of understanding of the complexity involved in providing sound pedagogic language teaching; the lack of understanding by the community at large of the value of languages which one may argue is also a problem for governments with their sole emphasis on trying to 'cut costs'.

The present 'corporatist' climate in which universities are now seemingly forced to operate has taken a serious toll on language departments. It is questionable that without some alleviation of under-funding, understaffing and under-resourcing which is the lot of most language departments in the late 1990s, that the broadly based teaching of languages will remain viable as a significant part of offerings at many universities.

The pressure that many language departments find themselves under is beginning to affect research and the ability to work with languages other than English. In the long term this will reduce the ability of Australian scientists to work in other languages and will reinforce the notion that English is the sole language of science.

7. Survey of Australian university science staff

It is estimated that Australia produces about three percent of the world's published research. The University of Sydney is the largest tertiary provider of languages in Australia, offering in 2000 some 34 languages[15] through its Faculty of Arts and over a hundred languages on a self-study basis through its Language Centre. The University has a multilingual and multicultural staff, many of whom are involved in cross-national research or have international connections. It is one of the major research Universities in Australia in terms of the amount of research funding it is able to attract each year and keeps a register of languages that staff are able to speak/use as a research tool. As such, one might expect that if Australian scientists were using languages other than English in their research work, evidence of this would be found here.

To examine the question of pro-active language use, as opposed to passive reading or citation use, by University Staff for research communication purposes, University of Sydney's *Research Reports* for 1981 and 1988 were examined to see what languages were being used for which purposes – to develop hypotheses about language other than English use. The five Faculty/Departmental groupings most similar to the areas examined in Section 3 (i. e. Chemistry/Biochemistry, Biological Sciences/ Agriculture, Engineering, Medicine and Mathematics) plus the Faculties of Arts (in two groups – languages and other) and Law were examined. These results are tabulated in Table 3.

Table 3: Number of citations by faculty group with non-English (NE) sources by year

Year	1981 Total[1]	1981 NETitle[2]	1981 NE Source[3]	1988 Total	1988 NETitle	1988 NE Source
Chem/Biochemistry	73	0	1	208	0	11
BioScience/Agricul	136	0	0	213	0	0
Faculty Engineering	150	0	1	300	0	3
Faculty Medicine	548	2	13	1277	3	20
Mathematics/Stats	43	0	5	62	0	1
Faculty of Law	27	3	4	114	3	3
Faculty Arts-Langs	45	11	15	122	28	36
Faculty Arts-Other	198	14	16	455	10	14
Totals	1220	30	55	2751	44	88

[1]The number of published language events (books, chapters, articles, papers, reports) found in the University of Sydney *Research Report* for 1981 and 1988. No *Reports* have been published since 1990.
[2]Number of non-English titles listed.
[3]Number of non-English sources. These may be underestimated as it is not always possible to tell from the title whether the source is non-English. It may also include some English language journals published in non-English speaking countries. For example, Ammon (1998: 48) indicates that most science journals – using science in the broadest sense – published in Germany are now (almost exclusively) in English with an English title.

The data indicate that only a relatively small number of publication events[16] make use of a language other than English (LOTE). For 1981 only 30 titles and 55 Sources out of 1220 publication acts were in a LOTE. While the total number of publication acts had more than doubled by 1988 to 2751, the numbers of titles (44) and sources (88) remained constant except for those working in the Faculty of Arts where the languages departments had increased research and use of languages publication. Overall, German and French were the languages most frequently used with publication titles also in Chinese, Dutch, Finnish and Polish. In a further examination of the literature, 362 publications were identified as having a University of Sydney author in the 1973–1998 *Linguistics and Language Behaviour Abstracts* database. Of these only nine were written in a LOTE (7 in French and 2 in Italian). The seven authors involved (one of the authors published three articles) were members of either the French or Italian Departments. For only one of the seven was the LOTE a second language.

An examination of the language-use events represented by these publications, suggests the following hypotheses about proactive non-English language use by Australian scientists:

- Personal language background (mother-tongue) may account for the use of a LOTE in some circumstances (Chinese, French, Italian and Polish examples)
- Co-author expertise (non-Sydney) may facilitate publication in a LOTE (German, Italian and Dutch examples)
- Discipline expertise in languages may encourage scientists to write in a LOTE (French, German and Italian examples)
- Some disciplines are more internationally and interculturally oriented than others (e. g. in the Faculty of Law, no papers from the Department of Law (Australian focus) were in a LOTE whereas a number of papers in Jurisprudence were (international focus))
- Some grant or project requirements may require the work to be published in a LOTE (a Chinese language example – possibly translated in the Peoples Republic of China)
- Transcripts of new language materials (e. g. publication of Aboriginal songs in the original language with Aboriginal co-authors)
- Articles about work in language, where the article and the source are primarily in English, but analyse a language text (e. g. a Welsh text; reviews of books in a LOTE)
- Research completed on a local topic and published in a local journal (e. g. on Pompeii, published in Italian)
- A few papers given overseas are in a LOTE (pre-print versions, e. g. ISA Papers)
- As journals become more English language oriented, it is increasingly difficult to tell whether a journal is solely in English or includes other language materials (e. g. Sudan Medical Journal, Pflügers Archiv, European Journal of Physiology, Electrochimica Acta)

Three additional language-related activities were noted in the data.

- Papers were presented all over the world in countries where a LOTE was spoken, with an apparent increase from 1981 to 1988, but in English (increased language contact)
- German and Dutch (in particular) publishers publish a lot of scientific material in English although their own language is not English

– There were a number of non-University of Sydney authors with non-typical Australian names (whatever that may be!) suggesting there may be a mentoring or English language facilitation process occurring with some overseas co-authors.

The results of this examination of publications parallel the database information discussed in Section 3, i. e. the number of publications and the use of English are increasing. For English language milieu speakers (at least at this site), the pro-active use of a language in scientific publication is very limited and becoming increasingly so in terms of the total number publications produced. The fact that science is cumulative and the major databases are in English and favour "international" journals means that this trend is likely to continue in the print media unless something is done to change the language ecology. However, the increased overseas (direct cross-linguistic and cross-cultural) contact may make English-speaking scientists more aware of language issues.

From a language planning perspective, these decisions represent 'unplanned' micro-language planning decisions (see Baldauf 1994; Kaplan and Baldauf 1997). While such decisions may not be as immediately interesting as the planned macro-language planning in Québec or the Baltic states, never-the-less when taken as a totality, they do have a major impact on language communities. In a world where individual rights and responsibilities are increasingly being stressed, it may be appropriate for language planners to look more to individuals to drive language change.

8. Implications for English speaking language communities

The studies cited in this paper indicate that the universality of science which is often taken for granted is undoubtedly segmented by language and language selection. While ELS has created an apparent communication channel for scientists, the data presented in this paper also indicates that the channel comes at a cost and that a language barrier for the communication of science information still exists. The language ecology of scientific communication is changing and this raises the question of "What can we do as scientists, sociologists, applied linguists or TESOL teachers to reduce the effects of this barrier – to make scientific communication more effective and more accessible to all scientists?"

First, the critical approach to language teaching, language planning and literacy is increasingly providing an important critique of the role that particular language use and English play in creating language problems and language inequality. They suggest that modernist planning approaches to language planning and TESOL teaching may be inadvertently creating more problems and inequality than they solve. Luke, McHoul and Mey (1990), Tollefson (1991), Phillipson (1992) and Pennycook (1996) raise many of the issues central to the problem and the latter provides field exercises which may help English speakers to experience the language barrier. Being aware and sensitive to the problem is an important first step.

Second, for those working with TESOL students, especially those who aspire to become scientists, materials about the language barrier need to be explicitly made part of their curriculum. They need not only to understand how academic discourse works, but the politics of it as well. Swales (1990) provides a good introduction to these issues, while Barnes (1982) is more language teaching oriented.

Third, in a world that is becoming increasingly multilingual, there is a need for monolinguals, particularly monolingual English speakers because they symbolise the language barrier, to learn and actively use a language other than English, particularly in their role as scientists. One way to support the non-English literature is to subscribe to a LOTE journal or to publish an article in a LOTE (cf. Conseil de la langue français 1986; L'Heureux 1986)

Fourth, journals and their editors need to have a greater mentoring role generally, but especially in the case of second language authors. As this takes time and effort, it may mean that journals will need a different level of support from their sponsoring bodies and institutions, and that editorial boards will need to take on different functions. The need for language editors (e. g., *Scandinavian Journal of Psychology*) may need to be explored (*NAVF* 1976).

Fifth, journals and presentations need to consider publishing abstracts of their articles or providing resumes of their presentations in languages other than the original (Lewin and Jordan 1981; Touraine 1998). For non-English language journals, having an English abstract available may be the difference between making it into a database or not. Ikushima and Tenopir (1988: 118) report that "language of periodicals or summaries seems to be the most important single factor that determines whether Japanese scientific and technical literature is included in major English language databases". For English language journals, there is an obligation to our colleagues to provide access to our work.

Sixth, journals might also consider encouraging and publishing more reviews, especially those that collect their information cross-linguistically.

Good reviews provide better access to materials than abstracts. Gordon and Santman (1983) discuss the role of reviews and the language barrier. Larsen and Magnussen (1984) introduce a policy of in depth reviews as a way to bring Scandinavian psychology to the attention of psychologists world-wide.

Seventh, there is need for scientists, to be more collegial, to provide more mentoring, to do more collaborative work with authors who would not otherwise be able to publish, either through lack of experience, or a lack of linguistic skills (cf. Subramanyam 1983).

Eighth, English speakers in particular need to recognise the limitations of English bound databases (Smith 1979; Byrne 1983) and go beyond them to seek out materials in LOTEs. Not all important work is in English. This may mean using translation (van Bergeijk and Risseeuw 1980; Harris 1983).

Finally, there is a need be sure that our science information is being communicated, not just published (Manten 1980). This may mean finding more appropriate ways and outlets for our science communication. Chatelin and Arvantis (1989) document the changing habits of French soil scientists, who continue to publish in "local" journals, while presenting papers more frequently at meetings and congresses. Authors may need to consider a wider variety of publication outlets to take the problem of language into consideration. In this context, English speakers need to be sure when working in a non-English speaking situation, that appropriate and accessible "publication" of findings is available to that community.

There are undoubtedly other ways to create a more supportive language ecology for scientists in the era of ELS, but most important are the changes we make ourselves as scientists to foster more equity and access within the community of scholars.

9. Notes

1. Paper presented at the XIV World Congress of Sociology RC 25 Sociolinguistics symposium on the "Effects of the Dominance of English as a Language of Science (ELS) on the non-English Language Communities", Montreal, 29 July-3 August 1998.
2. For a variation on this model and an example of its use as a tool for understanding scientific communication, see Swales (1990: 104–105).
3. Jernudd and Baldauf (1987: 165) indicate that:
   The sociolinguistic type determines dominant language selection in local, (normal) national and (sometimes) regional institutions, based on the conditions of the communication system in the society at large. It determines what languages in-

dividuals normally acquire in their life cycle and what languages they normally use.

Many of the late 1980s examples given in that chapter have changed, but it might be said that the standard English-speaking type (e. g. the United States, Australia, New Zealand) is that English is seen to function at all levels (local, national, regional and international) with only specialists (diplomats, some academics) learning other languages and with migrant languages supported or tolerated on a laissez-faire basis. In southeast Asia there is a growing 'English-knowing bilingualism' type developing, exemplified by Singapore, where English is increasingly becoming a mandatory second language for a wide segment of the population and where English is taking on some national (e. g., university education) functions.

4. While this paper focuses on the language problems that exist for individual scientists in their work, Tollefson (1991) points out some of the wider implications for ELS and ELWC. In particular, that the spread and dominance of English creates inequalities 1) between 'developed' and 'developing' societies and 2) within 'developing' societies. Luke, McHoul and Mey (1990) have raised some of the same issues, but in the broader context of language rather than English alone. (Also see Phillipson 1992; Pennycook 1995)
5. The online data for each of these years is that "acquired" during that year and this matches the printed database. The number of citations in a database for any particular year is variable, as additional items are added to the database.
6. It will be interesting to see if Russian has declined significantly in the 1990s with the break up of the Soviet Union and the Russian Empire in Eastern Europe.
7. A comparison of the data presented for the five databases presented in Tables 2 and 3 suggests that we must be careful not to over generalise and to treat databases individually. For instance, the MATHSCI database is the only one not to show a growth in Japanese language publication.
8. For example France's prestigious Institut Pasteur, which has published *Les annales de l'Institut Pasteur* since 1887 (in French) decided in 1987 to henceforth publish only in English. It is now known as *Research in Microbiology* and has a circulation of about 1000. On the other hand, *Médecine-Sciences* (in French), which was founded in 1985, has gone from 200 to 4,200 paid subscribers. Thus journals, like individuals are making "market" decisions about how best to distribute their information (Anonymous [JC] 1989). Ammon (1996) notes that *Archiv für Kreislaufforschung* has changed into *Basic Research in Cardiology, Zeitschrift für Kinderheilkunde* into *European Journal of Pediatrics* or *Psychologische Forschung* into *Psychological Research.*
9. In 1999 the University of Sydney adopted a policy of preference buying from publishers using journals from providers over the internet with no hard copy being held. Such holdings are cheaper to buy and store, and are now readily available in the hard sciences (but less so in the humanities and social sciences).
10. Byrne (1983) in a study of the Australian literature demonstrates that much of that literature is not referenced in any overseas database. Thus, even English language literature may be lost if it is view as being on only local significance. Many local journals therefore have overseas editors and may even include "international" in their title in their quest to stand out, and be included in the major databases.
11. There are of course databases like PSYNDEX, a German-English database of German-language psychological literature containing over 20,000 references

(1977–1984). Only about ten percent of these are found in PsychINFO (Becker 1984). See Ammon (1988) for a review of these issues related to German language publication.

12. Some of the material in this section is adapted from Kaplan and Baldauf (1997: 171–180).
13. A collation and critique of this literature (20 studies) can be found in ALLC (1994: 33–50).
14. NAATI is a long established accreditation body for interpreters and translators with considerable national standing (see Ozolins 1993).
15. These include Arabic, Akkadian, Aramaic, Balinese, Chinese (Mandarin), Chinese (Classical), Middle English, Old English, French, Old French, German, Modern Greek, Classical Greek, Classical Hebrew, Modern Hebrew, Hindi, Old Icelandic, Indonesian and Malay, Modern Irish, Old Irish, Italian, Japanese, Javanese, Javanese (Old), Korean, Latin, Russian, Sanscrit, Spanish, Syriac, Ugaritic, Thai, Welsh, Yiddish. (Portuguese and Vietnamese may be added in 2001.)
16. As we are not concerned with publication length, but with whether a language other than English was used or not, no weightings were given to books when compared to articles, papers, reports, etc.
17. The costs of participating in science are very real for some scientists. Chris Candlin reports being told by a colleague at the AILA Congress in Finland in 1996 that it would cost her a month's pay to get her article translated into English so it could be published (*AILA News*, 4, 2000, pg.1).

10. References

Ammon, Ulrich

1998 *Ist Deutsch noch internationale Wissenschaftssprache? English auch für die Lehre an den deutschsprachigen Hochschulen*. Berlin/New York: de Gruyter.

1996 The European Union (EU – formerly European Community): Status Change of English During the Last Fifty Years. In Fishman, J. A./Conrad, A. W./Rubal-Lopez, A. (eds.) *Post-Imperial English: Status change in Former British and American colonies, 1940–1990*. Berlin: Mouton de Gruyter, 171–184.

1988 Deutsch als Publikationssprache der Wissenschaft: Zum Umfang seiner Verwendung im Vergleich mit anderen Sprachen. *Germanistische Mitteilungen* 28: 75–86.

Australian Language and Literacy Council (ALLC)

1994 *Speaking of Business: The Needs of Business and Industry for Language Skills*. Canberra: Australian Government Publishing Service.

Baldauf, Richard B., Jr.

1994 'Unplanned' Language Planning. In Grabe, W. et al. (eds.) *Annual Review of Applied Linguistics* 14. Cambridge: Cambridge University Press, 3–12.

1993 Fostering Bilingualism and National Development Through School Second Language Study. *Journal of Multilingual and Multicultural Development* 14: 121–134.

1986 Linguistic Constraints on Participation in Psychology. *American Psychologist* 41: 220–224.

–/Djité, Paulin G.
2000 An Australian Perspective: Second Language Teaching and Learning in University. In Rosenthal, J. W. (ed.) *Handbook of Undergraduate Second Language Education*. Mahwah, NJ: Lawrence Erlbaum, 231–251.

–/et al.
1995 *The Viability of Low Candidature LOTE Courses in Universities*. Canberra: DEET and National Languages and Literacy Institute of Australia.

–/Jernudd, Björn H.
1987 Academic Communication in a Foreign Language: The Example of Scandinavian Psychology. *Australian Review of Applied Linguistics* 10 (1): 98–117.
1986 Aspects of Language Use in Cross-Cultural Psychology. *Australian Journal of Psychology* 38: 381–392.
1983 Language of Publication as a Variable in Scientific Communication. *Australian Review of Applied Linguistics* 6 (1): 97–108.

Barnes, Gregory A.
1982 *Communication Skills for the Foreign-Born Professional*. Philadelphia: ISI Press.

Becker, Joachim H.
1984 German-Language Psychological Journals: An Overview. *German Journal of Psychology* 8: 323–344.

van Bergeijk, D./Risseeuw, M.
1980 The International Translation Centre: The Language Barrier in the Dissemination of Scientific Information and the Role of the ITC. *Journal of Information Science* 2: 37–42.

Bettoni, Carmilla/Leal, Barry R.
1994 Multiculturalism and Modern Languages in Australian Universities. *Language Problems & Language Planning* 18 (1): 19–37.

Byrne, Alex
1983 How to Lose a Nation's Literature: Database Coverage of Australian Research. *Database* 6 (3): 10–17.

Chatelin, Y./Arvanitis, R.
1989 Between the Centres and Peripheries: The Rise of a New Scientific Community. *Scientometrics* 17: 437–452.

Clyne, Michael
1997 Multiculturalism in Australia. *Current Issues in Language and Society* 5: 1–48.

Commonwealth of Australia
1992 *Australian Language and Literacy Policy*. Canberra: Australian Government Publishing Service.

Conseil de la langue française
1986 Agir ensemble en faveur du français dans le domaine scientifique. *Bulletin du Conseil de la Langue Française* 3 (3): 5.

Council of Australian Governments (COAG)
1994 *Asian Languages and Australia's Economic Future*. Brisbane: Queensland Government Printer.

Crystal, David
1997 *English as a Global Language.* Cambridge: Cambridge University Press.

Djité, Paulin G.
1994 *From Language Policy to Language Planning: An Overview of Languages Other Than English in Australian Education.* Canberra: NLLIA.

Garfield, Eugene
1985 Journal selection for *Current Contents*: Editorial Merit vs. Political Pressure. *Current Contents* 11: 3–11.

Gordon, M. D./Santman, A.
1983 Language Barriers, Literature Usage and the Role of Reviews: An International and Interdisciplinary Study. *Journal of Information Science* 3: 185–189.

Grabe, William
1988 English, Information Access, and Technology Transfer: A Rationale for English as an International Language. *World Englishes*, **7**: 63–72.

Harris, Brian
1983 Translation, Translation Teaching, and the Transfer of Technology. *META* 28: 5–16.

Ikushima, Keiko/Tenopir, Carol
1988 Availability of Japanese Scientific and Technical Periodicals in Major English Language Databases. In Williams, M. E./Hogan, T. H. (eds.) *Online Meeting: Proceedings 1988.* Medford, NJ: Learned Information, 115–119.

Jernudd, Björn H./Baldauf, Richard B., Jr.
1987 Planning Science Communication for Human Resource Development. In Das, B. K. (ed.) *Language Education in Human Resource Development.* Singapore: SEAMEO Regional Language Centre, 144–189.

Kaplan, Robert B.
1987 English in the Language Policy of the Pacific Rim. *World Englishes* 6: 137–148.

–/Baldauf, Richard B., Jr.
in press Not only English: "English only" and the World. In Gonzalez, R. A./ Melis I. (eds.) *Language Ideologies: Critical Perspectives on the Official English Movement.* Champaign-Urbana, Ill.: NCTE.
1997 *Language Planning From Practice to Theory.* Clevedon: Multilingual Matters.

Kipp, Sandra/Clyne, Michael/Pauwels, Anne
1995 *Immigration and Australia's Language Resources.* Canberra: Australian Government Printing Service.

Lange, L.
1985 Effects of Disciplines and Countries on Citation Habits: An Analysis of Empirical Papers in Behavioural Sciences. *Scientometrics* 8: 205–215.

Large, J. A.
1983 *The Foreign Language Barrier. Problems in Scientific Communication.* London: André Deutsch.

Larsen, S. F./Magnussen, S.
1984 Reviews of Scandinavian Psychology: Editorial Introduction. *Scandinavian Journal of Psychology* 25: 1–4.

Leal, R. Berry/Bettoni, Camilla/Malcolm, Ian G.
1991 *Widening our Horizons: Report of the Review of the Teaching of Modern Languages in Higher Education.* 2 vols. Canberra: Australian Government Publication Service.

Lewin, Ralph A./Jordan, David K.
1981 The Predominance of English and the Potential Use of Esperanto for Abstracts of Scientific Articles. In Kageyama, M./Nakamura, K./Oshima, T./Uchida, T. (eds.) *Science and Scientists.* Tokyo: Japan Scientific Societies Press, 433–441.

L'Heureux, Robert
1986 Écrire la science en français. *Bulletin du Conseil de la Langue Française* 3 (3): 1.

Lo Bianco, Joseph
1990 Making Language Policy: Australia's Experience. In Baldauf, R. B., Jr./Luke, A. (eds.) *Language Planning and Education in Australasia and the South Pacific.* Clevedon: Multilingual Matters, 47–79.
1987 *National Policy on Languages.* Canberra: Australian Government Publishing Service.

–/Freebody, Peter
1997 *Australian Literacies.* Canberra: National Languages and Literacy Institute of Australia.

Louttit, C. M.
1957 The Use of Foreign Languages by Psychologists, Chemists and Physicists. *American Journal of Psychology* 70: 315.

Luke, Allan/McHoul, Alex W./Mey, Jacob L.
1990 On the Limits of Language Planning: Class, State and Power. In Baldauf, R. B., Jr./Luke, A. (eds.) *Language Planning in Australasia and the South Pacific.* Clevedon: Multilingual Matters, 28–54.

Manten, A. A.
1980 Publication of Scientific Information is not Identical With Communication. *Scientometrics* 2: 303–308.

Mühlhäusler, Peter
1996 Ecological and Non-Ecological Approaches to Language Planning. In Hellinger, M./Ammon, U. (eds.) *Contrastive Sociolinguistics.* Berlin/New York: Mouton de Gruyter.

NAVF (Norges almenvitenskapelige forskningsråd)
1976 *Publisering og formidling av samfunnsvitenskapelig forskning.* Oslo: En konferanserapport.

Ozolins, Uldis
1993 *The Politics of Language in Australia.* Melbourne: Cambridge University Press.

Pennycook, Alistair
1995 English in the World/The World in English. In Tollefson, J. W. (ed.) *Power and Inequality in Language Education.* Cambridge: Cambridge University Press.

Phillipson, Robert
1992 *Linguistic Imperialism.* Oxford: Oxford University Press.

Senate Standing Committee on Foreign Affairs, Defense and Trade (SSCFADT)
1992 *Australia and Latin America*. Canberra: Senate Printing Unit.

Smith, Nick L.
1979 The Limitations of ERIC in Reviewing Evaluation Literature. *Educational Evaluation and Policy Analysis* 1: 68–73.

Stanley, John/Ingram, David/Chittick, Gary
1990 *The Relationship Between International Trade and Linguistic Competence*. Canberra: Australian Government Printing Service.

Stanton, P. J./Aislabie, C. J./Lee, J.
1992 The Economics of a Multicultural Australia: A Literature Review. *Journal of Multilingual and Multicultural Development* 13: 407–21.

–/Lee, J.
1996 Australian Cultural Diversity and Export Growth. *Journal of Multilingual and Multicultural Development* 16: 497–511.

Subramanyam, K.
1983 Bibliometric Studies of Research Collaboration: A Review. *Journal of Information Science* 6: 33–38.

Swales, John
1990 *Genre Analysis: English in Academic and Research Settings*. Cambridge: Cambridge University Press.

Thogmartin, C.
1980 Which Languages for Students in the Social Sciences. A Survey to Help Academic Advisors. *Anthropology Newsletter* 21 (2): 6.

Tollefson, James W.
1991 *Planning Language, Planning Inequality: Language Policy in the Community*. London: Longman.

Touraine, Alain
1998 Report of the Bureau of International Sociology. *Social Knowledge and the Multiplicity of Languages and Cultures*. Montréal: International Sociological Association, 7–15.

Tsunoda, Minoru
1983 Les langues internationales dans les publications scientifiques et techniques. *Sophia Linguistica*, 144–155.

Valverde, Estella
1992 *Language for Export: A Study of the Use of Language and Language Related Skills in Australian Export Companies*. Canberra: Office of the Prime Minister and Cabinet.

Wallerstein, Imanuel
1995/1998 The Language of Scholarship. *Letters from the President*. No. 3. Montréal: International Sociological Association.

Wood, D. N.
1967 The Foreign-Language Problem Facing Scientists and Technologists in the United Kingdom: Report of a Recent Survey. *Journal of Documentation* 23: 117–130.

White, Peter/Baldauf, Richard B., Jr./Diller, A.
1997 *Languages and Universities: Under Siege*. Canberra: Australian National University.

The Penetration of English as Language of Science and Technology into the Israeli Linguistic Repertoire: A Preliminary Enquiry

Bernard Spolsky and Elana Shohamy

1. The Revitalization of Hebrew

In two earlier papers (Spolsky 1996; Spolsky 1998) and in a recent book (Spolsky and Shohamy 1999), we have continued in a path set by Fishman, Cooper, and Conrad (Fishman et al. 1977) to trace aspects of the spread of English in Israel and its penetration into the linguistic repertoire of the country. Here, we want to look in more detail at the domains of science and technology, and trace the partially successful but increasingly challenged efforts of supporters of revitalized Modern Hebrew to resist incursions into the domain by English and other languages.

The revitalization (Spolsky 1991) and revernacularization (Fishman 1982) of Hebrew were processes that started towards the end of the 19th Century as part of the development of Jewish territorial nationalism. Starting in Jewish agricultural settlements in Ottoman Palestine around 1890 with the use of the language in schools, the movement was bolstered by zealous support from the pioneers who founded the first communal settlements and by acceptance in the new Jewish town of Tel Aviv established in 1906 (Harshav 1993).

During these early years, it was the secular schools in Palestine that were the center of the struggle and the sites of the first victories. In the settlement schools, there had been two rivals with Hebrew. For a while, in the settlements supported by the Rothschilds, French had been favored for teaching modern subjects. More generally, Yiddish was the language for teaching Jewish subjects, just as it was the most common mother tongue of Ashkenazi Jewish migrants. In the town schools, German became a strong candidate for teaching modern secular subjects, and this led to the first battle for linguistic dominance over the field of science and technology.

2. The Language War

This conflict between German and Hebrew as the language for scientific and technical education won the popular title of the Language War. A full but somewhat partisan account of the affair is given in Cohen (1918), published in London as part of a successful campaign to persuade the British Government to order General Allenby to support the Hebrew language in newly conquered Palestine.

The path of those who fought for the revitalization and the revernacularization of the Hebrew language had not been smooth. Apart from the normal forces of inertia working against such a drastic change in speech practice – Hebrew had been a religious and literary language but had not been spoken as a vernacular for at least 1500 years – there was ideological opposition on a number of fronts. One was the hostility of the *haredi* (ultra-orthodox) Jews in Palestine and in the Diaspora who believed that the sanctity of the Hebrew language must be preserved by not sullying it with secular use (Poll 1980). Another was the competition of an alternative Jewish nationalist ideology, the secular cultural proponents of the standardization of Yiddish as a Jewish national language (Fishman 1980; Pilowsky 1985). The third front was fought by the advocates of the value of two Western European world languages, French and German, as logical and necessary media for scientific, cultural and educational modernization.

Both French and German as languages of modern science and culture were originally proposed and buttressed by Jewish philanthropic organizations in their respective metropolitan countries. The French-based *Alliance Israëlite Universelle* was the first to do this. It was set up in the late nineteenth century to work for "emancipation and moral progress of the Jews" through teaching French language and culture. It was more active in other Middle East countries, but the *Alliance* did also establish a few schools in Palestine, some of which still exist and encourage their pupils to learn French as well as English.

The German-based *Hilfsverein der Deutschen Juden* was modeled in many ways on the French *Alliance*. It was founded in 1901 in furtherance of the German language diffusion efforts begun by Bismarck (Ammon 1992). The *Hilfsverein* provided the funds and initiative to establish a strong system of education in Turkish-ruled Palestine, including elementary and secondary schools and teacher training seminaries. Quite early, though, the teachers and pupils in their schools became committed to the new movement for Hebrew revival, and as a result, Hebrew was firmly established as language of instruction in *Hilfsverein* schools in Palestine by 1910.

This state of affairs was challenged when, in 1913, a proposal was submitted to the Governing Board to create a tertiary technical institute, to be named the *Technikum*. Teachers in existing *Hilfsverein* schools wrote to the Governing Board proposing a language policy for the planned institute. They recognized that it would be difficult to use only Hebrew as language of instruction in the *Technikum* both because of the lack of Hebrew terms and Hebrew texts on scientific subjects and because of the shortage of qualified Hebrew speaking teachers. They accepted then that, temporarily, languages other than Hebrew (including German) might have to be used, but they insisted on a policy of transferring to Hebrew as soon as possible. The Governing Board ignored their plan, and determined instead that German would be the only language for scientific and technical courses. They did agree that Hebrew could continue to be used in the High School section and also in some general courses.

The Governing Board's proposal was strongly resisted by teachers and pupils in the schools, and there were a number of public demonstrations and petitions (Cohen 1918). The pro-Hebrew campaign was successful, and in February 1914, the Board reversed its position, agreeing that Hebrew would be used for mathematics and physics when the school opened, and that other subjects should follow. Thus, when the *Technion* (as it came to be called) finally opened in 1924 after the end of the First World War, it was as a Hebrew-only institution.

The Technion was not in fact the first university to be founded in Palestine. In the event, it was preceded, shortly after the end of the First World War, by The Hebrew University of Jerusalem. Its name was not accidental. Partly it represented the policy of the secular Zionist movement to replace the word "Jew" with "Hebrew" in its rhetoric and ideology. But the name also proclaimed clearly a commitment to the Hebrew language, which was quickly entrenched in the university's constitution and regulations.

3. The hegemony of Hebrew in the universities

The Hebrew University's language policy applied not just to teaching but also to all written work. All student papers and dissertations were required to be written in Hebrew. This strict ideologically backed policy continues in place, and only in exceptional circumstances will the Hebrew University accept written work from students in any other language. As other Israeli universities were founded over the next forty or so years, they adopted the same policy as the Hebrew University and the Technion, but they implemented the policy somewhat less stringently.

The struggles in support of revernacularized Hebrew in the 1920s and 1930s were focused on demands for its use in all public domains, and were aimed in particular at two proclaimed enemies. One was the Yiddish that had been the principal Jewish vernacular in Europe and the second was English of the Mandatory government of Palestine. The British conquest of Palestine at the same time as the defeat of Imperial Germany effectively ruled out support for German even before Nazi policy made it unthinkable. Similarly, the Sykes-Picot agreement setting British and French zones of influence in the Middle East meant that English and not French was the colonial language for Palestine.

Supporters of Hebrew at the Hebrew University successfully resisted proposals to establish a chair of Yiddish until 1951. Ussishkin, speaking at the 1923 national conference of the Legion for the Protection of the Hebrew Language, also criticized those who used English to assert an elite status and closeness to the government. Partially reflecting this prejudice, it was not until the 1960s that the Hebrew University had a full professor of English, although the English department had grown in size in the intervening years.

The other universities were and are less strict than the Hebrew University. Most extreme in the opposite direction is The Weizmann Institute of Science, established in 1934 as a center for advanced scientific research and renamed in 1949. Weizmann is widely assumed to accept and adopt casually the English of its many visiting and foreign-born staff members.

The newer universities established since the creation of the State of Israel in 1948 – Bar-Ilan University, Tel Aviv University, Ben Gurion University and Haifa University – have the same written policy as the Hebrew University for use of Hebrew. With some fairly restricted exceptions, almost all university courses are taught in Hebrew. The exceptions are obvious. All the departments of English Language and Literature teach in English. Some non-Hebrew literature departments (such as

French at Bar-Ilan and the Hebrew University and Yiddish at Bar-Ilan) prefer to teach in the language. In addition, there is a parallel medical school taught in English at Tel Aviv University, and English language MBA programs at Ben-Gurion and Bar-Ilan. In some cases, new staff members might use their native language in other courses for a year or so. But as a whole, the language of the lecture rooms, the seminar and the laboratories, as well as of the corridors and cafeterias is predominantly spoken Hebrew.

In comparison to the Hebrew University, however, the newer universities are more tolerant of English and more ready to accept the writing of theses and dissertations in scientific subjects in English in order to speed up the process of international publication. On this, we have empirical evidence. Looking over 100 dissertations in scientific fields at the Hebrew University between 1940 and 1997, Drorit Ram (personal communication) found only a dozen, and those in the last 30 years, written mainly in English. Of these, eight of the writers had English surnames and may well have been native speakers of English who were given special permission to use their native language. In contrast, at Tel Aviv University, dissertations in English in the sciences are much more common than this. Checking 102 recent dissertations in the fields of physics, biology, engineering and chemistry, Ram found 62 that were written entirely or mainly in English. The other 40 followed the published requirement of being written in Hebrew with only the abstract and bibliography in English.

4. English reading requirements

Control over language choice in written theses and dissertations is clearly easiest to enforce, as all pass through central university committees. It is in reading requirements that the penetration of English has been most effective and even accepted. In spite of the continued resistance of an entrenched Hebrew ideology on the one hand and the accompanying regulatory requirement for Hebrew use, there has been growing encroachment of English into the academic and technological domains. The main driving forces are the mass of scientific and technical written sources in English which it is impractical to translate into Hebrew and the need for written and oral communication in a world language with non-Hebrew speaking academics.

A first factor is terminology. While the Hebrew Language Academy has been working energetically to develop new scientific and other modern terminology (Fellman 1977; Fellman and Fishman 1977; Saulson 1979), the process has generally been of limited success (Allony Fainberg 1977; Fellman and Fishman 1977; Hofman 1974). Lectures given in Hebrew are thus likely to be thickly spattered with English (or rather anglicized, international) technical terms. While high school textbooks are more prone to attempt to use the approved Hebrew words, seeing that most advanced textbooks used at the universities are in English, the English words easily slip into academic scientific discourse.

Even if the terminological innovations of the Hebrew Language Academy were to be more widely adopted, there would still remain the issue of the shortage of scientific and scholarly material published in Hebrew in the rapidly expanding fields of knowledge. The cost of translating specialized material into national languages is usually well beyond the economic possibilities of smaller countries. Only a nation like China can make this a matter of general policy.

As a result of the recognition of the need for English reading ability for advanced study, all Israeli universities have a requirement for English proficiency for admission. Candidates for admission must pass the *Bagrut* (matriculation) examination in English at the four or five point level. Israeli universities also use the score on the nationally administered Psychometric test, an important section of which is an English test, to determine admission to popular departments and to decide whether a student needs extra hours of remedial English while at university. As a result, the English requirement comes to serve not just as a justifiable prerequisite to university study, but also as a means of filtering out students from those schools, neighborhoods, and social and ethnic groups which have less adequate access to good English teaching.

The assumption behind English requirements is that while courses are taught in Hebrew, students will need to do a fair amount of their reading in English. Students who do not meet minimal standards of English proficiency must take remedial courses. However, the fact that some of them are still taking them in their last year suggests that the requirement is in some cases purely formal.

Certainly after the first year, and especially in science subjects, a large amount of the students' reading lists will consist of items published in English. Most university teachers will give this as the first and sometime the only justification for the requirement. In practice, especially in less demanding subjects, students often manage to find ways to get around the English reading. As it is a common practice for lecturers to repeat orally in their Hebrew lectures the material included in the required English reading, students often choose to rely on this. In other cases, groups of students band together to divide up the work, with those with better English skills preparing Hebrew summaries or translation, the *quid pro quo* being help with homework or notes from class lectures. We have no detailed information of the extent of this method of avoiding the need to use English, but there are signs that it is quite widespread.

At more advanced levels, and particularly in scientific fields, a student who cannot read English easily is likely to be severely handicapped. At the advanced graduate levels, universities even sometimes ask for courses in English writing. It is at the most advanced levels, especially with requirements for research and publication, that the demand for English proficiency becomes more serious. Most Israeli academics in all but a handful of Hebrew-bound subjects (Jewish studies, for instance) are expected to publish internationally, which means publishing in English or another world language. Most are also assumed to do a large amount of reading in English. When doing their own writing in English (or other international languages), a number of lecturers use translators (especially for book-length manuscripts) or editors.

5. Scientific and technological need for speaking English

So far we have concentrated on reading and writing skills in English. To a certain extent, the Israeli university tends to that division of language functions where speaking is in Hebrew and reading and some writing. True, the normal framework for social life as well as professional, in laboratory and lecture rooms and common rooms, remains the dominant Hebrew, albeit with the code switching engendered by technical terminology. But oral skill in English is also felt to be of growing importance. The presence of a foreign staff member or a few foreign post-doctoral students will also increase the tendency for a laboratory to tend towards English use. International scientific communication is mainly in English, as even the French discovered when they tried to enforce a section of the Toubon Act requiring French in international scientific congresses. At international meetings, or when hosting or visiting non-Israeli scientists, both scientific and social communication tends to be in English, for few non-Israelis know enough Hebrew to use it. A good proportion of scientific meetings in Israel set English as language for lectures, especially when there are foreign participants.

Hebrew academic lectures do involve a good deal of code switching, with common use of borrowed rather than Hebrew Academy coined terminology. Transcriptions of Hebrew lectures will reveal a high proportion of *loazit* or foreign terms and expressions, most commonly from English. When writing in Hebrew, though, academics will probably be more careful in their choice of vocabulary, influenced my the common ideology of normativism. Given this pervasive acceptance of Hebrew hegemony, while English has succeeded in penetrating the scientific and academic domains, especially at the upper and more technical ends, most Israeli academics appear firmly committed to the need to maintain Hebrew use.

In commercial and technological fields outside the universities, it is the demands of international trade that leads to a growing value for English proficiency and use. One chemical engineer once assured me that it was his native English that accounted for his success as vice-president for sales. In the growing and vital Israeli hi-tech industries, the role of English as a language of work and selling is even more strongly apparent. The many Israeli computer start-up companies already have or are seeking US-based partners, with the result that, at all work levels, there is an advantage in being able to work in English.

A study of a single company for a doctoral dissertation revealed the linguistic and cultural difficulties faced by engineers forced to negotiate with American counterparts. In hi-tech fields, products must be closely tailored to suit user needs. Engineers designing equipment or computer experts developing software must be ready not just to provide adequate written support in English (a task that can be carried out by employing translators and technical writers) but to stay in close contact with non-Hebrew-speaking clients and customers both before and after selling their product or service. Because of the technical complexity, this is not a task that can be relegated to sales staff. As a result, Israeli experts must be able to work with other experts, usually in English. Israeli engineers selling hardware to American or other foreign buyers must therefore develop proficiency in the language and culture of their customers. As a result, proficiency in English is critical for advancement in hi-tech fields in Israel.

6. Summary

Summing up, it is our belief (to be further investigated in planned research) that in spite of the very strong ideological support for Hebrew,

English is managing to penetrate Israeli linguistic space not just via the mass media and tourism, but also through the demands of science and technology and the education associated with both. At the same time, the high level of terminological development and the even higher level of ideological loyalty to Hebrew means that this penetration is not in itself a serious threat to the continued hegemony of Hebrew. English is now seen as an important added skill, but as the attitudes of the Arabic-speaking minority and the huge immigrant population show (Shohamy and Donitsa-Schmidt 1997), Hebrew remains the first language that one must master for social, educational and economic success.

7. References

Allony, Fainberg, Yaffa
1977 The Influence of English on Formal Terminology in Hebrew. In Fishman, J. A./Cooper, R. L./Conrad A. W., 137–167.

Ammon, Ulrich
1992 The Federal Republic of Germany's Policy of Spreading German. *International Journal of the Sociology of Language* 95: 33–50.

Cohen, Israel
1918 *The German Attack on the Hebrew Schools in Palestine*. London: Jewish Chronicle and Jewish World.

Fellman, Jack
1977 The Hebrew Academy: Orientation and Operation. In Rubin, J./Jernudd, B. H./Das Gupta, J./Fishman J. A./Ferguson, C. A. *Language Planning Processes*. The Hague: Mouton, 97–109.

Fellman, Jack/Fishman, Joshua A.
1977 Language Planning in Israel: Solving Terminological Problems. In Rubin, J./Jernudd, B. H./Das Gupta, J./Fishman J. A./Ferguson, C. A., 79–96.

Fishman, Joshua A.
1980 Attracting a Following to High-Culture Functions for a Language of Everyday Life: The Role of the Tshernovits Conference in the 'Rise of Yiddish'. *International Journal of the Sociology of Language* 24: 43–73.
1982 The Role of the Tshernovits Conference in the Rise of Hebrew. In Cooper, R. L. (ed.) *Language Spread*. Bloomington, IN: Indiana University Press, 291–320.

Cooper, Robert L./Conrad, A. W.
1977 *The Spread of English: The Sociology of English as an Additional Language*. Rowley, MA: Newbury House.

Harshav, Benjamin
1993 *Language in Time of Revolution*. Berkeley: University of California Press.

Hofman, John E.
1974 Predicting the Use of Hebrew Terms Among Israeli Psychologists. *International Journal of the Sociology of Language* 3: 53–65.

Pilowsky, Arye L.
1985 Yiddish Alongside the Revival of Hebrew: Public Polemics on the Status of Yiddish in Eretz Israel, 1907–1929. In Fishman, J. A. (ed.) *Readings in the Sociology of Jewish Languages*. Leiden: Brill, 104–124.

Poll, Solomon
1980 The Sacred-Secular Conflict in the Use of Hebrew and Yiddish Among the Ultra-Orthodox Jews of Jerusalem. *International Journal of the Sociology of Language* 24:109–225.

Saulson, Scott B. (ed.)
1979 *Institutionalized Language Planning. Studies in the Sociology of Language*. The Hague: Mouton.

Shohamy, Elana/Donitsa-Schmidt, Smadar
1997 *Attitudes, Stereotypes and Priorities of Jews Towards Arabic and Arabs Toward Hebrew*. Tel Aviv: The Steinmitz Center for Peace in the Middle East, Tel Aviv University.

Spolsky, Bernard
1991 Hebrew Language Revitalization Within a General Theory of Second Language Learning. In Cooper, R. L./Spolsky, B. (eds.) *The Influence of Language on Culture and Thought: Essays in Honor of Joshua A. Fishman's Sixty-Fifth Birthday*. Berlin: de Gruyter, 137–155.
1996 English in Israel after Independence. In Fishman, J. A./Rubal-Lopez, A./ Conrad, A. W. (eds.) *Post-Imperial English*. Berlin/New York: Mouton de Gruyter, 535–555.
1998 The Role of English as a Language of Maximum Access in Israeli. Language Practices and Policies. *Studia Anglica Posnaniensia* 33: 377–98.

Shohamy, Elana
1999 *The Languages of Israel: Policy, Ideology and Practice*. Clevedon: Multilingual Matters.

Aspects of Scholarly Language Use In Malaysia: Switching Codes in Formal Settings

Rodolfo Jacobson

1. Introduction

It is a widely attested fact that code-switching, the alternation between two or more languages, occurs in most parts of the world and also does so in Malaysia, the target country of this study. It is in no way unusual to hear in informal conversation an utterance like the following:

(1) I feel *artis-artis veteran lebih berpengalaman untuk beri dorongan*
veteran artists [are] more experienced to give encouragements
'I feel that veteran artists have more experience encouraging

kepada artis-artis muda sekarang dan saya rasa, ini
to artists young now and I feel, this
young artists now and I feel that this

untuk ya, saya rasa dia orang tak perlu
[is] for, yes, I feel a person not need
is for, isn't it, I feel that a person does not need

tanya what the association can do for them,
[to] ask
to ask what the association can do for

they should ask *apa yang dia orang boleh buat untuk*
what a person can do for
them, they should ask what a person can do to

menaikkan tahap persatuan itu.
to raise [the] status association that
to raise the status of that association.' (OKS,2 [II A] in Jacobson forthcoming)

The Malay portion of the discourse is well integrated into the English part of the discourse and is rendered so smoothly that the observer gets the impression that no language change has occurred at all. Even though not all mixed sentences may exhibit this kind of balance between the two participating languages (see below), all the data of code-switched discourse found in Malaysia show that, in principle, the Malaysian has found a most gracious way of joining the resources of his two most important languages, that is, that of the national language with English, Malaysia's "second most important language" (Asmah 1992: 84).

Examples from a host of other languages would show the same phenomenon, that is, that in informal speech bilinguals tend to rely in their conversation with peers and family members on the resources from the languages with which they have grown up. One major question presents itself at this point and it is the consideration whether the switching between or among languages *only* occurs in informal interaction, that is in the chitchatting between friends and family members or whether it can be shown to occur in more formal situations, such as, the exchange of ideas among members of the faculty of a university, members of a professional staff or among students discussing academic issues. The evidence of current research had usually shown that the switching occurs when members sharing a similar ethnic background converse informally. In my own experience, however, this is not always the case. I have been exposed to formal discussions at Malaysian universities where the switching between languages is frequent and a meeting may start out in the national language to establish a framework for the language to use but as the meeting develops more and more insertions from English become noticeable. It is therefore the purpose of the present study to examine formal Malaysian speech, identify the occurrences of code-switching and try to conceptualize why in Malaysia code-switching can occur at the formal level, whereas elsewhere it tends to be restricted to informal settings alone.

2. The Malaysian language scenario

It would be unwise to proceed without offering an explanatory description of the language scene in Malaysia as this country obtained its independence from Britain. Under the condition that the new nation unify under one roof the three races residing in the territory – the Malays, the Chinese, the Indians – Britain agreed to granting Malaya its independence in 1957. The former British colony became a free nation as Malaysia and as such she became aware that, in order to comply with the condition of unification of her various races, she had to solve the language issue at once. Realizing that language must not be seen as a dividing factor but rather as a unifying element, the new Government decreed in Article 152 of the Constitution of Malaysia that *the national language shall be the Malay language*. There was actually little disagreement on this choice. English as the language of the former colonial government, was found inappropriate for the free nation, Chinese, a cluster of different lects of the broader family of Chinese languages (Cantonese, Hokkien, Hakka, Mandarin) and spoken by the second largest group was – in part, because of their mutual unintelligibility – equally found unfit by the politically dominant Malays to serve as national language and Tamil together with some other ethnic languages counted only on a small percentage of speakers and was not eligible for that very reason. Says Asmah Hj. Omar (1995: 160) in this respect that

> the Malay language was chosen as national language to give an identity to the people who called themselves, Malayans, later Malaysians. It was the language of the national anthem and the language most Malaysians, i. e., those who were not English-educated, use in intergroup communication. This is so, because even before it was given the status of national language Malay was already the common language for most Malaysians, manifested in various lectal forms: acrolect, mesolect and basilect (...) What its establishment as a national language did was to give it a *de jure* status to what was already *de facto*.

English shared with Malay the status of official language from 1957 to 1967 at which later time Malay or Bahasa Malaysia became the sole official language of Peninsular Malaysia even though the Government allowed Eastern Malaysia (Sarawak and Sabah) to continue with English as an official language for a few more years. As Bahasa Malaysia strengthened its hold on the entire country, it became the medium of instruction in all national schools. English was pushed into a low-key position for the number of years that it took Bahasa Malaysia to fully establish itself as *the* language of the Nation. The presence of English, on the other hand, never disappeared entirely but lingered on despite official pressure to speak Malay at all public places and is now reemerging as a result of a changing governmental policy. A guideline issued in August 1995 by the Ministry of Education followed suite to the Prime Minister's announcement in 1993 that allowed the use of English in universities and colleges for the teaching of sciences. The guideline goes far beyond an earlier announcement that allowed the use of English in tutorials, seminars, assignments, foreign language classes and other like activities. This resurgence of English is seen, not as a return to colonial standards – the colony is hardly remembered by most Malaysians today – but rather as a means to advance in scientific and technological matters and to assist Malaysia in its quest of becoming a global community, one that by the year 2020 hopes to have achieved the full status of an industrialized nation. The English language that has reemerged may however not be quite the same language that went underground some decades ago. It has shed some of its British elements and become nativized in the process and, by the same token, it may have become more acceptable to reappearing in the discourse of Malaysians, particularly in Malay-English mixed speech events.

3. Dual language strategies

3.1. Some general notions

The continued presence in the Malayan/Malaysian landscape of the English language – although be it at different levels of proficiency – may be in part responsible for the Malaysian's fascination with code-switching. We have suggested above three stages of the English presence that we may – with Asmah (1992: 67) – briefly refer to as *colony*, *low-key* and *re-emergence*. Even though the low-key stage has often been blamed for the decreased English language proficiency of the younger generations, the fact is that English has never fully disappeared. Malaysians are therefore tempted, depending on their language proficiency, to embed portions of English into their Malay discourse as well as insert portions of Malay into their English discourse or maintain a balance between the two languages where in the latter case it becomes quite difficult for the observer to determine whether the person is communicating mainly in English or mainly in Malay. The above cited example, (1), is a valid example of such a linguistic balance.

Sociolinguists have tried to cope with the different mechanisms of code-switching by labeling the dominant language *matrix language* and the subordinate language *embedded language*. The so embedded portions are often referred to as *islands*. Since in the language balance of (1) no portion can be said to be superordinate nor subordinate, some of us prefer to label such utterances, not code-switching but *language alternation* (Bentahila and Davies 1998; Jacobson forthcoming). Much theoretical debate has been carried on in regard to the matrix– embedded-language notion and one American sociolinguist has refined an earlier language frame concept and proposed the model known today as the *Matrix Language Frame Model* which has been adopted widely by scholars carrying out research in the field of code-switching. Since the major thrust of this chapter is not on theoretical constructs developed in the field but on the pragmatics of code-switching, the interested reader is referred to the professional literature (Myers-Scotton 1997; 1998).

3.2. *Switching in informal settings*

Switching languages is most common in informal settings. It has been attested in scores of language pairs, such as, English-Spanish, English-Swahili, English-Arabic, English-Turkish, English-Japanese as well as French-(Moroccan) Arabic, French-Alsatian, French-Dutch, French-Hebrew, Sardinian-Italian, Marathi-Sanskrit and many more. It is of course also very common in the language pair that is the object of this study. Note the following Malay-English examples:

(2) (…) the lecturer make[!] his own research, *buat rakamkan dekat*
makes record close to
'The lecturer makes his own research, makes a recording near

studio, dekat bilik rakaman, then we make a master, the lecturer
studio close to room
the studio, close to the recording room, then we make a master;

need everything*lah* [emphasis]
the lecturer needs really everything.' (J-22/sy in Jacobson and Safiah 1988)

(3) Women *pandai menseimbangkan tugasnya, di pejabat, di*
clever balance their duty, in office, in
'Women are clever balancing their duty, in the office, at

rumah sebagai ibu, sebagai isteri.
house as mother as wife
home as mother, as wife.' (S-05/sy in Jacobson 1996: 83)

Example (2) shows an English language matrix containing a Malay language embedded island. The next example (3), in turn, reveals the reverse, that is, a Malay matrix containing one single English word embedded into it. Utterances of this kind are the most common types of code-switching and the speakers seem to enjoy mixing the two languages in informal conversations, that is, when they are talking to peers of like linguistic competencies. This pleasure of switching has usually been interpreted as a sign of intra-ethnic identification, a demonstration of shared biculturality. As true as this may be for code-switchers like the Hispanics of the United States whose switching techniques Jacobson studied intensively in the past (1978 1983), this is not always true for other bilingual or multilingual speakers. As for Malaysia, Azhar and Bahiyah (1999: 133) argue that "[c]ode-switching is not only a means of social cohesiveness, but an effective means of production [as] it allows formal meetings to be concluded quickly." We will say more about this later in the chapter.

In an earlier study of Malay-English code-switching that has appeared in published form only in the Malay language (Jacobson 1993: 426), we attempted to conceptualize the difference between the switching of codes in the southwestern part of the United States, in particular South Central Texas, and the one occurring in Malaysia and this is what we then said about Malay-English code-switching:

> I have recognized a two-pronged motivation among Malaysian bilinguals, one of fusing the two languages together into a single linguistic manifestation and another, of maintaining the contrast between English and Malay in order to use the contrasting features of the two languages as lexical or stylistic devices (...) Malaysia seems to stress by means of this duality, the co-existence of a past with English as an official language and a present with Bahasa Malaysia as the language of the public domain. Be it as it may, these two languages are complementing one another most effectively in the creation of a colloquial variety to which today's bilinguals can relate. (From the conclusion in English of the paper entitled *Fusion and contrast in the alternation of two genetically different languages: the case of Malay-English code-switching* and delivered at the 1992 ISA-RC25 meeting in Nijmegen, The Netherlands)

Spanish-English code-switching is, in contrast, a means of ethnic identification, that is, a way to say, *we are peers and share the same linguistic-cultural background*. This peer recognition and bicultural sharing can only manifest itself in an informal setting. Formality, on the other hand, requires of the setting that a single language be spoken. Malay-English code-switching, in turn, does not involve ethnic identification but does allow the inclusion of a second language for such reasons as *quality education, appropriateness of expression, speed, idiomatic idiosyncracy.* Speed has recently been suggested as a reason of alternating between languages (Azhar and Bahiyah 1994) as a word may become available to a speaker sooner in one language than it does in the other. None of the cited reasons seems to suggest that code-switching must be restricted to informal settings. The use of code-switching in Malaysia may not even be representative of *bicultural* behavior at all, as there are no intimate ties for the Malaysian between the use of English and the membership in an Anglo (British or American) world. This was recently pointed out to the writer in a comment from the audience at a lecture. (Gill, personal communication) As a matter of fact, we may just have stumbled here on a powerful reason why code-switching is found admissible in Malaysia even when professional issues are being discussed.

3.3. Switching in formal settings

We will loosely define formal settings as those gatherings of professionals where scholarly topics are being discussed. This includes but is not limited to university or college reunions but also refers to gatherings at academic/professional institutions, such as *faculty meetings, departmental meetings, committee meetings, formal class discussions.* Language practices at such meetings can be witnessed by listening to the debates while they are in progress. For research purposes, it is easier, not to study live discussions, but to audio-tape and then transcribe them for the purpose of analysis. Permission to record official meetings is often difficult to obtain, particularly when outsiders are involved in the research process. This writer, however, was fortunate enough ' not only to identify an insider who had already accumulated data on the switching in formal settings but also to find a cooperating university official who would allow that an upcoming departmental meeting be recorded and later transcribed, so that two valuable sets of data have become available, one based on past data collection and the other, on the recording and transcription of a meeting conducted while the writer was still in the country, even though not present at said meeting so as not to affect adversely the normal behavior of the participants. The following subsections will discuss these two sets of data.

3.3.1. Holey English

The term *Holey English* is not a familiar term understood by an international audience but has recently been coined by some English language instructors in Malaysia suggesting that where *Wholly English* refers to the acrolectal variation of English that according to Platt and Weber (1980) is used for professional purposes even in Malaysia, Holey English is a kind of English full of perforations (or holes) that are filled with stretches of Malay speech. This playful juxtaposition of *wholly* vs. *holey* (Azhar and Bahiyah 1994) stresses an interesting point, namely that the approximation to a British English RP variety in formal discussion in Malaysia is a myth and that the variety actually used is a blend of mesolectal/acrolectal English and Malay. In other words, Azhar and Bahiyah talk about a very specific kind of code-switching where English is the matrix language and Malay the embedded language, – to use a more familiar terminology. They do not provide any example of the reverse situation, that is, Malay being the matrix language but containing English language islands. On the other hand, they provide some good examples of what this writer has called *Language alternation* where English and Malay alternate in such a way that it is difficult to determine which of the two languages is the matrix, as this can be appreciated in the following selection of sentences (Azhar and Bahiyah 1994: Table 3, 136):

4) B: *Siapa* the (…) as the head of this Accomodation Committee, *aar?*
Who eh
'Who is the (…) functions as the head of this Accomodation Committee, eh?'

(5) I: *Penginapan* *tak ada* head*lah.*
Accomodation Committee not have (…) emphasis
'The Accomodation Committee does not have any head at all.'

The reason why there are no clear examples of a Malay matrix with English embedding is obviously the fact that these authors' concern is the specific form of English used in formal discourse with the intent to stress their view that English language teaching programs in Malaysia are misleading in as much as they profess the unrealistic goal of an acrolectal dialect of English for all students. As a matter of fact, English-speaking Malaysians will always insert, even in formal discourse, words or phrases from their national language. The approach in this chapter, however, is broader as this writer is investigating the nature of discourse in a variety of different formal settings. To use Azhar and Bahiyah's terminology, it is not Holey English that the writer investigates, but *Holey Language*.

The reason why the language medium used by Azhar and Bahiyah's subjects is English can be best understood by examining more closely who the interlocutors are and what their academic affiliation is. As members of a TESL [Teaching English as a Second Language] department, they may be inclined to use English but switch occasionally to Malay to accelerate the speed of their task. Meetings of a more general nature are likely to be conducted in Malay with stretches of English discourse. In other words, many meetings will reflect the reverse situation, namely, the use of Malay as matrix language containing English embedded islands of varying complexities. In addition, there may be instances of the presence of Malay and English in a balanced situation where each language holds its own and neither dominates the other. These are the language uses that the writer will describe below in his analysis of the recordings made at a higher-level departmental meeting in the National University of Malaysia.

3.3.2. Malay-English code-switching in formal discussions

The following discussion of code-switched speech in formal meetings does not suggest that *all* formal meetings necessarily reflect this particular language routine, that is, they need not involve the use of the two languages Malay and English. In effect, many such meetings are conducted entirely in Malay. After all, Malay or Bahasa Malaysia is the national language. The potential bilingual discourse in formal gatherings depends on two crucial factors, (1) Does the nature of the meeting or the topic of discussion lend itself for a dual language performance? and (2) Is the facilitator conducting the meeting liberal enough to permit dual language use? As for the first factor, code-switching is likely to be avoided if the meeting is called by high-ranking Malaysian officers or by representatives of the Malay Studies Faculty and/or if the topic deals with strictly Malaysian issues. On the other hand, for meetings called by functionaries of a linguistic or language education entity of the university, switching between the two languages may be in order. As for the second factor, even a language-oriented meeting may shun the use of the two languages if the facilitator of the meeting, or the head of the department whose orders the facilitator must obey, firmly believes in the sole appropriateness of Malay for administrative matters. As for the data collected for the present discussion, both factors were resolved favorably in regard to dual language use, namely, the meeting was indeed called by a language department and the topic concerned departmental issues concerning the teaching of language education courses in English as well as in Malay. Furthermore, the head of the department as well as the meeting facilitator both felt positively toward the use of the two languages as this behavior, not only reflected the very essence of the department, but contributed to a more informal and natural atmosphere during the reunion.

With this overall philosophy in mind, this writer wishes to share with his readers some comments on the language behavior of a group of university instructors during a departmental meeting conducted by its chair. The cited meeting was recorded unobtrusively so as to allow for the most natural behavior of each of the participants. The meeting lasted exactly one hour and the recording was later transcribed to render a total of 102 utterances, some consisting of more than one sentence. A typical utterance would consist on the average of 45 words, so that there would be about 4,400 words in the transcription.

The data show an amazing range of code-switching episodes going from an incidental use of one other-language item in otherwise monolingual discourse to that of a completely balanced performance in the two languages:

(6) Cuti *sabbatical so far* tak ada permohonan?
Holiday not there is request
'Is there any request for a sabbatical holiday so far?'

(7) Projek *paper* ini akan dinilai semula oleh penilai kedua
Project this will be evaluated again by evaluator second
'This project paper will be evaluated again by a second evaluator

just to make sure that, you know, they are as bad or they are as
just to make sure that, you know, they are as bad or they are as

good as the grades given.
good as the grades given.' (UKM 1.3/14)

There is no hesitation detectable as the speaker moves from one language to the other. Poplack et al. (in Jacobson 1990: 188) would refer to this kind of uninterrupted performance as *unflagged* and this lack of flagging is called *seamlessness* by Azrah and Bahiyah. Because of this smooth transition from one to the other language, it is particularly difficult to pinpoint why the speaker would switch at a given point be it for one word or be it for a long stretch of lexical items. On the other hand, the researcher was able to identify some semantic fields that lend themselves more than others to prefer English words over Malay expressions, either because the speaker was more familiar through reading and study with the terms in English or because the rendering in Malay would have been too cumbersome and the English expression was just at the tip of the tongue as such would be the case for most idiomatic expressions. In effect, the semantic fields that did emerge seemed to suggest to the interlocutors that an English term was more profitable at this point than the Malay counterpart. They were as follow:

(a) The university's administrative units;
(b) Courses taught in the department;
(c) Fields of study;
(d) Departmental administrative matters;
(e) Computer and Internet-related notions;
(f) Social groups;
(g) Levels of university study;
(h) Academic degrees.

A discussion of each of these fields cannot be included here but the writer will select field (e) to demonstrate how English words are inserted into Malay speech when the speaker refers to matters concerning computer technology:

(8) (…) juga satu salinan *loose bound* sesama satu *diskette* (…)
(…) also one copy together with one
'(…) also a loosely bound copy and one diskette (…)'

(9) Muka surat yang sama, masalah *e-mail,*
Face letter which same problem
'On the same page, [there is] an e-mail problem;

kebanyakan masalah *e-mail* telah pun diselesaikan (…)
multitude problem even resolved
many e-mail problems had already been resolved.'

(10) Saya rasa pejabat TD-1 ada mengumpul *list websites* yang
I feel office is to gather which
'I feel that the TD-1 Office should gather a website list and

mungkin boleh dipertimbangkan supaya ianya boleh
perhaps can be considered so that it can
consider it so that we might

diterima, (…)
be received
receive it (…).' (UKM 1.7/ 15/ 16)

A close examination of the transcription shows some of the resources used to make several points of discussion clearer to the audience; for example, the repetition of a given phrase for reasons of emphasis plays an important part in the speaker's strategy. Examine the following:

(11) *For example 'School of Human Development',* Pusat Pengajian
Center Studies
'For example the 'School of Human Development,' the Study
Pembangunan Manusia.

Development Mankind
Center of Human Development.' or

(12) (...) agar program-program yang kita tawarkan berdasarkan
So that programs that we offer based
'So that programs that we offer are based upon language,

kepada basaha, *language-based,* (...)
to language
[are] language-based, (...).' (UKM 2.3/14)

As the reader can appreciate, the repetition goes both ways. If the first mention is in English, the second is in Malay; if, in turn the first is in Malay, the second is in English.

Another interesting point is the syntactic fusion that occurs between the two languages. Please note the following example:

(13) (...) pensyarah-pensyarah yang telah *in a way volunteer, in a way*
lecturers who already
'lecturers who in a way had volunteered, in a way were forced,

dipaksa, diajak, tetapi (...)
forced, invited, but
invited but (...)' (UKM 1.9)

There are no verb tenses in Malay but only adverbial expressions like *sedang, sudah, telah*, that convey the notion of the time when the action was performed, so that *sedang* 'now', *sudah* 'already' and *telah* 'completed' used before a verb serve respectively as indicators of present, past and past perfect tense formation. In example (13) the past perfect of the verb *volunteer* is denoted by the presence of *telah*, a Malay word, and *volunteer*, an English word, thus combining the two languages in the conveyance of a single verb form.

Not only is the presence of Malay speech in English discourse or the reverse situation, that of English in Malay, noteworthy but so is also the use of some English constructions that convey a notion of nativization of English. Phrases like *no problem, so far performing well, could be two different things, must be certain number of courses, don't think so?* all reveal, because of their briefness or lack of pronominal forms, the desire to be as curt in English as the speaker tends to be in Malay where much is left to the interlocutor to fill in, a kind of restricted code, in Basil Bernstein's terms (Bernstein in Gumperz and Hymes 1966). On the other hand, where both languages are joined together, it is often the Malay grammatical structure that determines the outcome as in *katagori interdisciplinary, kursus major, grade yang baik, thought ini* where speakers follow in all instances the Malay word order of adjectives following, and not preceding, nouns.

The writer can merely offer here a limited number of comments on a rather extensive corpus that must undergo further analysis. However, even these few comments show that the switching of codes also occurs in formal discourse, so that the view of limiting switching to informal settings alone does not apply to the Malaysian situation. Furthermore, the notion of biculturality that is so evident when one examines Hispanic language behavior in the United States does not seem to apply here fully. Although some cultural facts are learned as a new language is acquired, it now appears to be questionable whether this acculturation-through-language process accounts for the total cultural package. The remainder is learned by functioning in close contact with those who speak the language natively. Mexican-Americans live in such close contact with Anglo-Americans and can therefore acculturate effortless to the main culture of the United States. Malaysians, in turn, do not enjoy a similar close contact with native speakers of English, since – with the exclusion of Eurasians – English is for all Malaysians a second language which serves them well as a linguistic tool rather than a transmitter of British or American culture. Hence, the familiarity with English is more an inheritance from the colonial past and not a product of cultural awareness. In other words, there are apparently no close cultural ties between Malaysia – and probably many other countries in Asia – and, say, Britain or the United States, so that one may have to rethink, in regard to Malaysia, many of the views that he harbors when dealing with biculturalism in America.

4. Conclusion

The foregoing discussion has stressed the fact that studies concerning the presence of code-mixed speech in many parts of the world including Malaysia have mostly focused on situations where peers communicated among themselves in informal encounters. It has been the object of this study to explore the extent to which this language behavior also exists in formal situations. The writer has first provided a description of the Malaysian scenario with special reference to the way how the former colonial language English had lingered on after Malaysia's independence, even though Bahasa Malaysia had been chosen as the national language and how it was now re-emerging because of the country's desire to fully participate in the affairs of the world.

Some general notions concerning the study of code-switching have been included here in order to acquaint the reader with three mechanisms to produce mixed speech on the basis of either embedding one language into the other or of creating a balanced presence of the two participating languages. The differences in code alternation by Mexican-Americans and Malaysians were briefly explored and the presence of code-switching events, also in formal settings, was shown to exist in Malaysia. The evidence of switching in formal contexts was based on studies by Malaysian researchers as well as on recent recordings done at one of Malaysia's universities.

Holey English, a term created by Malaysian researchers was then correlated with similar notions familiar to most from international research in the field. The data of the Malaysian project were examined in light of who the interlocutors were and what their academic specialization was. In the later discussion of the writer's own data, he commented on the factors that control the options for bilingual discourse in formal settings. The data from a departmental meeting at a university showed a number of interesting aspects, in particular (1) the absence of flagging, (2) the semantic areas most favorable to the switching of codes, (3) repetitions in the other language as a technique for emphasis, (4) syntactic fusion through which elements from both languages would contribute to the formation of a single grammatical unit and (5) nativization of English reflecting the restricted code of Malay discourse.

The final remarks addressed once more the query as to whether or not the presence of English in Malay-English speech actually reflected biculturality as it did in the Mexican-American setting in the United States and speculated on the role of English merely as being a neutral linguistic tool rather than an instrument of acculturation.

5. Note

The writer wishes to express his gratitude to those who participated in the data collection process: to Marlyna Maros for her arrangements to record a departmental meeting, to Dean Imran Ho for his generous support to allow that such a meeting could be recorded and to Sandra Khor Manickam, Azean Abu Samad, Norhasliza Hassan for their valuable assistance in transcribing the data.

6. References

Asmah Haji Omar

1992 *The Linguistic Scenery in Malaysia.* Kuala Lumpur: Dewan Bahasa dan Pustaka.

1995 Language Policy and Management in Malaysia. *Asian Pacific Communication* 6 (3).

Azhar M. Simin/Bahiyah Dato Hj. Abdul Hamid

1994 English in Malaysia: The Case of Two Myths. In [name of editor and title of volume unknown]. Bangi: Universiti Kebangsaan Malaysia, 129–144.

Bentahila, Abdelâli/ Davies, Eyrlis

1998 Code-Switching: An Unequal Partnership? In Jacobson, R. (ed.), 25–49.

Bernstein, Basil

1966 Elaborated and Restricted Codes: Their Social Origins and Some Consequences. In Gumperz, J. J./Hymes, Dell (eds.) *The Ethnography of Communication* [Special publication of *American Anthropologist* 66].

Gill, Saran Kaur

2000 *Personal conversation.* Bangi: Universiti Kebangsaan Malaysia.

Hellinger, Marlis/Ammon, Ulrich (eds.)

1996 *Contrastive Sociolinguistics.* Berlin/New York: Mouton de Gruyter.

Jacobson, Rodolfo

1978 *The Social Implication of Intra-Sentential Code-Switching.* In Romo, R./ Paredes, R. (eds.), 227–258.

1983 Switches to English or Spanish: Does It Matter? Unpublished Paper Delivered at the South Central Modern Language Association (SCMLA) Meeting, Fort Worth, TX, Oct. 27.

1993 Penyatuan dan kontras dalam alternasi antara dua bahasa yang berbeza dari segi genetik: penukaran kod bahasa melayu-inggeris [Fusion and contrast in the alternation of two genetically different languages: Malay-English code-switching]. *Jurnal Dewan Bahasa* 37 (5).

1995 Penyampaian mesej yang lebih luas melalui wacana dwibahasa: kajian kontrastif dalam penukaran kod [Conveying a broader message through bilingual discourse: contrastive study in code-switching] *Jurnal Dewan Bahasa* 39 (5).

1996 In search of a deeper message: code-switching rationales of Mexican-Americans and Malaysians. In Hellinger, M./Ammon, U. (eds.), 77–102.

Jacobson, Rodolfo

1990 *Code-switching as a Worldwide Phenomenon.* Frankfurt a. M./New York: Lang.

1998 *Code-switching Worldwide*. Berlin/New York: Mouton de Gruyter.

Forthcoming
Language alternation: the third kind of code-switching mechanism. In Jacobson, R. (ed.).

Forthcoming
Code-switching Worldwide II. Berlin/New York: Mouton de Gruyter.

Osman, Safiah
1987 Gaining Insights into Malay-English Code-Switching. Paper Presented at the National Conference on Modern Languages, Kuala Lumpur (unpubl.).

Myers-Scotton, Carol
1997 *Duelling Languages: Grammatical Structures in Code-Switching*. (Paperback ed. with new *Afterword*). Oxford: Oxford University Press.
1998 A Theoretical Introduction to the Markedness Model. In Myers-Scotton, C. (ed.), 18–39.
1998 *Codes and Consequences: Choosing Linguistic Varieties*. Oxford/New York: Oxford University Press.

Platt, John/Weber, Heidi
1980 *English in Singapore and Malaysia – Status, Features, Functions*. Kuala Lumpur: Oxford University Press.

Poplack, Shana/Wheeler, Suzanne/Westwood, Anneli
1990 Distinguishing Language Contact Phenomena: Evidence from Finnish-English Bilingualism. In Jacobson, R. (ed.), 185–218.

Romo, Ricardo/Paredes, Raymund (eds.)
1978 *New Directions in Chicano Scholarship*. San Diego: University of California La Jolla.

English in Science Communication in Hong Kong: Educational Research Output

William Y. Wu, Dennis W. K. Chan and Björn H. Jernudd

1. Introduction

There might seem little reason for systematically investigating language selection in scientific communication in Hong Kong. People expect that English can appropriately be used in any communicative situation among participants in scientific communication. Texts of whatever format in English overwhelmingly dominate work in scientific organizations. Written communications with business interests and government offices are routinely in English.

The Hong Kong government has been promoting use of written Chinese since long before the political transition of Hong Kong into Chinese sovereignty in 1997, however, this process has taken on a greater urgency since 1997. The fact that the Chief Executive speaks in Cantonese strongly supports the use of Chinese in government and public affairs. He addresses the public in important speeches in Chinese from his position of executive power. This is something the British Governor did not do. Nevertheless, the use of spoken Chinese is hardly an innovation.

That two Cantonese speakers speak Cantonese between themselves was then and is now just as routinely expected as is their reading and writing in English on professional content. The latter (English in writing) dominates the former (Cantonese in speaking) because use of English carries over into discussing English texts or text-based content. The effect is especially strong for individuals who have been trained in English, and at graduate level overseas. Professionals and scientists active in today's Hong Kong who were born in Hong Kong have typically been educated in English-medium schools and universities; and potential authors of science articles have very likely been educated at graduate level in English overseas.

The presence at places of work and study of non-Cantonese speakers does not imply that foreigners acquire Cantonese (Chinese). Given the colonial socio-political situation the opposite was the case; the transition of sovereignty is only three years in the past and any change in Hong Kong's social communicational pattern will show only after more time has passed. There were of course exceptions, among whom were Christian missionaries in particular and many of whom were also educators. But in the normal case, English was further upheld and still continues to be upheld by the participation or the anticipation of participation of foreigners in a communicative situation. Foreigners were neither expected nor helped to learn Cantonese. While Cantonese speakers will unapologetically communicate in Cantonese in the presence of foreigners on personal matters or 'to chat', they will nevertheless switch to English for instrumental-transactional purposes. There is thus a sharp diglossic relationship between Cantonese/Chinese on the one hand and English on the other in Hong Kong in science-related contexts. For those with competence in Chinese, Cantonese is an oral, in-group medium. English is a written medium, and is routinely spoken to express text- or foreign-related instrumental purpose whether or not a foreigner is present.

2. Scientific field of investigation: hypotheses for language selection in the educational sciences

We have chosen to investigate when Chinese and not English is actually selected as language of publication in the field of the educational sciences. We rely on a straightforward proposition. Quite unremarkably, there is a greater likelihood to use the community's language in disciplines that help construct the community such as the study of the community language, or the community's history (Jernudd and Baldauf 1987: 159). We feel that study of publications in the educational sciences will allow study of this variable. We expect some use of Chinese for articles that address local educational policy issues and local educational practices, and we expect Chinese for articles that deal with Chinese language and other strongly Chinese-associated content in education, such as Chinese history.

3. Sources of data

There are seven leading journals in education which are based and published in Hong Kong: *Asian Journal of Counseling* (The Association of Psychological and Educational Counsellors of Asia), *Asia-Pacific Journal of Teacher Education and Development* (the Hong Kong Institute of Education), *the Curriculum Forum* (The University of Hong Kong), *Education Journal* (The Chinese University of Hong Kong), *Educational Research Journal* (Hong Kong Educational Research Association), *Journal of Basic Education* and *New Horizons in Education* (The Hong Kong Teachers' Association).

We selected to study *New Horizons in Education* and *Curriculum Forum. New Horizons in Education* has the longest history among all the journals and the *Curriculum Forum* has been published by the Department of Curriculum of the University of Hong Kong where English has been the dominant spoken and written language. By expectation of contrast, *Curriculum Forum* serves as an important reference for both teachers and students as it provides valuable information and insight about the Hong Kong subject teaching in general with attention to specifics of curriculum design development and implementation (Wu and Tse 1998).

The *Asia-Pacific Journal of Teacher Education and Development* was not chosen for this study because it was a new journal with a three-year history only. The youth of this journal brings up an interesting point. The authors have noted how colleagues in the academic community in Hong Kong in arts and humanities fields express interest in managing their own publication outlets. As a result, a number of new journals have started up in Hong Kong.

There is the *Journal of Translation Studies* published by the Chinese University of Hong Kong (number 4 in 2000), *East-West Dialogue* by the Hong Kong Baptist University (vol. 4 in 1999), and the *Hong Kong Journal of Applied Linguistics* by the University of Hong Kong (vol. 4 in

1999). Half of the articles in number 4 of the *Journal of Translation Studies* are in Chinese, the articles in the latest issues of the other journals all in English. The applied linguistics journal does not mention language of submission in its notes for contributors and *East-West Dialogue* announces that "papers may be written in English or Chinese". We expect an increase in Chinese language selection as the language of publication of articles because of the availability of journals such as these.

4. Data and methodology

Data

The data consist of

(a) 194 articles published in the *Curriculum Forum* over a nine year period from 1991 to 1999 and

(b) 625 articles published in *New Horizons in Education* over a thirty-eight-year period from 1961 to 1999. The analysis of data was conducted by re-categorization, cross-tabulation and interpretation by application of expert knowledge by the authors.

Journal content

Wu and Ho have examined the contents of journal articles in the *New Horizons in Education*. They "revealed the following developmental profile: an increasing number of authorship in journal articles, expanding scope of educational research, improving teaching practice attempts, emphasis on curriculum and instruction, enhancement of teaching Chinese, English and language, seeking educational administration efficacy, emerging special education, better educational assessment, and developing cross-cultural education" (Wu and Ho 1998).

The 194 articles in the *Curriculum Forum* concern a range of aspects of education and educational development ranging from teaching of mathematics to the environment and intrinsic motivation (Wu and Tse 1998). There are primarily two types of articles published in the *Curriculum Forum*, of which the majority are research-based and the others practice-oriented. There are 59 (30.41%) articles considered to be practice-oriented with no more than four references, with 39 having no references.

Journal language policy

The *Curriculum Forum* encourages the contribution of articles written in either English or Chinese. Of the one hundred ninety four articles already published, 143 are written in English (73.71%) and 51 in Chinese (26.29%).

New Horizons in Education is an annual publication of the Hong Kong Teachers' Association. With the production of two volumes in 1961, a total of forty volumes have been published thus far. Of the 625 articles, 367 are written in English (58.63%) and 259 in Chinese (41.37%). Although the majority of the articles are written by one author, starting in the 1980s there is a slight increase in the number of articles with multiple authors (Wu and Ho, 1998: 98 Table 2). This trend is also in accord with the international sector, particularly in the field of science (Bayer and Smart 1991). For Hong Kong educators, international collaboration provides cross-cultural understanding both in research and in educational practice to improve the quality of education.

Among authors of both journals, 232 educators present their research in Chinese, and 382 educators in English[1]. In *Curriculum Forum*, 64 local scholars wrote their articles in Chinese and 55 local scholars in English. Among the 51 non-local scholars-authors, all selected English as their medium[2]. In *New Horizons in Education*, 184 local scholars wrote their articles in Chinese and 125 local scholars in English[3]. All 55 non-local scholars used English[4].

5. Hypothesis 1: Chinese language selection favored for practical and not research-oriented articles

We have categorized the articles into practice-oriented versus research-oriented articles according to expert judgement. A distinction is made regarding the inquiry. Research-oriented articles are data-based or address research issues. In a practice-oriented article, inquiry is reflective and offers personal commentary without any substantive literature review. Thus, the number of citations and hence references become an important indicator. Practice-oriented articles have no more than four references. Appendix 1 offers some illustrations for both practice-oriented and research-oriented articles in the two journals.

A majority of articles, 452, are practice-oriented and the rest are 368 research-oriented in the two journals.

In *Curriculum Forum*, there are eleven (11) practice-oriented articles out of the total of 51 Chinese language articles, compared to 48 (out of 143) in English. The Chinese language articles reflect content which implies significant use of Chinese, such as methods for teaching the Chinese language and Chinese history subjects, and content that deals with issues of local educational policy, such as targets and target-related assessment and basic education in China. The majority of the 40 (78.43%) research-based articles in Chinese involve mostly teaching methods on Chinese-language-related subjects. Among them, there are five articles on other subjects are such as Economics and Mathematics, and General Education.

The reverse is found in *New Horizons in Education*. There are 200 articles in Chinese (77.22% of the Chinese language articles in this journal) in the practice-oriented category, compared to 193 in English (52.59% of the English language articles in the journal). Most of the Chinese-language articles published before the 1980's have no references and cover curriculum development for subjects such as the use of Chinese and Chinese culture. Some of them deal with topics such as local education policy, moral education, and ethics issues.

There are 59 articles in Chinese (22.78% of the total Chinese-language 259) in the research-oriented category in *New Horizons in Education*, compared to the corresponding 47.41% in English. Content of the Chinese language articles ranges from Chinese-related teaching methods to non-language based subject teaching, from local educational issues to international educational issues.

Table 1: Articles classified into research- and practice-oriented categories of content according to language of article in *New Horizons in Education* and *Curriculum Forum*

Content orientation:	Total	In Chinese				In English			
		Curriculum Forum		New Horizon in Education		Curriculum Forum		New Horizon in Education	
		No. of Articles	%	No. of Articles	%	No. of Articles	%	No. of Articles	%
Practice-oriented	452	11	21.57	200	77.22	48	33.57	193	52.59
Research-oriented	368	40	78.43	59	22.78	95	66.43	174	47.41
Total		51		259		143		367	

Curriculum Forum is more research-oriented overall and *New Horizons in Education* is more practice-oriented. In *Curriculum Forum*, the educators prefer to use English in writing both practice- (48, 81.36%) and research-oriented (95, 70.37%) articles. In contrast, in *New Horizons in Education*, the educators prefer to use Chinese on practice-oriented content (200, 50.89%) and to use English (174, 74.68%) on research-oriented content.

Table 2: Articles classified into research- and practice-oriented categories of content according to language of article in *New Horizons in Education* and *Curriculum Forum*

	Total	Practice-oriented				Research-oriented			
		Curriculum Forum		New Horizon in Education		Curriculum Forum		New Horizon in Education	
		No. of Articles	%	No. of Articles	%	No. of Articles	%	No. of Articles	%
In Chinese	310	11	18.64	200	50.89	40	29.93	59	25.32
In English	510	48	81.36	193	49.11	95	70.37	174	74.68
Total		59		393		135		233	

The hypothesis that practice-orientation of article content favors publication of the article in Chinese since practice would involve significant use of Chinese which therefore would be reflected also in writing about the practice is only weakly if at all supported by our data. It may be weak because prescribed practice in secondary education until last year was use of English as medium of instruction. Whether strictly implemented or not, that fact undermines this hypothesis. This has not been the case for primary education which is conducted in Cantonese/Chinese. Can a refinement be made to sort out those articles among the practice-oriented ones that deal with primary level education and will the hypothesis then be upheld?

6. Refinement: Articles dealing with primary education content favor Chinese language selection

We sorted out data first according to the local or non-local authorship of articles on primary level educational content. The most striking finding is that overall 56 articles are published in Chinese compared to 15 in

English. As for authorship, almost all authors are Chinese. However, a few nevertheless published in English. The relatively few articles (17) printed in the *Curriculum Forum* were equally divided between Chinese and English ones. Despite the small numbers, this is not at all surprising in Hong Kong. English is the default language in scientific academic publishing and the use of Chinese would either signify an emerging new norm or special motivation for its use. Our surmise here is that the use of Chinese in primary education is a motivating factor for selection of Chinese as the language of the article. Interestingly, in *New Horizons in Education* only 11% of the articles are in English.

Table 3: Authorship of articles with primary level education content by language of publication in *Curriculum Forum* and *New Horizons in Education*

	Articles in Chinese		Articles in English	
Authorship	Local	Non-Local	Local	Non-Local
Curriculum Forum	9	0	8	1
New Horizon in Education	47	0	5	1
Total	56		15	

The practice- *versus* research-oriented sorting reveals that most articles are practice-based, which we now associate with Chinese language selection of publication. We also found that as many as 16 research-oriented articles are printed in Chinese compared to eight in English, between the two journals. Primary education content apparently favors Chinese language selection.

Table 4: Articles associated with primary level education in *Curriculum Forum* and *New Horizons in Education* according to practice- and research-oriented content by language of publication

Content orient.:	Total	In Chinese				In English			
		Curriculum Forum		New Horizon in Education		Curriculum Forum		New Horizon in Education	
		No. of Articles	%	No. of Articles	%	No. of Articles	%	No. of Articles	%
Practice-oriented	41	4	80	33	68.75	2	33.33	2	33.33
Research-oriented	24	1	20	15	31.25	4	66.67	4	66.67
Total		5		48		6		6	

7. Hypothesis 2: Chinese-language selection favored for articles dealing with Chinese-content subjects

Curriculum Forum is primarily a teaching subject-based refereed journal. A majority of articles are dedicated to discussion of teaching of subjects such as English, Chinese, History, Geography, Chemistry, Mathematics and Science. In it, a full two-thirds (31, 60.78%) of the 51 articles in Chinese concern Chinese-content learning. There are only two articles in English on Chinese-content learning. The one deals with the subject Composing in Chinese and the other is an article for Chinese language teachers. The Chinese-content articles range over Chinese writing, Chinese History, Putonghua, Chinese language and culture, and also China education.

In *Curriculum Forum*, another 10 articles in Chinese (19.60% of the 51 Chinese-language articles) deal with local educational issues; five articles (9.80%) with other non-language subjects such as Economics, Mathematics, and General Education; and five with other topics altogether. The articles in English are distributed over these other content categories.

New Horizons in Education is a journal with a majority of articles dedicated to subjects such as Curriculum, Mathematics, Science Education, Physical Education, Languages, and Teacher Education. The most often discussed topics include: particular areas of education, specific subjects such as Biology, Mathematics, Physics, Language Education, Curriculum Development and Medium of Instruction, Educational Psychology, Teaching, Educational issues, Assessment, Educational Administration, Special education for handicapped and minorities, and education across regions.

In *New Horizons in Education*, there are only 95 articles in Chinese (36.68% out of a total of 259 Chinese-language articles) on Chinese-language content subjects. One-third (87, 33.59%) of the Chinese-language articles are concerned with local educational issues. There is also quite a large number of Chinese-language articles in the content category of non-language subjects (39, 15.05%) and other special educational issues or teaching methods (38, 14.67%). There are significantly more English-language articles proportionately in this latter 'other' category (201, or 55.01% of all English-language articles).

The hypothesis that Chinese-content of an article favors Chinese as the language of publication cannot be rejected.

We conclude that the hypothesis that articles with a Chinese-language-oriented content (Chinese-language learning, *etc*) attract being published

Table 5: Number of articles by specific content and by language of publication in *New Horizons in Education* and *Curriculum Forum*

Topic	Total	In Chinese				In English			
		Curriculum Forum		New Horizon in Education		Curriculum Forum		New Horizon in Education	
		No. of Articles	%	No. of Articles	%	No. of Articles	%	No. of Articles	%
Language Content									
Chinese Content	129	31	60.78	95	36.68	2	1.40	1	2.74
English Content	62	0	0	0	0	19	13.29	43	11.78
Other Subject Content	137	5	9.80	39	15.05	53	37.06	40	10.96
Local Education Issues	196	10	19.60	87	33.59	19	13.29	80	21.92
Others	294	5	9.80	38	14.67	50	34.97	201	55.01
Total		51 310		259		143 508		365	

in Chinese deserves elaboration. One explanation could be that the author as well as the intended audience necessarily both know and practice Chinese in dealing with the content matter of the article. Furthermore, in sociolinguistic practices in Hong Kong, attention to Chinese is distinctly marked, still sharply demarcated from the norm of use of English in discourse about educational and scientific topics. This diglossic reality allows Chinese language publication for Chinese-language and other Chinese-associated content which prompts the use of Chinese. At the same time, however, these articles are also accessible to the generalist and the broader audience of educators in Hong Kong, an increasing majority of whom command Chinese as well as English.

8. Notes

1. In the case of multiple authorship, only the first author was included in the data set and subsequent analysis. An educator may produce more than one article in either Chinese and/or English.
2. Four Chinese-language articles are articles that were translated into Chinese by local educators. The distinction between local and non-local authorship is a matter of institutional affiliation in most cases and/or whether the name of author indicates Chinese ethnicity.

3. An additional six articles, four in English and two in Chinese, report only institutional affiliations as authorship.
4. To accommodate contributions from teachers who could present their work to a larger audience, the *Curriculum Forum* began publishing articles concerning curriculum from China, the Asia-Pacific region, and the rest of the world starting in November 1994 with vol. 4, issue 1.

9. References

Bayer, A. E./Smart, J. C.
1991 Career Publication Patterns and Collaborative Styles in American Academic Science. *Journal of Higher Education* 6 (6): 613–636.

Jernudd, B. H./Baldauf, R. B., Jr.
1987 Planning Science Communication for Human Resource Development. In Das, B. K. (ed.) *Language Education in Human Resource Revelopment*. Singapore: SEAMEO Regional Language Centre, 144–189.

Wu, W. Y./Ho, K. K.
1998 A Developmental Profile of "New Horizons in Education". *New Horizons in Education* 39: 96–147.

Wu, W. Y./Tse, S. K.
1998 An Examination of the 1991–1997 Cumulative Index of Curriculum Forum. *Curriculum Forum* 7 (2): 112–133.

English as the Medium of Instruction for Science and its Effects on the Languages of the Philippines

Jerzy Smolicz, Iluminado Nical and Margaret Secombe

1. Introduction

The Philippines enjoys particular linguistic advantages in the South-East Asian region, as a country where English is not only an official language but also widely spoken among the population. Historically, English was introduced into the Philippines in 1898 at the time of the imposition of American rule on the islands. Over the twentieth century, which established the domination of English as a world language, it has preserved its status as one of the two official languages of the Philippines and has continued as a language of schooling, although in the last quarter of the century, it has been used as a medium of instruction alongside the national language, now known as Filipino.

Despite this access to the learning of English and its powers of attraction as the language of American affluence, global influence, scientific development and technical advance, the Philippines has remained a multilingual country. In the course of their daily living, every Filipino speaks at least two, often three, and sometimes even four languages. According to Gonzales (1998: 489), at least 84% of the present population can speak Filipino, the national language developed from Tagalog, and 56% speak some form of English, while 74% report being able to understand spoken English.

The indepth small scale study reported in this paper sought to investigate the emerging generation's use and evaluation of the English language vis a vis Filipino/Tagalog and Philippine languages other than Tagalog (referred to in this paper as PLOT). The participants were groups of Filipino young people, in their last year of secondary school, who came from three non-Tagalog speaking communities in the Philippines.

2. The multilayered linguistic heritage of the Philippines

Throughout its history, the Philippines has been a country of cultural and linguistic complexity with a heritage which includes the infusion of a variety of cultures upon the Indo-Malay base (Bowering 1863; Zialcita 1995). Although the many islands which make it up are small in terms of geographical area (116, 220 sq. miles in all), the Philippines is a pluralistic and multilingual society. The multi-ethnic population speaks ten major and many other numerically smaller languages, mainly belonging to the Austronesian family. Although they constitute separate languages, in that they are not mutually intelligible and have many subordinate dialects, they are clearly related to one another and regarded as belonging to the "Philippine type" (Gonzales 1998: 493).

The indigenous pre-colonial past of the Filipino cultural stream had been infiltrated by other peoples, long before the country was conquered by Spain in the sixteenth century. Cultural tributaries came originally from India, and later there arrived Chinese traders, many of whom settled in the Philippines, mainly after the coming of the Spaniards. In the South, Islamic influences had begun to spread, bringing with them Arabic and Persian infusions. Over the three centuries of hispanic rule, Spanish became established as the language of government and the small ruling elite. When the Spaniards were finally forced to leave the Philippines, the Americans who replaced them overpowered the emerging indigenous independence movement and introduced English as the language of government and education (Bautista 1981).

In the Philippines today Chinese racial and cultural influences abound (Sin 1995), but Spain and United States can be seen to have left the most striking heritage in terms of religion and language. As an independent republic since 1946, the Philippines has successfully transmuted all the cultural inputs of its neighbours and former imperial powers to constitute a cultural blend that is uniquely Filipino, while still showing the imprint of the various imported cultural values (Smolicz 1986; 1990). It can be argued, however, that the double exposure to colonial linguistic domination, first in Spanish and then in English, has delayed the literary development of all the major indigenous languages of the country.

The Spanish members of the religious orders in the Philippines learned the indigenous languages in order to spread the Catholic faith to the masses. Their approach was successful to the extent that they succeeded in making the Philippines one of the most Christian countries of Asia. At the same time, their efforts helped the indigenous languages of the Philippines to acquire their first written records (De la Costa 1961). Their literary development was, however, slow due to the restriction of literacy to a small elite, with literature mainly confined to religious subjects and many aspects of indigenous culture excluded as pagan. In practice, the Spanish colonizers showed much more preoccupation with the inculcation of Christianity than with the imposition of Spanish and, in fact, limited education at higher levels to Spaniards alone until 1863. Decrees of the Spanish court advocating the more widespread use of Spanish were largely ignored by the members of the four religious orders which virtually ran the country and whose members preferred to use the indigenous languages in their missionary work. Only the elite was linguistically hispanised over the latter half of the nineteenth century. Hence, the great Filipino patriot and hero, Jose Rizal, executed by the Spaniards in 1896, wrote his call to Filipino independence in Spanish.

With the imposition of American rule, however, came the promulgation of compulsory education in English for all Filipinos. Three factors can be seen to have favourably predisposed the Filipino people toward the learning of this new colonial language: the positive attitudes of Filipinos towards the Americans who were widely seen as liberating them from rigid Spanish rule; the system of public instruction in English which they established; and the incentives given to those Filipinos who successfully learned English, in terms of career opportunities, government service and participation in politics.

While the American English language initiatives were initially directed mainly against Spanish, they were almost equally hostile to the indigenous languages of the country, with penalties imposed upon pupils using their home languages on the school premises (Manhit 1980; 1981). Throughout the period of American rule, indigenous languages were excluded from schools and universities and most forms of public life. Yet they continued to thrive in the homes and hearts of the people until just before the Second World War, when the emerging demand for political independence was paralleled by a movement demanding the recognition of the right of Filipinos to their own national language(s).

Independence was conceptualised in terms of the European model of the monolingual nation-state (Smolicz and Nical 1997). When the search for one national language through the fusion of the ten major indigenous languages failed (Gonzales 1974), the adoption of Tagalog, the dominant language of the Manila area, as the basis for developing a national tongue, was perceived as the only way to prevent total domination by the colonial language, English. In general, this move was interpreted by native Tagalog speakers, as an advantageous and inevitable outcome, while most of the elite members of other language groups were eventually reconciled to accepting such a compromise, provided that English remained dominant in government, universities and business life. Those who supported the adoption of an "intellectualised" version of Tagalog as the national language entitled "Pilipino", and eventually renamed Filipino, often pointed to the negative impact of English on Philippines society (Sibayan 1994). The imposition of English by the colonial regimes in Asia and Africa, it has been claimed, led to the colonised people internalising the norms and ideology of the colonisers and becoming alienated from their own linguistic and cultural heritage (Phillipson 1992: 27). In the context of the Philippines, education became negatively affected in that the time spent learning English most often meant not only that the standard reached in other subjects was inadequate but that the learning of indigenous languages was neglected (Constantino 1982).

3. The introduction of bilingual education

The struggle between the advocates of English and Filipino was resolved in 1974 through the adoption of the Bilingual Education Program, which aimed to develop a nation competent in both English and Filipino (Manuel 1974). Despite difficulties in its implementation over the next decade, the Bilingual Education Program was re-affirmed in the democratic transformation that succeeded the Marcos regime. The 1987 guidelines of the Department of Education, Culture and Sport (DECS) stated that English and Filipino were to be taught in all grades of elementary and secondary schools. Filipino was to be the medium of instruction in Social Studies/Social Science, Character Education, Work Education, Health Education, and Physical Education; English was to be the medium of instruction in all other areas, in particular, Science and Mathematics. The 1987 guidelines contained an additional provision in relation to Muslim regions of the country, in that "Arabic was to be used in areas where it was necessary". Some allowance continued to be made for schools to use the local non-Tagalog "vernacular" or regional language of the area "as auxiliary to the media of instruction, but only when necessary to facilitate the understanding of concepts being taught in English, F(P)ilipino or Arabic" (Quisumbing 1989: 300).

Bilingual education remains a controversial issue in the Philippines. Reports of the failure of bilingual policy appear constantly in the press (in such Manila dailies as The Philippine Star and The Inquirer) with reference to the perceived decline in the standard of English in the schools. This is usually attributed to the time which needs to be allocated to Filipino and the influence of Filipino linguistic structures upon English usage. The learning of Mathematics and Science is said to have become consequently more difficult, as these subjects are taught in a language that is not fully comprehensible to the students. The Congressional Commission on Education (1991: xii) expressed concern about the decline of educational standards in the country as a whole, when it bluntly stated: "Our elementary and high schools are failing to teach the competence the average citizen needs to become responsible, productive and self-fulfilling."

Public criticism, however, rarely speaks about the handicaps experienced under the present Bilingual Education Program by native speakers of non-Tagalog languages, especially those who come from areas of rural poverty or low socio-economic status. In upper and middle class homes throughout the country, some English is usually spoken, so that children have at least a background knowledge of the language when they start school at the age of six. But the rural poor with very limited or no English, as well as only peripheral knowledge of Tagalog/Filipino learned from television, films or comics, face a double linguistic barrier in learning in the school context, so that it is not surprising that only the most able and dedicated achieve success (Gonzales 1998: 520; Smolicz 1986).

4. An investigation of language use and attitudes

4.1. Methodology

After over 25 years of English as the medium of instruction in Science and Mathematics under the Bilingual Education policy, preceded by more than 75 years of compulsory education in English, it might be expected that English would have become the dominant language used by people in the Philippines. This study sought to investigate the activation of, and attitudes toward English, as compared to Filipino and other indigenous languages, among young people drawn from three different linguistic communities of the Philippines (Nical, 2000). One final year class of secondary school students was selected from the laboratory school attached to the state college/university in the three regions under investigation. A total of 152 students were asked to participate and all agreed to be involved. The regional breakdown was 62 respondents from the Cebuano speech community in Cebu City; 55 from the Ilocano speech community in Nueva Viscaya; and 35 from the Waray speech community in Tacloban, Leyte. According to the 1995 census figures, Cebuano speakers constituted 21% of the total Philippines population, Ilocano speakers 9% and Waray speakers 4% (Gonzales 1998: 490–492).

Data on language usage were collected through a questionnaire in which the respondents were asked to indicate : (1) how frequently they used English and any other languages in the three communication activities of speaking, reading and writing; and (2) how often they activated English and any other languages in different domains, as indicated by specific interlocutors. The analysis of responses to these questions was carried out through the calculation of mean values which were interpreted as follows:

4.50 – and above	– Always (in the non-exclusive sense of "all the time")
3.50–4.49	– Often
2.50–3.49	– Sometimes
1.50–2.49	– Seldom
0.50–1.49	– Never

In order to compare the language usage of students from the three speech communities, F-ratio and Scheffe results were calculated.

The data on the activation of languages with different interlocutors in a trilingual situation were analysed through a procedure developed by Ammon (1989: 73–76). His concepts of sole activation and co-activation of languages were expanded to include the particular patterns shown by the respondents in co-activating Philippine languages. The patterns identified for use in this classification were: dominance of any language over two equally less used; dominance of one in an hierarchical order; equal dominance of two over one less used; and balanced use of three languages.

Attitudes to the languages concerned and the meaning which the respondents attached to each language were derived from two essays which the students were asked to write. The first topic which the students were given was, My feelings about the languages used at home. The second essay was on the topic, My feelings about the languages used at school. This method of using personal statements and writings as a source for identifying individual attitudes is based on humanistic sociology (Znaniecki 1969; Smolicz 1979, 1999). The basic principle is that all social and cultural activities need to be interpreted from the perspective of the participants rather than the researcher. The focus is on individuals as cultural beings and as conscious agents in a given social system, with the aim of understanding how they themselves view their current cultural reality.

On the basis of what they had written in the essays, the students were judged to have a positive or negative attitude to the languages being discussed. When the attitude was positive, it was further classified as *high*, *moderate* or *low*. To be assigned a high positive attitude, respondents needed to demonstrate their appreciation of the intrinsic qualities of the language, as well as recognition of the advantages that were assumed to be derived from its mastery. Closer reading of the essays also led to the identification of a number of distinct meanings which the writers associated with each of the languages. These meanings were then grouped under three categories – autotelic, instrumental and negative – and a frequency table for the citations of these meanings in each language was drawn up, so that the pattern of meanings for respondents across the three language communities could be compared.

4.2. Discussion of results: usage and domains of activation

The data on the frequency with which the students said they utilized English and other Philippine languages in the three communication activities of speaking, reading and writing are summarized in Table 1.

The pattern of usage for English across the three groups of respondents proved to be quite complex, with the differences among them being statistically significant. The Ilocano group revealed the lowest level of English usage for all three activities and the Cebuano students the highest. Among the Cebuano and Warray respondents, English was the language predominantly used for reading and writing (at the "often" to "always" level), but the Ilocano respondents revealed that Filipino was the language they used most frequently for reading and writing.

Table 1: Comparison of language usage of Filipino students

Communication activities in different languages	Mean frequency of language usage			F-ratio	Scheffe results at .05 level
	Waray N=35	Ilocano N=55	Cebuano N=62		
PLOT					
Speaking	4.57	4.33	4.63	2.32	
Reading	1.86	2.80	2.05	20.99**	M2>M1,3
Writing	1.80	2.53	1.89	11.74**	M2>M1,3
FILIPINO					
Speaking	3.32	3.91	3.22	14.01**	M2>M3,1
Reading	3.62	4.38	3.37	19.34**	M2>M3,1
Writing	3.50	4.27	3.35	18.37**	M2>M3,1
ENGLISH					
Speaking	3.48	2.98	3.87	25.02**	M1>M2; M3>M2,1
Reading	4.48	3.94	4.79	23.09**	M1,3>M2
Writing	4.46	3.73	4.66	28.86**	M1,3>M2

Legend – The calculation of the mean was related to the following scale of usage:
1.00 – Never
2.00 – Seldom
3.00 – Sometimes
4.00 – Often
5.00 – Always (in the non-exclusive sense of "all the time").

**Significant at .01 level.

In relation to speaking ‘ respondents in all three communities reported that the language they used "often" to "always" was their regional PLOT. The level of PLOT use for reading and writing was very much lower, ranging from "never" to "sometimes". In the case of Filipino, the main feature of the results, revealed by all three groups of respondents, was its consistent usage for all three communication activities at the "sometimes" to "often" level.

The information indicating which languages the respondents used in speaking with a range of different interlocutors is summarised in Table 2. In the home and peer group domain, which included parents, brothers, sisters, relatives and fellow students as interlocutors, the use of PLOT

Table 2: Comparison of Filipino students' mean usage of languages in different domains

Languages and Interlocutors	Communities			F-ratio	Scheffe Test Results
	Waray	Ilocano	Cebuano		
PLOT					
a. Parents	4.46	4.31	4.58	1.36	
b. Relatives/friends	4.26	4.22	4.40	.96	
c. Brothers/sisters	4.53	4.37	4.66	4.66	
d. Fellow students	4.17	4.18	4.35	1.03	
e. Market vendors	4.66	3.75	4.48	13.52**	M 1&3 > M 2
f. Sales people	3.88	3.64	3.58	1.09	
g. Office people	2.97	2.83	2.83	.32	
h. School teachers	2.57	3.85	3.13	9.55**	M 2&3 > M 1
FILIPINO					
a. Parents	1.88	3.42	2.54	29.87**	M 2 > M 3&1; M 3 > M 1
b. Relatives/friends	3.00	3.38	2.80	6.57*	M 2 > M 3
c. Brothers/sisters	2.59	3.52	2.29	24.87**	M 2 > M 1&3
d. Fellow students	2.83	3.48	2.83	11.69**	M 2 > M 1&3
e. Market vendors	1.74	3.05	1.78	29.59**	M 2 > M 1&3
f. Sales people	2.74	3.27	2.38	14.14**	M 2 > M 3&1
g. Office people	2.89	3.81	2.42	32.81**	M 2 > M 3&1
h. School teachers	3.28	4.04	3.09	19.66**	M 2 > M 3&1
ENGLISH					
a. Parents	2.86	2.36	3.23	17.05**	M 1&3 > M 2
b. Relatives/friends	2.74	2.56	3.37	17.94**	M 3 > M 2&1
c. Brothers/sisters	2.82	2.56	3.02	3.79*	M 3 > M 2
d. Fellow students	3.11	2.84	3.62	15.57**	M 3 > M 2&1
e. Market vendors	1.60	1.71	2.02	3.18*	M 1&3 > M 2
f. Sales people	3.00	2.14	3.61	31.78**	M 3&1 > M 2
g. Office people	3.54	3.02	4.02	18.60**	
h. School teachers	4.26	3.51	4.02	12.84**	M 3&1 > M 2

Legend:
*Significant at .05 **Significant at .01

The calculation of the mean is related to following scale of usage:
1.00 – Never 2.00 – Seldom
3.00 – Sometimes 4.00 – Often
5.00 – Always (in the non-exclusive sense)

predominated (in the "often" to "always" range) for all the language groups. The use of Filipino with these interlocutors was reported by the Ilocano students at the "sometimes" to "often" level, whereas the Cebuano respondents used English to much the same extent in communication with relatives, friends and fellow students.

The domain of the market place also revealed a dominant usage of PLOT. Among the Waray and Cebuano students, PLOT was spoken to market vendors "often" to "always", with both Filipino and English being used "seldom" to "never". The Ilocano respondents, however, revealed a more balanced usage of PLOT and Filipino at the "sometimes" to "often" level, with English being "seldom" or "never" spoken in this context. These results are consistent with a number of earlier statistics reported by Gonzales and Bautista (1986: 7–26)

In the business domain of sales and office people, respondents from all groups indicated a usage of English, alongside the other two languages, which were activated in the "seldom" to "sometimes" range. Here, too, the Ilocano students showed a preference for the more frequent usage of Filipino, while the Waray and Cebuano respondents indicated that they were more likely to use English.

Communicating with teachers in the school domain also highlighted a pattern of trilingual usage for all respondents. The Ilocano students most frequently used Filipino and PLOT rather than English, while the Cebuano respondents and even more, the Warays revealed a usage of English at the "often" to "always" level. All the student respondents, however, reported some usage of PLOT with teachers, despite the fact that it had no official place as a medium of instruction or as a subject in the curriculum.

Overall, the most striking feature of these results was the fact that in no domain was the exclusive use of English or any other language evident. The procedure developed by Ammon (1989: 73–76) for analysing activation in multilingual settings was therefore applied in order to understand the patterns of co-activation revealed by the respondents and to determine whether any language proved to be dominant. The results, which are presented in Table 3, highlighted the great range of patterns of activation apparent among the individual respondents.

Table 3: Patterns of language activation to different interlocutors among Filipino students (All figures in percentages of the number of respondents in the group concerned.)

Speech Community/ Interlocutor	Patterns of language use																	
	Sole activation			Co-activation														
				One dominant over two equally less used			Dominance of one in an hierarchical order						Equal dominance of two over one			Balanced
	P	F	E	P	F	E	P FE	P EF	F PE	F EP	E FP	E PF	PF	PE	FE	Equal use
Waray N=35																
a. Parents				40		3	11	28	3				3		3	9
b. Brothers/ sisters	3						3	24	3				3	3		6
c. Relatives/ friends							18	15	3				9			3
d. Fellow students						8	3	20					3	8	3	14
e. School teachers				3		28		3			37				14	14
f. Market vendors	35			24			29	12								
g. Sales people	6			34		11	11	9			3	3	6	9		9
h. Office people				3		11		11		3	20	3	6	14	11	17
Ilocano N=55																
a. Parents	2	2		24	2		40	2	5	5			13			5
b. Brothers/ sisters				30	6		35	4	7	2			9		4	4
c. Relatives/ friends				15	7		33	7					27			11
d. Fellow students				24	2		25	2	9				33		2	4
e. School teachers				7	31	11			2	5			5	5	13	20
f. Market vendors	4			5	2		33	4	22				25	2		4
g. Sales people				9	4	2	25	2	11				34	2	4	7
h. Office people		2		5	25	4	9		11	7			7	5	11	13
Cebuano N=62																
a. Parents	8			29		2	18	27					5	5		6
b. Brothers/ sisters	6			26			6	46				3	5	5		2

Table 3 (Continued)

Speech Community/ Interlocutor	Patterns of language use															
	Sole activation			Co-activation												
				One dominant over two equally less used			Dominance of one in an hierarchical order						Equal dominance of two over one			Bal-anced
	P	F	E	P	F	E	P FE	P EF	F PE	F EP	E FP	E PF	PF	PE	FE	Equal use
c. Relatives/ friends	2			29	2	2	13	26	2				6	8		11
d. Fellow students				24			3	24			2	8	2	24		10
e. School teachers				2	2	27		2			18	6	2	24	13	5
f. Market. vendors	40			23			10	15				3	2	3	2	2
g. Sales people	5		2	18		5		18			8	19		14		11
h. Office people			6			20		8			18	21	2	18		6

Legend:
P–PLOT F–Filipino E–English

The sole activation of one language occurred only rarely – and mainly in the context of the market place. The tendency for the PLOT to predominate in the home domain, but with varying patterns of lesser activation of the other two languages, was evident in all three communities. However, a pattern of balanced usage of PLOT and Filipino, which specifically excluded English, was indicated by a proportion of the Ilocano respondents. In the business domain, too, the co-activation of languages was most often dominated by PLOT and Filipino among the Ilocano students, while English along with PLOT was more frequently used by the Cebuano and Waray groups.

The school was clearly the domain where English was most often activated – but even there, it was always together with Filipino and PLOT. Among the Waray and Cebuano students, English most often pre-dominated over the other two, but over half the Ilocano respondents indicated either the dominance of Filipino or the balanced use of all three languages in communicating with teachers.

The more frequent activation of English with teachers, as compared to other interlocutors, which was found overall, is not surprising, given the status of English as the medium of instruction in Mathematics and Science. From the observation of one of the researchers, it could also be regarded as either the result of school enforcement, or the students' own sense of responsibility for their learning. There are rules and regulations which require students to use English with teachers. There are schools in the Philippines which impose some form of punishment if students are caught using the local PLOT in the context of the school. At the same time, the students themselves feel a responsibility to use English with teachers in order to develop their competence in the language and to show proper respect, since the usage of English with teachers and other school authorities is generally associated with courtesy and politeness. Despite this press for English, there was evidence in the data that PLOT was not being completely excluded from the school domain.

4.3. Language attitudes and meanings

According to the overall summary of of data on student attitudes to languages (see Table 4), English was the language toward which the respondents most often expressed moderately positive or low positive attitudes, with only a quarter evincing high positive attitudes. In contrast, almost half the respondents expressed high positive attitudes to Filipino and about the same percentage were moderately positive, while the majority indicated moderately positive attitudes to their respective PLOT.

Yet the above summation of the attitudinal responses concealed some very important differences across the linguistic communities. The Waray students were clearly most positive to English, revealed moderately positive attitudes to their PLOT and a range of attitudes, including a comparatively large number of negative attitudes, to Filipino. Among the Cebuano respondents, high positive attitudes were evident equally for English and PLOT, while their attitudes to Filipino were more often at the moderate and low level. In contrast, the Ilocano respondents most often expressed high positive attitudes to Filipino, moderately positive attitudes to PLOT and mainly low and negative attitudes to English.

The analysis of the meanings which the students attributed to the various languages in their essays pointed to a different cluster of meanings for each language. English was associated with the global, international community, with business and progress, and with access to knowledge. In contrast, PLOT was seen as a language of regional significance, related to the home and local community, and the language which the students felt most comfortable with as a vehicle for communicating ideas and feelings. Filipino was regarded as the language of national significance and a source of pride and identity. Some illustrative examples of the meanings given in the essays are listed below.

Table 4: Filipino students' attitudes towards languages
(All figures represent percentages of the number of respondents from the linguistic group concerned)

ATTITUDES to	PLOT	FILIPINO	ENGLISH
Waray (N = 35)			
Positive			
High	10	37	50
Moderate	79	30	50
Low	10	13	–
Negative	–	20	–
Ilocano (N = 55)			
Positive			
High	26	62	7
Moderate	58	32	27
Low	16	6	47
Negative	–	–	18
Cebuano (N = 62)			
Positive			
High	43	16	40
Moderate	36	58	45
Low	21	22	14
Negative	–	5	–
Overall (N = 152)			
Positive			
High	29	40	29
Moderate	54	41	39
Low	17	13	24
Negative	–	6	7

– "My feelings about the Ilokano language being used at home is great. It represents people who are sturdy in spirit and having their own identity." (Source of pride /identity)
– "I feel comfortable if I use Waray as a language at home and in the neighbourhood. We understand each other well and we feel pleased ev-

erytime we speak to each other in Waray." (Feeling comfortable + Home and local community)
– "Ilokano is our lingua franca at home and in the neighbourhood. There exists no gap and barrier among ourselves." (Home and local community)
– "Being in Cebu City, my family and I always speak Cebuano at home. I was born here, thus I learned to speak Cebuano. I feel great, relaxed and happy every time I am with my family talking to each other during meals, free time and other family gatherings in Cebuano." (Home and local community)
– "Waray language is rich enough to be able to accomodate any and all nuances pertaining to language use and applicability. In fact, there had been (and they still continue to flourish) great compositions (both musical and poetic) done in Waray. It gives us great pride to know about these." (Aesthetic)
– "Educated people find it more interesting [to talk] in English." (Mark of good education)

As the Table 5 indicates, differences in the distribution of these meanings were apparent among the three linguistic communities. The Ilocano respondents attributed positive meanings to Filipino much more frequently than the other students, while the Cebuanos expressed positive meanings toward the PLOT more often than the other groups. For their part, the Waray students proved most likely to indicate positive meanings for English.

Table 5: Meanings given to languages by Filipino students (All figures represent percentages of the number of respondents from the linguistic group concerned)

Meanings given to languages	Languages and communities											
	PLOT				FILIPINO				ENGLISH			
	W*	I*	C*	T*	W*	I*	C*	T*	W*	I*	C*	T*
Autotelic												
1. Aesthetic	11	7	15	11	3	15	5	8	11	2	8	7
2. Socio-cultural:	17	15	16	16								
regional significance												
national significance					11	38	23	26	9	2	13	8
international significance									43	13	27	26
link with origins		11	15	10	8	9	8	9				
mark of good education									3		8	4
sign of respect					6		2	2	9			2

Table 5 (Continued)

Meanings given to languages	Languages and communities											
	PLOT				FILIPINO				ENGLISH			
	W*	I*	C*	T*	W*	I*	C*	T*	W*	I*	C*	T*
3. Personal identificational:												
source of pride/identity	6	7	24	14	20	55	18	32	3	5	11	7
sign of loyalty/nationalism			5	2	17	4	3	7				
indicates smartness/confidence	3	2	2	2					6	4	6	5
4. Affective												
feeling comfortable/relaxed	63	25	35	38	6	22	16	16	11	11	29	18
full expression of feelings	31	15	24	22			2	1	3		3	2
easy to understand	26	18	37	28	3	7	3	5	11	4	31	16
Instrumental												
5. Economic:												
used in business/employment		4		1					6	5	18	10
key to progress/bright future					11		5	5	26	4	16	14
6. Political:provides harmony												
– in home & local commmunity	20	13	27	20								
– in national community					31		10	11				
– in global community									26			6
7. Social:												
making friends/entertaining			6	3			5	2	23	2	18	13
8. Educational												
appropriate for all subjects						2		1	26		10	10
appropriate for some subjects					11	2	6	6	3	2	3	3
access to knowledge			2	1			3	2	23	33	18	24
Negative												
9. of limited use	11	2	18	11	14	2	5	6				
10. excludes other languages	8		8	5								
11. has social stigma	6			1	3		3	2			5	2
12. causes difficulties:												
in speaking			2	1	6		2	2		4		1
in reading	3	4	2	3			2	1				
in writing	6	4	5	5								
confusion at school					23		5	7	3	62	5	25
13. is not developed					3		2	1				
14. useless/hated							2	1		13		5

* W = Waray; I = Ilocano; C = Cebuano; T = Total.

The negative meanings expressed were much more limited and often specific to a particular individuals or group of respondents. For example, although many Cebuano respondents discussed the positive meanings they associated with English, there was a few comments such as the following:

- "The disadvantage in using English is that many people dislike you for using the language." (Social stigma)
- "Speaking in English or in Tagalog [at home] made me feel so awkward. I was afraid that friends and neighbours would think I am acting strangely."

A number of the Waray and Cebuano students pointed to the limitations of their PLOT, while a proportion of the Waray respondents was critical of Filipino on the grounds of its limited usefulness and the confusion it caused at school. The strongest negative meanings, however, – overwhelmingly higher than any other – were revealed by the Ilocano students who considered that English caused confusion at school.

The comments of one of the Cebuano respondents illustrated the negative impact of the trilingual situation in the school context.

> "In our home, we oftentimes use the Cebuano and Filipino languages...
> But when I go to shopping malls, I start feeling uncomfortable with it. Usually teenagers, like myself, like to speak in English so I use the English language, too. Speaking these languages in and out of the house has disadvantages. Some of us students can hardly speak, read, and write very well. I, for one, can not perfectly speak or use the English language. Although I am used to speaking the Filipino and Cebuano languages, I can not perfectly read the Cebuano writings. I am having a difficulty in reading and in using it."

For the majority of the respondents, however, the trilingual context of their day to day lives seemed a taken for granted reality that presented no special difficulties, despite the predominance of English in the school setting. Another Cebuano student wrote :

> "In school, we use three languages – Cebuano, Filipino and English. But the medium of instruction used is actually English. Honestly speaking, I really feel comfortable."

4.4. Conclusions

In considering the conclusions to be drawn from this study, it is important to recognise its limitations. As the data were collected in three regional linguistic communities, the paper does not consider the effect of English on other Philippine languages and particularly upon the mainly Tagalog speaking areas of Luzon, which includes the "melting pot" of Manila. Furthermore, the respondents were senior secondary school students attending what were regarded as leading schools in their region; they could thus be said to have benefited from some of the better opportunities which the schooling system could provide. The experiences of young people from poor rural areas where schools were often lacking basic resources and adequately trained teachers, or those from low socio-economic background who were forced to leave school early before they had mastered literacy in any language, have not been included in this paper.

One important feature of the results was the differences in activation and attitudes revealed among respondents from the different linguistic groups. English, for example, was more often activated among the Cebuano and the Waray groups of students than among the Ilocanos. This variation might initially be assumed as complementary to the consistently greater usage of Filipino revealed by the Ilocanos. This preference, in turn, could be explained by the proximity of the Ilocos region to the Manila area, and the heartland of Tagalog, the language on which Filipino is based. In contrast, Leyte and Cebu are geographically isolated from Manila so that people in these two provinces have naturally had less contact with the national language.

Other political and historical factors can also be seen to have influenced these language differences. The Cebuano preference for English over Filipino could be rooted in memories of the struggle to have Cebuano endorsed as the national language of the country in preference to Tagalog, on the grounds that Cebuano was the indigenous language which at that time had the greatest number of native speakers. In the Visayan regions of Cebu and Leyte, therefore, Filipino has often been regarded as the language imposed by the central authorities, while English has been seen as opening up possibilities for interaction and employment. Conversely, the high levels of negative evaluation of English and relatively low levels of English activation among the Ilocanos may be traced to their perception of English as the language of colonial domination and to the bitter struggle of Ilocano nationalists against Americans at the turn of the nineteenth century.

Overall, it can be seen from the detailed discussion in the results section that the statistical measures of language usage and domains of activation and the more humanistic cultural data from student essays on language attitudes and meanings provided strong confirmation of one another. The conclusion to be drawn from these results was that the use of English as the language of instruction for Science and Mathematics in schools throughout the country had not significantly affected the use of Philippine languages as a means of oral communication in the home, the neighbourhood and the local community, according to the reported experience of the young participants in this study.

In reading and writing activities, however, English was clearly the dominant language, although its use had not precluded the development of Filipino as a language of literacy as well. The low to almost non-existent levels of reading and writing in PLOT, however, could be attributed to the entrenched status which English, supplemented by Filipino, has in these communication activities.

This situation may be partly explained by the different opportunities to learn the languages. In the years before they go to school, children learn to speak their PLOT informally through their everyday interactions in the home, neighbourhood and local community and the language is maintained through the natural on-going interaction of daily life. The literacy skills of reading and writing, on the other hand, are formally learned at school – not through the language of the home and local community which they have learned orally, but in what is for many, two new languages – English and Filipino – which are then used as the media of instruction in schools. There is thus no context in which children specifically learn to read and write in PLOT.

At the same time there is only a very limited amount of reading materials available in PLOT. Outside the school, the print and broadcast media remain dominated by the English language. Most of the newspapers and magazines that circulate in each of the localities included in this investigation are in English as opposed to PLOT. At the most, a couple of tabloids and a variety of illustrated "komiks" published in Tagalog are available in the sidewalks of the city streets. These could help to account for the comparatively higher levels of reading and writing in the Filipino language, compared to PLOT.

In conclusion, this small-scale investigation in three non-Tagalog speaking communities revealed a complex but stable trilingual situation which the emerging generation of young people gave evidence of taking for granted or accepting as normal. While the majority of respondents seemed to interpret this situation in positive terms, there was a minority who expressed negative views towards one of the languages concerned or explained the confusion it engendered for them. Thus this study provided only some limited evidence of conflict between the languages or of respondents who were antagonistic to a specific language. The evidence of the resilience of the Philippine languages indicated by this empirical research would appear to be consistent with the judgement of Gonzales

(1998: 519) that "none of the major Philippine languages and hardly any of the minority languages" are faced with extinction at the present moment.

In the present study, the data on linguistic activation to different interlocuters and the meanings assigned by the students to the various languages pointed in particular to a pattern of what Sibayan (1978) called "complementary distribution", where each of the languages used was recognised by the participants as having its own particular on-going role. These findings closely reflect the assessment by Gonzales (1998:519–520) that a trilingual situation is being maintained in the non-Tagalog speaking regions of the Philippines. He maintained that PLOT (or the "vernacular", as it is often referred to in the Philippines) was still "the language of the home and the neighbourhood", while Filipino was the national language, which functioned as "a symbol of unity and linguistic identity". English, in contrast, was "the language of academic discourse, especially for business, science and diplomacy". In the experience of the respondents in this study, access to English as the literary language of education, science and international communication had not resulted in a general shift away from Filipino as the national language or from the spoken use of their PLOT, as a core value of their community life (Smolicz 1981; Smolicz, Hudson and Secombe 1998) and the language closest to their homes and hearts.

5. References

Ammon, U.

1989 Towards a Descriptive Framework for the Status/Function (Social Position) of a Language Within a Country. In Ammon, U. (ed.) *Status and Function of Languages and Language Varieties.* Berlin/ New York: de Gruyter, 21–106.

Bautista, M. L. S.

1981 An Explanatory Note: Round-Table Conference on Philippine National Language Development. In Gonzales, A./Bautista, M. L. S. (eds.) *Aspects of Language Planning and Development in the Philippines*. Manila: Linguistic Society of the Philippines.

Bowering, J.

1863 *A Visit to the Philippine Islands in 1858*. Manila: Filipiniana Book Guild Inc.

Congressional Commission on Education

1991 *Making Education Work*. Manila/Quezon City: Congress of the Republic of the Philippines.

Constantino, R.
1982 *The Miseducation of the Filipino*. Manila: Philippines Foundation for Nationalist Studies.
De la Costa, H.
1961 *The Jesuits of the Philippines*. Cambridge, M. A.: Harvard University Press.
Gonzales A.
1974 *The 1973 Constitution and the Bilingual Education Policy of the Department of Education and Culture*. Philippine Studies 22 (4): 22–31.
1998 The Language Planning Situation in the Philippines. *Journal of Multilingual and Multicultural Development* 19: 487–525.
–/Bautista, M. L. S.
1986 *Language Surveys in the Philippines (1966–1984)*. Manila: De La Salle University Press.
Manhit, B. J.
1980 The Case for Reading: A Socio-Political Perspective. *Education Quarterly* 26 (4): 32–41.
1981 Alternatives for Functional Literacy: A Socio-Pyschophilosophical Perspective. *Education Quarterly* 27 (4): 1–25.
Manuel, J. L.
1974 Department of Education and Culture, Department Order No. 25, s. 1974, Implementing Guidelines for the Policy on Bilingual Education. Repr. in Sutaria, M. C./Guerrero, J. S. Guerrero/Castano, P. M. (eds.) 1989 *Philippine Education: Vision and Perspectives*. Manila: National Bookstore, 350–352.
Phillipson, R.
1992 *Linguistic Imperialism*. Oxford: Oxford University Press.
Nical, I. C.
2000 Language Usage and Language Attitudes Among Education Consumers: The Experience of Filipinos in Australia and in Three Linguistic Communities in the Philippines. Adelaide: PhD University of Adelaide.
Quisumbing, L.
1989 The DECS Bilingual Education Policy. In Sutaria, M. C./Guerrero, J. S./Castano, P. M. (eds.) *Philippine Education: Vision and Perspectives*. Manila: National Book Store Inc., 309–315.
Sibayan, B. P.
1978 Bilingual Education in the Philippines: Strategy and Structure. In Alatis, J. (ed.) *Georgetown University Round Table on Language and Linguistics*. Washington DC: Georgetown University Press, 302–329.
1994 Philippine Language Problems. In Acujia, J. E. (ed.) *The Language Issue in Education*. Manila: Congressional Oversight Committee in Education, Congress of the Republic of Philippines, 47–86.
Sin, J. L. (Cardinal)
1995 Sin Lectures on Chinese Values. *The Academia* 22 (2):5.
Smolicz, J. J.
1979 *Culture and Education in a Plural Society*. Canberra: Curriculum Development Centre.

1981 Core Values and Cultural Identity. *Ethnic and Racial Studies* 4 (1): 75–90.

1986 Multilingual Policy in the Philippines. In Spolsky, B. (ed.) *Languages and Education in Multilingual Settings*. Clevedon: Multilingual Matters, 97–116.

1990 Polarity of Filipino Values. *Asian Migrant* 3 (1).

1999 *Education and Culture*. Melbourne: James Nicholas Publishers.

–/Nical, I. C.

1997 Exporting the European Idea of a National Language: Some Educational Implications for the Use of English and Indigenous Languages in the Philippines. *International Review of Education* 4 (5/6): 1–21.

–/Hudson, D. M./Secombe, M. J.

1998 Border Brossing in 'Multicultural Australia': A Study of Cultural Valence. *Journal of Multilingual and Multicultural Development* 19 (4): 318–336.

Zialcita, F. N.

1995 State Formation, Colonialism and National Identity in Vietnam and the Philippines. *Philippine Quarterly of Culture and Society* 23:72–117.

Znaniecki, F./Bierstedt, R. (ed.)

1969 *Florian Znaniecki on Humanistic Sociology: Selected Papers*. Chicago: University of Chicago Press.

III. Countries with a History of Foreign Languages for Science Other than, or in Addition to, English

The Impact of English as a Language of Science in Finland and its Role for the Transition to Network Society

Harald Haarmann and Eugene Holman[1]

1. Introductory remarks

This presentation deals generally with the manner in which English has attained the status of a dominant language for scientific discourse in Finland during the 20th century, and, more specifically, with the impact which scientific English has had on Finnish usage and norms.

Finnish is one of the three non-Indo-European languages of Europe to have achieved the status of a developed language with the functional capability of being used for all spheres of national life, including scientific discourse and university instruction. This notwithstanding, the language is little known outside of Finland, for which reason Finns have always understood the importance of mastering other languages for both scientific and everyday communication with the world beyond its borders, in addition to which many occasional texts somehow regarded as noteworthy, such as toasts, orations, and nowadays, popular songs, advertisements, and restaurant names, have a long tradition of being given in foreign languages in Finland.

Under Swedish rule from the early 12th century until 1809 when it fell under Russian rule, Finland did not emerge as an independent nation until 1917. Within this framework Swedish established for itself a secure if sometimes modest role as a language for the country's scientific discourse, something that Russian never achieved. The Finnish tradition of learning did not, as had been the case in France and Germany, produce a distinct set of literati who represented the educated elite of their time and who professed their skills in belles-lettres and non-literary texts which served as monuments in the development of the native language. Learning in Finland depended, for most of its history, on individual personalities who accommodated themselves to the internationally recognized educational standards of their times and demonstrated their status as scholars of world rank by publishing their most important works in whichever exoglossic language happened to be the current lingua franca of scientific discourse.

From the Middle Ages, when Finnish scholars are known to have studied at the universities in Paris, Bologna, Prague, and elsewhere on the continent, until the late nineteenth century, the preferred language of scientific discourse was Latin.[2] Although centuries of Swedish rule had left Finland with a significant and influential Swedish-speaking minority, and Russian rule saw the establishment of a small Russian one, neither Swedish nor Russian, despite serving in various official functions and thus enjoying the status of superimposed endoglossic languages, ever attained a dominant role in scientific discourse. The late 19th century began to see the dominant role of Latin yield to German[3] and, to a lesser degree, French, in scientific discourse, in Europe as well as Finland, although efforts to cultivate a standard which could be the language of an independent nation also saw Finnish begin to be used for scientific discourse in specific fields. Finnish independence saw the country fall under the cultural influence of the German-speaking world, thus providing the conditions which allowed German to establish itself as the preferred medium for scientific discourse during the 1930s, a position which it retained throughout WW II, during which Finland was in a limited alliance with

Germany, as well as during the immediate postwar period, when German-educated scholars held key academic posts, despite the fact that Finland, like Germany, came out on the losing side of the war and was under various kinds of political and cultural pressure to loosen its ties with Germany and central Europe.

Only with the postwar reorientation of Finnish academic life towards the English-speaking world, particularly the United States, did English establish itself firmly as the dominant academic language in Finland.[4] This, coupled with school reforms of the early 1970s which made English an all but compulsory primary and secondary school subject have resulted in a Finland in which English is not only a widely spoken language but also a national resource which is widely used both for special effect, as in advertising, as well as for the primary production of texts belonging to an increasing number and variety of genres. As recent as this trend is, it continues the tradition of Finland preferring exoglossic languages for scientific and other highly valued texts.

2. Aspects of the dominance of English in modern Finnish society

A person who wants to consider the impact of English as a language of science has to take its polyfunctional role in Finnish society into account, since language use in the scientific sector is heavily dependent on, and supported by, other facets of the complex multilingualism that characterizes modern Finland as a whole.

As has been the case elsewhere in Western Europe, Finland has also experienced the trend toward an increasing influence of English in various domains of public and private life. In addition to the educational impact of English as the first foreign language taught in primary schools and the Anglo-American influence in the world of business and entertainment which has continuously strengthened its grip on Finnish culture since the 1950s, English also enjoys the role of the promoter of globalization and of the network society and, in this role, is leaving its imprint in life-styles and verbal behavior.

There is more behind the mere practical use of English which is noticeable in many domains, and this is the fact that English also has numerous symbolic functions. Since the 1980s, English has ranked high in the strategies of giving fashionable names to Finnish companies and products. This trend of conveying a feeling of modernity with English names has been an element which is widely spread in the business world of many countries (Haarmann 1989: 249 ff).

In commercial advertizing of Finnish companies, English is used for key-words or slogans together with the two official languages of the country, Finnish and Swedish. Among the representatives of the younger generation, there has been a strong tendency toward accepting English as an icon of modernity. This is even apparent in spoken language usage which is spiced with pragmasyntactic elements of English such as *jees* 'yes', *pliis* 'please', *abaut* 'about, approximately', *okei* 'okay', etc. (Haarmann 1991).

In recent years, Finnish society has experienced changes which favor the passive "consumption" and the active use of English even more today than a decade ago.

a) Trade relations with the West. – Despite the flourishing of Finnish-Russian trade relations after the demise of the former Soviet Union, the economic equilibrium underwent a downturn as a consequence of the recession which struck the Russian economy during the summer of 1998. Russia has not yet recovered from that collapse, and Finnish-Russian trade is now less important than Finnish-Estonian trade. Consequently, Finnish companies look westward for business partners, something which strengthens the role of English as a vehicle of commerce.
b) EU membership. – Finland has been a member state of the European Union since the beginning of 1995. Although the staff of Finnish officials who work in the EU bodies (Brussels, Luxembourg) and in the European parliament in Strasbourg is multilingual, the main contact language at the continental working-place has been English.
c) Economic boom. – The Finnish economy has seen a sharp upturn in its electronic industry in the 1990s. Among the top performers is the Finnish firm Nokia which has grown to become the biggest electronic company in Europe. As regards its trade and marketing, Nokia has gone global for years and has established relations with countries all over the world. In Nokia's global network, English is the main vehicle of business negotiations and contracts. Since electronic development, in a global comparison, is concentrated within the countries of the OECD, establishing and maintaining contacts with English-speaking countries having high electronic standards are an essential asset of Finland's globalization.
d) Growing shares on the financial markets. – With international financial markets expanding, Finland also participates in the global net-

work of financial economy. Dealers at the stock exchange in Helsinki and Finnish bankers are involved in the global flow of financial resources, as are also foreign investors with an interest in the booming Finnish economy. English is by far the most important means of communication in this domain.

e) Increasing the density of global intercommunication. – In the 1990s, the means of global intercommunication have spread rapidly among the Finns, among businessmen and scholars as well as private users. In the spring of 1999, Finland was the first country in the world to cross the magical boundary of 60 per cent mobile telephone density. With 37 per cent of the population using the Internet, Finland also holds a world record. Communication via the Internet which is dominated by English is a factor that further promotes contacts between Finland and the world. As a consequence, English is becoming all the more essential.

In a world-wide comparison of the density of information flow through the media (i. e. the press, radio, television, telephone, computer network) the traditional ranking of so-called "leading" countries has dramatically changed (Vartanova 1999: 21 f). The USA is the only country which has retained its first rank. However, formerly leading countries such as Germany or Japan which were still high-ranking in the 1980s, have lost their privileged status. Since the late 1990s, Finland has occupied second rank, with the other Scandinavian countries following.

f) Finland's public relations in the field of music. – Finns are known in the world for their high standards in music proficiency, especially in the field of opera singing, orchestra conducting and piano as well as violin playing. Many Finns have become famous for their professional skills and for their successful performances outside Finland. In Finland itself, international competitions such as the Sibelius competition for younger musicians and international festivals such as the Savonlinna Opera Festival have gained ever more attention abroad resulting in an fever increasing number of foreign visitors. The unmarked language for communication among Finns and foreigners in this field has been English.
g) Finland's public relations in the field of sports. – Finns are known in the world for their strong representation in international championships and competitions in various fields of sports, of car racing, icehockey, skiing and light athletics in particular. Sporting events draw

thousands of Finnish fans abroad, and the need for mobilizing English as a language of everyday contact with foreigners has been growing.

h) Expanding tourism. – As an expanding business tourism is developing in two major directions. On the one hand, the number of foreigners visiting Finland has been steadily increasing, partly resulting from the growing attraction of Finland as a EU country. On the other hand, Finns spend ever more money on travelling. Although the proportions of today's tourism have not yet reached the prime time of Finnish tourism in the late 1980s, there has been a steady growth in long distance travelling (to Far Eastern countries, to South East Asia, to the Americas). In the context of this mutually growing tourism, English is the dominant contact language. Even when travelling to Tallinn, the one city in the world where a language closely related to Finnish dominates and many of the inhabitants of the city have a mastery of Finnish and are eager to practice it, many Finns visiting the city still prefer to speak English.

Resulting from the processes which have been highlighted under (b), (c) and (e) are impulses which promote the use of English in the fields of science. International co-operation on the educational level and in the domain of scientific research has been increasing. Visits of foreign scholars to Finland have increased with Finland's participation in the European integration movement. The intercommunication among Finnish and foreign experts has especially intensified in various domains of electronic technology.

3. Parameters for the study of English as a language of science in the Finnish context

In a world-wide comparison, English is by far the most widely used language of science. This is a truism which has become a matter of stereotype mentioning in writings about English as a world language. In this connection, it is noteworthy that, in a study that broadly covers the global functions of English (i. e. Crystal 1997), the role of this language as a means of science is not specified or highlighted. This may be explained by the author's understanding that this role is such a self-evident property of English that it falls into the category of truisms and, therefore requires no explicit mentioning.

The dominant role of English is a fact that can no longer be questioned but has to be acknowledged as such (Ammon 1998: 205 f). The lesson that we are supposed to learn from this truism cannot reasonably be the launching of a counter-attack against the dominant role of English, but rather the insightful understanding of how local cultures react to the global challenge. The present contribution is mainly concerned with the response of the Finnish world of learning to the generally accepted role of English as a language of science.

The ecological conditions of the various languages which have been used in Finland's multilingual society differ from those in many other countries, with regard to the proportions of the speech communities, concerning the sociopolitical status of its endoglossic languages and, with respect to the sociocultural range of the languages involved in the world of learning.

Since Finland is officially a bilingual country, with Finnish and Swedish as languages used in state affairs, in public life, and in education, the two languages automatically function as the languages of academic teaching and scholarly writing. This means that, in Finland, there is not only one endoglossic language of science but two, a fact which makes discussion of the distribution and competition of languages in the world of learning a complex issue.

From an overall perspective, the situation has become even more complex since a third endoglossic language has recently also been introduced as an instrument of learning, and this is Saamic. Of the six written varieties of Saamic, Northern Saamic is the one which was used for the first doctoral dissertation ever written in that language (Hirvonen 1998). Her original work, in which the author investigates the biographies of Saamic writers, was subsequently translated by the author into Finnish (Hirvonen 1999). Whether this high-level academic use of Saamic will remain occasional or eventually become routine practise is yet to be seen.

A feature of the sociolinguistic profile of Finland which is exceptional in the European context is that official bilingualism is institutionalized in a country where the great majority of the inhabitants is nominally monolingual, with Finnish as the mother tongue of 93 per cent of the population, and Swedish that of 5.5. Regarding the distribution of mother tongues, Swedish is the language of a small minority (i. e. of Finland Swedes, the majority of which live in the western and southern part of the country). Functionally, the situation is more complex. Finnish is the dominant language of the country and, with the exception of the monolingually Swedish-speaking Åland Islands, even speakers of Swedish, Saamic, Russian, and other minority languages take in most of their public information in Finnish.[5] Private life, in turn, requires taking in an increasing amount of information in English. Subtitled Anglo-American television and motion pictures dominate entertainment, many Finns prefer to use English rather than localized versions of computer operating systems and software, and English language periodicals of a general and specialized nature are sold and read widely. During the 1990s English established itself as the unofficial "third domestic language" of Finland and documents intended to inform the public ranging from Helsinki Transit Authority maps to Internet websites are regularly prepared in Finnish, Swedish, and English, usually, but not always, in that order.

Swedish occupies a distinctive place in the overall communicative resources in Northern Europe. Its role as a spoken language may be delimited to Sweden and to Finland, as a passive means of communication, however, as a written language it reaches far beyond state boundaries, into Norway and Denmark as well. In an historical perspective, Swedish has played the role of an inter-Scandinavian lingua franca. From the standpoint of the use of Finnish as a language of science, Finland's world of learning is secluded. And yet, since Finland's scientists have access to Swedish, the world of learning opens toward an inter-Scandinavian dimension.

This particular situation with the world of learning being polarized between a regional vehicle (Finnish) and a language of interregional communication (Swedish) emphasizes the uniqueness of the Finnish situation in a European perspective. The fact that many other languages of learning have been in use in Finland (i. e. Latin, German, English, French, Russian) makes scholarly multilingualism in Finland an intriguing field of study.

The general overview of the languages of science in Finland as outlined in the foregoing now has to be specified by looking at the different sectors where English is passively consumed and/or actively used by students and scientists. From the standpoint of practical language use, there are four areas where the impact of English is noticeable, each with specific functional implications. A major difference is that between the consumption and the production of English literature, a term which is used here to cover texts in any form, printed (e. g. books, articles in journals) or unprinted (e. g. data collected from the Internet).

3.1. The passive consumption of academic texts in English by students

In all sections of university education there has been a growing need for students to read English texts. The situation of university studies in Finland differs significantly from the situation in larger countries such as

Germany, France or Spain in that the pressure on students to read foreign languages is not as strong in the larger countries. A German or French student would expect to read texts in his/her mother tongue since many texts essential for studying are available in German or French. In smaller countries such as Finland, translation of scientific texts or textbooks originally written in English into the mother tongue of students is, as a rule, commercially unprofitable, for which reason the same university textbooks are used by Finnish students as are used by their North American and British counterparts.

3.2. The passive consumption of academic texts in English by scientists

In the context of using literature for teaching and/or research, the situation of passive consumption of English texts may be the same as with students (see under 3.1), however the ends for which the information thus acquired is used differ considerably.

Acquiring information for the purpose of study means accumulating knowledge for two main purposes: passing the exams and becoming trained for professional activities. On the other hand, the reading process for teaching or research purposes produces a higher density of information flow through the medium of English as well as a greater dynamics in the accumulation of knowledge. In addition, the range of literature in English which has to be covered by university teachers and researchers for their purposes is naturally wider than the selected amount of literature required for ordinary study.

The steadily increasing demands for the consumption of English texts by researchers has, the widening of communicative horizons notwithstanding, produced some negative side-effects. English has become such a strong attractor that, in the scientific writings produced by Finns (judging from the display of bibliographical references), one can notice a phenomenon which has been termed the "Anglo-Saxon bias". This bias is feasible in many sectors of scientific research, in social psychology, genetics, biochemistry, information technology, and others. Research efforts concentrate and become dependent on English literature to such an extent that information in other foreign languages is practically ignored or even voluntarily discarded.

3.3. The production of oral English texts for teaching and training at the university level

There has been a growing need for the organization of teaching classes and courses in English at university level in recent years. In addition to his/her professional skills the teacher has to mobilize his/her language proficiency. As a precondition for the use of English as a language of instruction by Finnish or Finland Swedish teachers, a high-level proficiency of English as a foreign language is required. This level of language skills is much higher than the elementary level of everyday usage ("getting along in English").

Raising the level of language skills among teachers and preparing courses in English necessarily results in an even closer connection with the English-speaking world.

3.4. The production of written scientific texts (i. e. non-literary prose) in English

According to the typology of written texts as presented by Kloss (1978: 47), the greatest sociocultural potential of any language is reflected in its scientific literature. English-language scientific texts produced by non-native speakers represent the highest level of their specialized language skills, the highest density of their accumulated knowledge, and the highest grade of readiness for global intercommunication.

The choice of English as an active vehicle of science by Finns is an effective means for keeping up with the pace of global scientific development and, in addition, a challenge for actively influencing this development.

4. A historical outline on the languages of science in Finland

The historical view opens to the world of learning in general since, before the nineteenth century, the concept of *scientia* covered learning and scientific research indiscriminately.

As a rule, those Finns who were able to make a name for themselves in the world of learning before the 19th century were members of the Swedish-speaking population which had settled and evolved in western and southern Finland since the Middle Ages. Since Finland was, until the beginning of the nineteenth century, a part of the kingdom of Sweden, the ruling and intellectual elite of Finland came from among the Swedish population. Before the Academia Aboensis (later Åbo Akademi – Academy of Turku) was founded in 1640, higher education was only available in Sweden (Uppsala) or in Western Europe. It has been shown only recently that the great majority of those Finland Swedes who went abroad for studying were not aristocrats, but rather members of the urban bourgeoisie or even ordinary people (Nuorteva 1997: 473 ff).

During the Middle Ages Paris, Bologna, Marburg, Vienna, Prague and Krakova were among the preferred cultural centers for Finland Swedes to acquire learning. With the rise of Protestantism which was adopted as state religion in Sweden in 1523, Wittenberg in Germany became attractive as a center of learning. The most famous representative of Finnish learning in the older tradition, Mikael Agricola (ca 1510–1557), received his higher education in Wittenberg. Agricola is among the few exceptions in the world of Finnish learning since he was not of Swedish descent, but an ethnic Finn, the son of a farmer in Pernaja. In Wittenberg, Agricola's main tutor was Philipp Melanchthon who wrote a letter of recommendation to the king of Sweden, Gustav Wasa (Nuorteva 1997: 163).

For centuries, the world of learning in Northern Europe was dominated by Latin. This was also true after Protestant ideals of favoring the mother tongue as a vehicle of religious education had gained ground. Agricola may have even learned German to a certain degree during his stay in Wittenberg, but Latin was the exclusive means of his academic discourse and writings. The mother tongue – Finnish for Agricola – was meant for practical use. It served as the means for translating biblical texts and for authoring elementary schoolbooks (e. g. Agricola's *Abckiria* of 1543).

Even after the founding of the Academia Aboensis Latin continued to be used in Finland as a means of academic discourse, as in all parts of the Swedish kingdom. Things changed when, in 1677, O. Rudbeck introduced Swedish at the University of Uppsala. In the course of the eighteenth century, Swedish competed with Latin which still retained its dominant status as a language of learning in the context of Finland's society. Throughout the whole period during which the old university at Åbo (Turku) functioned, that is from 1640 to 1827, Latin was the favored language of science in Finland.

There are no statistics available as to the distribution of languages of the ca. 20,000 documents which were printed in Finland between 1488 and 1827. And yet, there is a corpus of thoroughly archived and investigated literature (so-called "Old Fennica") in the State Library of Lower Saxony in Göttingen (Germany) which may serve as a cross-section of older literature produced in Finland (Heininen 1988: 121 f). The 328 titles show the following distribution according to languages:

Latin (211), Swedish (84), Finnish (15), German (11), French (3), English (1), Russian (1)

It is noteworthy that, before 1827 – the year when the old town of Turku burnt down and the university was devastated by the fire –, Finnish ranked third (before German) as a language of academic writing. It is also noteworthy that Swedish experienced its breakthrough as the dominant language of science – at the cost of the rapidly receding Latin – only after the annexation of Finland by Czarist Russia in 1809. By the middle of the nineteenth century, the ranking of the three major languages of academic writing in Finland was the following: Swedish, German, Finnish.

Already in the latter half of the eighteenth century, German had gradually gained in importance in the academic world. Finnish students had been enrolled at the university of Göttingen since 1747, and there had been continuous contacts between German and Finnish scholars. Academic literature which was produced in Finland was almost regularly reviewed in contemporary German journals (see Kunze 1988 for these activities).

When Finland was separated from the Kingdom of Sweden and granted autonomous status as a Grand Duchy within Imperial Russia, this radical political change had no significant influence on the world of learning. Swedish retained its favored status, and also German continued in its academic function. Academic people in Finland became attracted to a new cultural center where German flourished, and this was St. Petersburg. The Russian Academy of Science had been dominated by German scholars since the times of Catherine the Great (ruled: 1762–1796), and German dominated the scholarly writings of the Academy well into the first half of the nineteenth century.

5. Scientific multilingualism and the role of English in modern times

Throughout the nineteenth century, English played a marginal role, in the Annals of the Academy of Science in St. Petersburg as well as in the production of scholarly texts in Finland. Finnish was the actual winner in the race for dominance in the world of learning. Around 1900 Finnish was competing with Swedish, which successfully retained its recently acquired position as the major language of science, with German ranking third. After Finland had gained its independence in 1917, a significant change in the traditional trend of language use in the fields of science occurred: Finnish advances to first rank, leaving Swedish behind.

In the 1920s and 1930s, the importance of German as an international language of science in Finland increased. Finnish and Finland Swedish scholars wrote entire texts in German, and summaries in German were added to many dissertations. Given the fateful alliance of Finland and Germany against the Soviet Union during the Second World War (i. e. between 1941 and 1944) one would have expected a dramatic break in the tradition of German as a language of high culture in the political atmosphere of the late 1940s. Then, Finland's politics started out on its cautious course of neutrality, the so-called Paasikivi line, and there were no close ties with either German state.

And yet, German continued for some time in its role of first foreign language in school education, and it was used in scholarly writing. The shift to English evolved gradually, independent of politics which was no longer oriented toward central Europe. In the humanities, where German had been a preferred means of scholarly discourse it retained this status also in the first decades of the post-war era.

In the field of Uralic studies (i. e. the study of Finno-Ugric and Samoyedic languages and cultures), for example, German was dominant in the pre-war decades, and it ranked second after Finnish well into the 1970s. This is the general picture when placing the material of the *Bibliographia studiorum uralicorum* (Länsimäki 1991) in a chronological order. Since the early 1980s, a drastic change can be observed. English is constantly gaining ground and has widely replaced German. This is well symbolized by the fact that the primary journal for Uralic studies in Finland retains its German name, *Finnisch-ugrische Forschungen*, even though the tendency during the past twenty years has been for its research articles, as opposed to book reviews and the chronicle of scientific life, to be in English.[6]

There is a practical reason for this trend of favoring English. Those scholars who have participated, since the 1980s in scientific discourse no longer learned German as their first foreign language at school. Instead, English was their main foreign language. If they learned German at all (as a second foreign language) their proficiency is likely to have remained at a lower level, as compared to their English skills. The members of the younger generation who prefer to write in English may be able to read some German, but language skills do not suffice for using the language actively.

English owes its continuous growth of influence in Finland to the general importance of the Anglophone countries in world economy and in the world of learning. Since the 1970s English has ranked third after

Finnish and Swedish in scholarly writings. This is true for the humanities. In many other domains English has challenged Swedish, and in some even Finnish. In many domains of the natural sciences, Finnish has been practically excluded from scientific writing.

Among the examples illustrating the latter trend are chemistry, biotechnology and engineering. In the domain of biochemistry, English dominates almost exclusively as the language of doctoral dissertations[7]. Finnish or Swedish are never used. The only exception to the rule is German in which one dissertation was published in 1972. English also dominates in the domain of engineering where doctoral dissertations have been published in that languages for decades.

And yet when inspecting the whole range of scientific production in Finland, Finnish still ranks first, and English second. However, the impact of English is growing.

5.1. Courses in English at Finnish universities

Although the university system in Finland is of a much more recent origin than in most other European countries, it provides today one of the most comprehensive networks in this continent. For Finland's 5 million inhabitants, there are 20 universities and 32 polytechnics. Of the 147,000 students who have enrolled in the academic year 1999–2000 more than 3,000 have come from abroad to study for a degree. About half of them study at polytechnics (Vento 1999: 15 f).

With the exception of students coming from continental Scandinavia (i. e. Sweden, Norway, Denmark) who can follow courses in Swedish, foreigners who want to study in Finland usually have no knowledge of either of the endoglossic languages of instruction in Finland. Courses in English have been offered for several years at most of the universities and polytechnics (see Table 1).

The range of scientific disciplines where English courses are organized is fairly wide, covering economics, natural sciences, social sciences as well as the humanities. The greatest number of offerings, however, is found in fields other than the humanities. It becomes apparent from the overview that certain fields of study are obviously more attractive for foreign students than others. English courses in economics and business administration as well as in environmental sciences are offered at many institutions at various locations while other fields of study are restricted to one location only (e. g. sport sciences or hypermedia).

Table 1: Courses in English at universities and polytechnics in Finland (1999–2000)

Field of study	Location of university and/or polytechnic
(1) Agriculture and forestry	Helsinki, Joensuu
(2) Architecture	Helsinki, Oulu
(3) Art, dance, design, music and theater	Helsinki, Rovaniemi
(4) Economics and business administration	Helsinki, Turku, Lappeenranta, Jyväskylä, Kuopio, Tampere, Vaasa
(5) Education	Joensuu, Jyväskylä, Rovaniemi, Turku
(6) Engineering and technology	Helsinki, Lappeenranta, Tampere
(7) Environmental sciences	Helsinki, Joensuu, Jyväskylä, Kuopio, Oulu, Tampere
(8) Health care and social services	Helsinki, Kuopio
(9) Information technology	Tampere, Vaasa
(10) Law	Helsinki, Rovaniemi, Turku
(11) Mathematics and natural sciences	Helsinki, Jyväskylä, Kuopio, Oulu, Tampere
(12) Medical sciences	Helsinki, Kuopio, Tampere
(13) Psychology	Tampere
(14) Social sciences	Helsinki, Kuopio, Rovaniemi, Tampere, Vaasa
(15) Sport sciences	Jyväskylä
(16) Theology	Helsinki
(17) Applied language studies, hypermedia, interactive media design	Oulu, Tampere

In addition to the courses in specialized fields of studies, English is the language of instruction in a number of interdisciplinary courses which are offered at all universities. The term "course in English" is not yet a clearly defined organizational entity. In the prospectus which contains information about courses in English at the University of Helsinki it says that it "deals not only with courses taught entirely in English, but also with courses that are taught only partly in English or that may be offered in English if demand arises" (Tammivaara-Balaam 1999: 1).

For the past ten years the University of Helsinki has published a booklet entitled *Instruction in English*. The 1991–1992 version was a modest booklet containing 26 pages and including more than 200 offerings ranging from courses in "Finnish" for foreigners to "Cell Biology" and "Environmental Aspects of Social Change". The most recent edition, 110 pages in length and covering the academic year 1999–2000, offers more than 400 courses and instructional events in English, including such specialties as "The Finnish Legal System" and "Finnish Folk Religion in Uralic Context" to "Tandem Mass Spectrometry" and "The Political Systems of the Baltic Countries."[8] Roughly, this instruction falls into four groups:

1. Instruction offered by the native-speaker and Finnish staff of the Department of English in various fields dealing with the history and structure of English and the literatures and cultures of the English-speaking world.
2. Instruction offered in English by visiting teaching staff. This can range from the instruction in an 'exotic' language such as Babylonian to a course in a topic such as "Christian-based Ecological Activism in the United States", taught by a visiting American professor.
3. Instruction offered in English by Finnish scholars intended primarily for foreign non-degree students spending a year in Finland to learn something about the country. This includes courses in the "Finnish Legal System", "Finnish Social Policy", and, of course, "Finnish Language and Literature".
4. Instruction offered in English by Finnish or foreign scholars desiring to make the results of their research known to the international element of the university community. Thus, for example, two professors, one Finnish, one German, offer a course in English entitled "Bible, Theology and Internet".

5.2. The use of English as an active language of scientific writing

Here, we refer to the use of English as the medium of scientific texts, not to the extent that English literature is mentioned or quoted in texts of this genre. There is not yet any monograph available in which this active role of English in comparison with other languages would have been empirically specified for the Finnish context. The current lack of such an investigation is certainly linked to the unavailability of information.

Despite the facts that we live in the information age and the data banks in Finland are well organized, no comprehensive set of data exists for studying the use of English in Finland in different fields of science. There is no data bank which would inform the researcher about the selection of languages used by Finnish scholars for their articles in national and/or international journals. Nor is there any comprehensive listing of books written by Finnish scientists. This is also true for the most specialized academic studies, for doctoral dissertations. Lists of dissertations are available for the faculties of individual universities, however not for all fields and not for all periods.

There are only very few exceptions to this unfortunate state of organizing the scientific information flow: one is the field of juridical studies, the other natural sciences (engineering and technology, in particular). Only recently have complete lists of dissertations been published, covering the period from at least 1900 to the late 1990s. On the basis of those lists the distribution of languages can be investigated. A comparison of the language choice for publications in these fields of science – one in the humanities, the other in technology – is significant since it illustrates the great difference between the still marginal role of English in a fairly secluded section of scientific research (juridical science in Finland) and the explosion of English-language use in the rapidly expanding technical fields.

5.2.1. The language choice of juridical dissertations

The language distribution for Finnish juridical doctoral theses is illustrated in Table 2. The language choice reflects the change in dominance from Swedish before independence to Finnish after it, and the selective participation of foreign languages.

The Finnish legal system did not attract much attention among foreign experts before Finland became a member of the European Union. The scholarly discourse about internal legal problems has been predominantly associated with the Finnish language which is the main working language in all legal institutions of the country. As an example, all laws are written in Finnish first and then are translated into the country's second official language, Swedish.

Regarding the language-subject correlation in the list of dissertations it can be observed that Finnish is used indiscriminately and its choice is not dependent on a specific subject. As far as inter-Scandinavian aspects of the Finnish legal system are concerned the choice of the language for the scholarly discourse is likely to be Swedish. This shows in the contents of the subjects of juridical dissertations.

Table 2: The languages of doctoral dissertations in the field of juridical sciences in Finland (1900–1997)
The data have been collected on the basis of the comprehensive listing of doctoral dissertations in Kangas (1998: 411 ff).

Period	Languages				
	Finnish	Swedish	German	French	English
1900–1909	5	7	1	0	0
1910–1919	5	4	0	2	0
1920–1929	11	2	1	2	0
1930–1939	16	7	0	0	0
1940–1949	13	2	0	0	0
1950–1959	17	4	1	0	1
1960–1969	27	6	2	0	0
1970–1979	40	3	1	1	3
1980–1989	43	6	0	0	7
1990–1997	50	3	0	0	5

The choice of a foreign (i. e. non-indigenous) language for a juridical dissertation has so far always been subject-bound. All dissertations in a foreign language treat pertinent subjects of international law. If Finns want to contribute to the international academic discussion they have to choose an internationally acknowledged language of science. The first juridical dissertation in German was published in 1907, the first in French in 1913 and the first in English as late as 1959.

The choice of French for juridical dissertations (and other juridical writings) in Finland reflects a common European tradition. In the nineteenth century, French was considered the language par excellence of international treaties and conventions. This was a function associated with its status as a preferred language in the world of diplomacy. For instance, the original version of the Berne Convention of 1886 on copyright is written in French.

The situation for juridical science has changed almost dramatically since the beginning of 1995. As a consequence of its official participation in the European integration process, Finland had to accommodate with European standards which, in practise, means it had to reconcile its repertory of internal laws with EU regulations and directives. The construction of the European dimension in the Finnish legal system makes Finland more attractive for foreign experts and, at the same time, stimulates Finnish lawyers' interest in European legal affairs.

The distribution of languages for juridical dissertations in Finland may be considered representative of a rather secluded domain of science. While, in other fields, English has been an indispensable active means of communication for many years, Finnish juridical science has only recently begun to open up to international circles. It is most likely that the language choice of doctoral dissertations, and of writings on legal problems in general, will tilt to give English the edge among the foreign languages in the future.

5.2.2. Language choice of dissertations in the fields of engineering and technology

The development of scientific research in the fields of engineering and technology is illustrative of the progress of technical sciences in recent decades and also of the fundamental shift toward a dominating medium: English (see Table 3). Of the more than 1200 doctoral dissertations which have been published since Finland's independence more than half have been produced in the last two decades[9]. There is no other field of science where the increase of dissertations is as striking as in technology.

Language choice of dissertations in the technical fields has never been subject-bound. The preferences for indigenous languages of science such as Finnish or Swedish, on the one hand, or for foreign languages such as German, English, etc., on the other hand, have been motivated by trends in scientific orientation. The first dissertation in German was published in 1917, the first in English as late as 1934. French was first used in 1968.

As regards the position of Swedish as compared to all other languages of science, this has always been less used than any of the others. German has been more important than Swedish for writing doctoral dissertations. Until the late 1940s, Finnish and German were equally important as languages of science, and English was still marginal. A shift in the importance of English can be observed in the 1950s and, during the 1960s, the trend of a preference for English becomes marked. From the early 1970s onwards, the immense increase of doctoral dissertations coincides with the growing dominance of English.

There are several specialized fields of technology where the endoglossic languages of Finland have been excluded from scientific writing and where exoglossic English dominates in absolute terms. This is true for analytical chemistry, industrial chemistry, the paper industry, and astronomy. In some other fields, English predominates, with only one or two exceptions of dissertations published in German, Finnish or Swedish. This is the situation in physics, geology, applied mathematics, construction engineering, etc. (Ala-Tuuhonen et al. 1998: 139 ff).

Regarding the publication of scientific literature in the fields of engineering and technology other than doctoral dissertations (Karjalainen et al. 1998), the picture of language choice is more varied. From the end of the 19th century until 1917, Swedish was predominantly used, with Finnish ranking second. The first article to be written in German appeared in 1882, the first book in 1893. French enlarged the range of languages in 1911. In the period between the two World Wars there were four major languages used for publications in engineering and technology: Finnish, German, Swedish and English. Since the late 1940s, Finnish and English have been in use. However, the impact of English is not as dramatic as in the case of doctoral dissertations.

Table 3: The languages of doctoral dissertations in the fields of natural sciences in Finland (1911–1997)
The data have been collected on the basis of the comprehensive listing of doctoral dissertations in Ala-Tuuhonen et al. (1998).

Period	Languages				
	Finnish	Swedish	German	French	English
1911–1917	2	0	1	0	0
1918–1929	6	1	6	0	2
1930–1939	14	3	14	0	2
1940–1949	14	3	14	0	2
1950–1959	3	6	6	0	20
1960–1969	16	0	4	1	76
1970–1979	35	4	9	0	200
1980–1989	17	1	2	1	257
1990–1997	24	0	0	2	456

6. The impact of scientific English on Finnish language usage

At present, no other language in Finland influences the domains of public and private life as does English. The impact of English in the various fields of science is not restricted itself to the competition with Finnish and Swedish, but it also extends to the actual linguistic structures of the languages with which English is in contact.

Since the scientific terminology and phraseology of both Finnish and Swedish have been exposed to intense influence from English for many years, it is only natural to expect some degree of linguistic interference to become apparent. The following remarks focus attention on some of the ways in which English has been influencing the structure of scientific Finnish, the genre of the language used for producing texts to be consumed by the community of Finnish scholars working within various fields which require the production of texts the purpose of which is to provide information for peers, these including the next generation of scholars, about the results and consequences of scientific research.

6.1. The phonological system

The Finnish phonological system differs from that of most European languages in having relatively few qualitative oppositions and a relatively large number of quantitative oppositions. This is particularly important in the obstruent system. The traditional Finnish obstruent system lacks the voiced stops /b/, /d/, or /g/, the voiceless bilabial fricative /f/, a voiced alveolar hissing sibilant /z/, nor has the palatal hushing sibilants /S/ and /Z/. There are no dental fricatives or phonemic affricates. Consequently oppositions between /p/ and /b/, /t/ and /d/[10], /f/ and /v/, /k/ and /g/, or /s/ and /z/, for example, do not exist. Indeed, the sounds /b/, /g/, /f/, and /z/ are not part of the native phonological inventory at all. In addition to having a small number of consonants, Finnish phonology places considerable constraints on word shape: native Finnish words are not allowed to begin or end with consonant clusters, and word-internal clustering is severely restricted. Finally, Finnish has various synharmonic constraints limiting the possible inventory of vowels that can occur in a single word. Thus a word like *kylä* ['kyla] 'village' is possible, while a word like **kyla* ['kyla] where a high front rounded vowel and a low back unrounded vowel co-occur in the same word is excluded.

The vast influx of scientific terminology from international sources coming mainly via a primary stimulus from English but going through a filter that can be understood as the accrued tradition by which Greek and Latin have been pronounced in Finland, provides scientific Finnish with literally thousands of English-motivated international words that violate the traditional phonological constraints of the language, e. g. *stereoskopia* 'stereoscopy', *plastiikka* 'plastic' ' both with otherwise impermissible word-initial consonant clusters, *konvergenssi* 'convergence', *stigmatisoitu* 'stigmatized', both with otherwise impermissible word-internal consonant clusters, *analysoida* 'to analyze', *polymorfi* 'polymorphy', both with violations of synharmonic restrictions, *binakulaarinen* 'binocular', *fotogrammetria* 'photogrammetry', *fetisismi,* 'fetichism', *šamanismi* 'shamanism', all containing non-native consonants.

6.2. *The morphological system*

The influence of English on the morphological system of scientific Finnish is more that of strengthening already existing tendencies, many of them resulting from the once more pronounced influence of Swedish, than of any specific innovation that can be solely attributed to English. In this section an introduced type of allomorphic alternation with unexpected and English-like orthographic consequences will be discussed, as will certain non-Finnish types of word building strategies which are shared by Swedish and English.

6.2.1. Borrowed allomorphic alternation patterns

Even before English began to exert an influence on Finnish, Latin and Greek words in their Swedicized pronunciations representing a type of minor morphological irregularity resulting from the manner in which Swedish dealt with certain classical vowel alternations had been introduced: Swedish *matematik* [matema'ti:k] 'mathematics' but *matematisk* [mate'ma:tisk] 'mathematical'. Finnish, which has a fixed initial accent that is less dynamically centralizing than that of Swedish, re-interprets the qualitative and quantitative changes undergone by the Swedish word as differences in the quality and quantity of word-internal syllables: *matematiikka ['matemati:kka] 'mathematics', but matemaattinen* ['matema:ttinen] 'mathematical'. This already-existing tendency has resulted in a situation where many technical terms entering the language through English have two allomorphic variants, one is used when the term occurs as a noun, the other when it is used as an adjective: *logiikka* 'logic', cf. *looginen* 'logical', *fysiikka* 'physics', cf. *fyysinen* 'physical', *fysikaalinen* 'physical'.

The Finnish orthographical system, often regarded as a model example of a spelling system that accurately reflects the phonological structure of the language, does not always indicate these allomorphic differences, thus *kromi* ['kro:mi], *kromaattinen* ['kroma:ttinen] 'chromatic', *alkoholi* ['alkoho:li] 'alcohol', *alkoholisti* ['alkoholisti] 'alcoholic'. Even if the innovation in question, ignoring a predictable allomorphic variation in spelling, is the result of the interaction of several factors, the exceptional manner in which it is dealt with by Finnish orthography is, if not necessarily influenced by English, certainly consistent with the manner in which English orthography deals with similar problems, e. g. *atom* ['ærəm] vs. *atomic* [ə'tɔmɪk], where the invariant written form conceals major and ono-trivial allomorphic variation.

6.2.2. Borrowed word-building strategies

Finnish has an extraordinary range of strategies for augmenting its vocabulary by derivational suffixes. Nevertheless scientific Finnish often produces derivatives by ignoring the resources offered by Finnish and calquing the structure of the Swedish or English term: *painovalmis* 'print-ready', cf. *valmis painettavaksi* 'ready to be printed', *bakteerivapaa* 'bacteria-free', cf. *bakteeriton* 'bacteria-free', *happiköyhä* 'oxygen-poor', cf. *vähän happinen* 'little-oxygenous', *käyttäjäystävällinen* 'user-friendly', cf. *helppokäyttöinen* 'easy-usable'. The synonyms given after the words in question represent the same concept using more traditional Finnish word-building strategies. What they have in common is a tendency to use a lexically empty derivational formative rather than a devaluated lexical morpheme as the derivational operator: *veroton* = *vero* 'tax' + *ton* 'LESS', cf. *verovapaa* = *vero* 'tax' + *vapaa* 'free', on the model of Swedish *skattfri* and English *taxfree.* Handbooks of Finnish written usage stress the importance of using the Finnish derivational apparatus for the purpose for which it was intended, but they concede that many of these words have established themselves as the preferred symbol for the concept.

Related to the above problem is the issue of compound verbs. Finnish has traditionally avoided compound verbs, although contacts with other languages have allowed a few to establish themselves, e. g. *ristiinnaulita* 'to crucify' lit. 'into-cross + nail', *ympärileikata* 'to circumcize' lit. 'around + cut'. Borrowings from the scientific and commercial genres of English and other languages have allowed compound verbs such as *aivopestä* 'to brainwash', *ilmajäähdyttää* 'to air-cool', *kauko-ohjata* 'to remote control', to become firmly enough established to form a strategy which, during the past generation, allowed Finnish to go further in this respect from the languages from which the pattern was originally borrowed: *äkkirikastua* 'to suddenly become rich' lit. 'sudden + become rich', *muovisuojata* 'to protect with plastic' lit. 'plastic + protect', *kaukolämmittää* 'to heat from a centralized heating plant' lit. 'distance + heat'. This change is not trivial: older Finnish compound verbs tended to be strongly indexical and include a case ending which expressed the relationship between the two parts of the compound quite explicitly, e. g. *ristiinnaulita* 'to crucify' both symbolizes the concept and explains it: 'being nailed into (= on to) the cross', *hyväksikäyttää* 'to abuse' lit. 'for good + use'. The newer compounds, while still indexical, have more of the nature of pure symbols: the two semantic elements are named, but the relationship between them is not made explicit: *henkivakuuttaa* 'to ensure a person's life' lit. 'life + insure', *etsintäkuuluttaa* 'to put out a search warrant' lit. 'search + announce'.

Both of these strategies taken together can be seen as part of a process by which an element of indeterminacy is being introduced to the extremely concrete and explicit Finnish morphological system.

6.3. *Syntax*

Perhaps the most striking manner in which Finnish syntax differs from English syntax is in its means for expressing definiteness.[11] This difference has far-reaching consequences for such seemingly unrelated issues as the structure of noun phrases, the use of different verb voices, and word order.

6.3.1. The structure of noun phrases

Finnish has neither a definite nor an indefinite article, the definiteness or indefiniteness of a noun phrase being indicated by its position in the sentence, by the use of specific pronouns or quantifiers, or by its case ending:

Timantinkaltaisten kalvojen valmistus on kuitenkin hidasta ja epätaloudellista isotoopinerotina käyttäen.[12]
'Nevertheless, the preparation of diamond-like tissues is slow and uneconomical using an isotopic separator.'
NP['this-kind + PLU + GEN' 'tissue + PLU + GEN' + 'preparation' + NOM + SG']NP + 'is' + 'nevertheless' 'slow + SG + PART' and 'un-economical + SG + PART' NP[isotope + separator + SG + PART]NP using.

This sentence contains two noun phrases, one of which, the subject, is definite, this being indicated by its position at the beginning of the sentence, the lack of any indefinite pronoun, and its occurrence in the nominative, the second, the direct object of a non-finite verb construction, is indefinite. Any text in scientific Finnish will exhibit numerous sentences with approximately the same structure; it is so much dependent on its own internal links that it is highly impervious to influence from other languages, one reason being that the monitored nature of text production when producing scientific texts would ensure that the basic rules for using the resources of the language of science are going to be observed.

The same is not true for the more spontaneous varieties of spoken Finnish. In them, emotional and other considerations often result in the above strategies not always being available or not amenable to the type of processing needed to produce or understand a sentence of the above type, meaning that alternative strategies will be available. Once again, these are often English-like, even if their source is more in Swedish than in English.

Ja sillon se kundi katto suoraan muhun, niin musta ainaki tuntu.[13]
And then $_{NP}$[the guy] $_{NP}$ looked directly at me, well, that's how it seemed to me.
‘and’ ‘then’ ’the’ ‘guy’ ’looked’ ‘directly’ ’me+ILLATIVE’, ‘well’ ’me + ELATIVE’ ‘at least’ ’it seemed’.

Here, the definiteness of the noun phrase is indicated by the article *se*, a weak form of a demonstrative pronoun originally neutral between proximative and non-proximative deixis.

Summarizing, the grammatical rules underlying the construction of the noun phrases used in Finnish scientific discourse are complex and well enough defined to preclude direct influence from English. On the other hand, the much simpler noun phrases used in the most informal genres of Finnish can be and have been subjected to outside influence, some of it analogous to what one might expect from English even if the source is more likely Swedish.

6.3.2. The use of the passive and word order

Finnish has a morphosyntactic category which is traditionally called the passive, but this is a misnomer. It is actually an impersonal voice used to indicate that the action designated by the verb is associated with a human but undefined agent:

> *Ovi avattiin.* ‘The door was opened [by somebody]’.
> *Ovi avautui.* ‘The door opened [perhaps due to the wind or a sensor]’.

In English we can say:

The window was broken by a falling brick.

The Finnish impersonal, in turn, could not be used here, the idea would have to be expressed in the active:

Putoava tiiliskivi särki ikkunan.
'A falling brick broke the window.'

In older Finnish, as a consequence of influence from Swedish and other Indo-European languages with true passives, various structures were used as equivalents of the passive. These included the ablative, as in:

Ovi avattiin mieheltä.
'The door was opened by [lit. 'away from'] the man.'

as well as postpositions:

Ovi avattiin miehen taholta/toimesta.
'The door was opened by virtue of/by the offices of the man.'

The ablative construction is no longer encountered in Finnish, but the constructions with postpositions are occasionally found in poorly written English-influenced scientific Finnish. Once again, this has arisen in Finnish as a result of language contact and has then continued as a consequence of an input from another language.

The difference between the active *The man opened the door* and the passive *The door was opened by the man* would normally be expressed in Finnish by contrasting word orders:

Mies avasi oven. 'The man opened the door.'
Oven avasi mies. 'The door was opened by the man.'

6.4. Lexicon

English has affected the scientific Finnish lexicon in three ways:

1. It has introduced thousands of new words, and, along with them, the elements of the international Graeco-Latin lexicon needed to coin a virtually limitless number of international terms. A text in scientific Finnish will contain words easily recognized by the international community of scholars such as *analyysi* 'analysis', *energia* 'energy', *ioni* 'ion', *materiaali* 'material', *mekanismi* 'mechanism', *metalli* 'metal', and *orgaaninen* 'organic'.

2. It has provided the already existing and largely native-based lexicon with synonyms, thus allowing for the development of further stylistic differentiation and associative links. The native neologism *muovi* 'plastic', from the verb *muovata* 'to form', is the generic term for the substance. Nevertheless, the term *plastiikka* is used in the term *plastiikkakirurgia* 'plastic surgery', as well as for the specialized meaning that has developed within photometry and dramaturgy. The native neologism *tietokone* 'computer' lit. 'knowledge + machine' is the generic term for that concept. Nevertheless, the English loanword *kompuuttori* exists too, but primarily as an informal synonym.

3. It has provided a motivation for Finnish scientists to invent or redefine already existing Finnish words so that they would more closely parallel their English counterparts in meaning, even if they remain totally different in form. The word *suihku* 'shower' has acquired new meanings influenced by English, e. g. *suihkukone* 'jet aircraft', *mustesuihku* 'ink-jet'; *näyttä*, which formerly had the more abstract meaning of display, as in *näyttäikkuna* 'display window', has not acquired the additional meaning of 'computer display'. As is the case in many other languages, the everyday words *hiiri* 'mouse' and *ikkuna* 'window' have acquired special meanings in computer terminology.

Thus, the impact of English on the Finnish lexicon is functionally similar to that of earlier periods of mass borrowing. According to Hakulinen (1979:364), the influx of Germanic loanwords into Finnish more than two thousand years ago "increased the expressive ability of our language even when they just introduced a synonym to parallel an already existing indigenous term; they have specifically offered opportunities for semantic differentiation and welcome word bases to serve as the basis for new derivatives."

6.5. Pragmatics

During the 19th and early 20th century Finnish was a language, the most prestigious speakers of which believed was best developed by expanding upon already existing resources. Such attitudes were motivated by two factors: the desire of the intellectual leaders in a newly-independent and self-confident nation to elaborate a distinctly Finnish identity, and the conviction that keeping the language as explicit and indexical as possible would ensure the "lexical democracy" by which a small child of average intelligence had no difficulty understanding even abstract terms such as *ajankohtainen* 'current', lit. 'time + point + ish', *johdonmukainen* 'consistent', lit. 'lead + along + -ish', or *lukemisto* 'anthology', lit. 'read + collection'. This resulted in normative attitudes which rejected the use of sounds, words, and constructions which were regarded as displaying obvious foreign influence. Thus, for example, the older Swedish borrowing *telefoni* 'telephone' was marginalized and eventually replaced by the neologism *puhelin*, lit. *puhel(e)*- 'to chat' + INSTRUMENT'. The nationalistic ideal of lexical democracy is thus facing a serious challenge from the internationalistic ideal of lexical uniformity, with unforeseen consequences.

Since Finnish has been widely used for scientific discourse both in the past and present, there are ready-made chunks available for referring to recurrent concepts, e. g. *satunnais*- 'random', *kaksois*- 'dual', *-oida* 'verbal formative', *-ointi* 'verbal noun formative', thus *satunnaispistestereogrammi* 'random point stereogram', *kaksoisklikata* 'to double-click'. Such "bastard words" are no longer regarded as linguistic hybrids to be avoided under all circumstances.

Thus, a page of scientific Finnish, tends to be full of English-inspired words of international origin, which differ radically from the structure and inventory of other genres of texts.

7. Conclusion

Finland's decision to favor English as its primary vehicle for scientific research has enabled the country, perhaps unexpectedly, to assume a major role, both active and passive, in the process of globalization.

Thus Finland is undergoing a transition to a networked society more rapidly and intense than in most other countries. In this sense, Finland is, in fact, playing a role as a producer of standards, innovative products and methodologies, and information, which is deeply connected with the ability to use English. With respect to the movement towards integration in Western Europe, a process which is definitely leading to the institutionalization of a "network state" (Castells 1998: 330 ff), Finland has been a forerunner.

The Finnish society has been under increasing influence from English, as the cultural vehicle *par excellence* of modernity. English has penetrated many "niches" of public and private life in Finland, including entertainment, advertising, marketing, and education. English has even more increased its already well-established role in information technology and the scientific sector.

In an interview given to the Finnish newspaper *Helsingin Sanomat* and published on January 2, 2000, the Spanish sociologist Manuel Castells emphasizes that Finland holds the edge in this evolution toward a network society (Castells 2000). According to Castells, technological modernization in Finland is evolving within the framework of high standards of living, whereas in the USA, modernization is intensifying at the cost of quality of life.

The particular strategies by which Finnish society is proceeding toward the stage of the network state have attracted the attention of experts around the world, some of whom speak of the "Finnish model" (e. g. Vartanova 1999; Mer 1999). The role played by English in the transition to an international, network-based society, particularly regarding the accumulation of know-how in Finnish companies and its utilization for the benefit of international marketing strategies, has only recently been noticed. (Huhta 1997). We hope we have been able to demonstrate that even if developments will have far-reaching consequences they are quite consistent with our knowledge about earlier periods of profound change instigated by intercultural contact. It is significant this time that Finland is an active rather than a merely passive participant, and that this would not have been possible without the knowledge and use of English.

9. Notes

1. Although individual parts of this article have been elaborated and written by different authors both authors sign responsible for the whole contribution.
2. See e. g. Toini Melander the occasional literature from the period 1562–1713 (both years included) which are registered in Toini Melander: Personskrifter hänförande sig till Finland 1562–1713 (Helsingfors 1951). This compilation includes approximately 800 orations, poetry and similar literature published at the event of births, weddings, deaths, funerals, namesdays, anniversaries, travels, academic degrees etc.
3. Typical of this period are works such as *Zur phonetik der finnischen sprache: untersuchungen mit Hensen's sprachzeichner*, by the phonetician Hugo Pipping, who otherwise wrote the majority of his publications, which were intended for a local readership, in Swedish, the linguist Emil Nestor Setälä, whose preferred language of academic publication was German, although he published scientific texts in Finnish, Hungarian, and French. Publications bearing his name and dating from the time he served as a diplomat and devoted to explaining Finnish issues to an international readership also appeared in English and Italian.
4. The extensive and intensive network of academic exchange programs between Finland and the United States that was begun in 1953 has allowed a greater number of Finnish academics to spend at least two months in the United States

doing research, teaching, or further study than is the case for any other European country. See Juha Vuorinen and Iris Heikkonen, 1983, "The Implementation of Finnish-American Academic and Professional Exchanges", pgs. 24–39 in W. Copeland et al. ed., *Finnish-American Academic and Professional exchanges: Analyses and Reminisces*. Espoo.

5. See M. Tandefelt 1988, 1996 for discussion of the concept of "mother tongue" as it applies to Finland Swedes.
6. See the *Finnisch-Ugrische Forschungen* website at http://www.helsinki.fi/jarj/sus/fuf.html for the contents of the journal by year.
7. I (HH) would like to thank Else Lindholm (Viiki Science Library, University of Helsinki) for her help in providing a list of doctoral dissertations in the field of biochemistry.
8. Available on-line at http://www.helsinki.fi/english/courses/.
9. I (HH) would like to thank Linnea Lindman (Helsinki University of Technology) for providing me with the list of dissertations edited by Ala-Tuuhonen et al. (1998).
10. Standard Finnish uses a phoneme /d/ which is of limited distribution and, in native Finnish words, is always morphophonemically derived from /t/, e. g. *madon* 'of the worm', cf. *mato* 'worm'. In this sense there is an opposition between /t/ and /d/, e. g. *maton* 'of the carpet', cf. *matto* 'carpet'. This /d/, which can only occur word-internally, is a different linguistic unit from the /d/ which has entered Finnish with recent loans, e. g. *diffuusio* 'diffusion', *dynamiikka* 'dynamics', *katodi* 'cathode'.
11. See A. Chesterman, 1991, for detailed discussion.
12. Source: the lecture on material physics in scientific Finnish at: http://beam.helsinki.fi/finnish/programme/materials.html
13. Source: the story in colloquial Finnish at http://www.matrix.fi/sivut/mekaks/2.97/tekstit/prinsessa.htm

8. References

Ala-Tuuhonen, Leena/Lindman-Sharma, Linnea/Märttinen, Eeva/Pasanen-Tuomainen, Irma (eds.)

1998 *Teknillisen Korkeakoulun väitöskirjat 1911–1997. Terpenikemiasta nanoteknologiaan*. Espoo: Teknillinen Korkeakoulu.

Ammon, Ulrich

1998 *Ist Deutsch noch internationale Wissenschaftssprache? Englisch auch für die Lehre an den deutschsprachigen Hochschulen*. Berlin/New York: de Gruyter.

Castells, Manuel

1998 *The Information Age: Economy, Society and Culture. Vol. III: End of Millennium*. Malden, MA/Oxford: Blackwell

2000 Castells tietää mistä tässä kaikessa on kysymys, *Helsingin Sanomat*. January 2.

Chesterman, Andrew

1991 *On Definiteness: A Study With Special Reference to English and Finnish*. Cambridge: Cambridge University Press.

Copeland et al. (eds.)
1983 *Finnish-American Academic and Professional Exchanges: Analyses and Reminisces*. Espoo.

Crystal, David
1997 *English as a Global Language*. Cambridge: Cambridge University Press.

Gelehrte Kontakte zwischen Finnland und Göttingen zur Zeit der Aufklärung (Katalog
1988 zur Ausstellung aus Anlass des 500 jährigen Jubiläums des finnischen Buches) Göttingen: Vandenhoeck & Ruprecht.

Haarmann, Harald
1989 *Symbolic Values of Foreign Language Use*. Berlin/New York: Mouton de Gruyter.
1991 Symbolic Internationalization: Beyond the Practical Use of Foreign Languages. In Sajavaara et al. 1991, 1–16.

Hakulinen, Lauri
1979 *Suomen kielen rakenne ja kehitys*. 4th corr. and suppl. ed. Helsinki: Otava.

Heininen, Simo
1988 Finnische Gelehrte in Göttingen während des 18. Jahrhunderts. In *Gelehrte Kontakte* … 1988: 47–77.

Hirvonen, Vuokko
1998 *Sàmeeatnama jienat. Sàpmelaš nissona bàlggis girječàllin*. Utsjoki: DAT.
1999 *Saamenmaan ääniä. Saamelaisen naisen tie kirjailijaksi*. Helsinki: Suomalaisen Kirjallisuuden Seura.

Huhta, Marjatta
1997 *The Dynamics of Language Training – From an Element of Cost to an Investment in Communication*. Helsinki: Opetushallitus.

Kangas, Urpo (ed.)
1998 *Oikeustiede Suomessa 1900–2000*. Helsinki: Werner Söderström Lakitieto.

Karjalainen, Kristiina/Lindman-Sharma, Linnea/Tuoresjärvi, Taija/Turkka, Helena (eds.)
1998 *Teknillinen Korkeakoulu kirjoissa ja kansissa. 150-vuotisbibliografia*. Espoo: Teknillinen Korkeakoulu.

Kloss, Heinz
1978 *Die Entwicklung neuer germanischer Kultursprachen seit 1800*. 2nd ed. Düsseldorf: Institut für deutsche Sprache.

Kunze, Erich
1988 Zur Geschichte der wissenschaftlichen Beziehungen zwischen der Georgia Augusta und der Academia Aboensis. In *Gelehrte Kontakte* … 1988, 79–120.

Länsimäki, Maija (ed.)
1991 *Bibliographia studiorum uralicorum 1917–1987. Uralistiikan tutkimuksen bibliografia/Bibliography on Uralic studies, III: Kielitiede/Linguistics*. Helsinki: Suomalaisen kirjallisuuden seura.

Mer, Jacques
1999 *La Finlande*. Paris: La documentation Française.

Nuorteva, Jussi
1997 *Suomalaisten ulkomainen opinkäynti ennen Turun Akatemian perustamista 1640.* Helsinki: Suomen Kirkkohistoriallisen Seuran toimituksia 177 (with an English summary: Finnish study abroad before the foundation of the Royal Academy of Turku (Academia Aboensis) in 1640).

Sajavaara, Kari et al. (eds.)
1991 *Communication and Discourse Across Cultures and Languages.* Jyväskylä: Jyväskylän yliopisto.

Tammivaara-Balaam, Irmeli (ed.)
1999 *University of Helsinki. Courses in English 1999–2000.* Helsinki: Helsinki University Press.

Tandefelt, Marika
1988 *Mellan två språk: en fallstudie om språkbevarande och språkbyte i Finland.* Uppsala.
1996 *På vinst och förlust: om tvåspråkighet och språkförlust i Helsingforsregionen.* Helsinki.

Vartanova, Elena
1999 *Finskaja model' na rubeže stoletij. Informacionnoe Obščestvo i CMI Finljandii v evropejskoj perspektive.* Moskva: Izdatel'stvo Moskovskogo universiteta.

Vento, Pauli (ed.)
1999 *Study in Finland. International Programmes in Finnish Higher Education 1999–2000.* Helsinki: Centre For International Mobility.

Vuorinen, Juha/Heikkonen, Iris
1983 The Implementation of Finnish-American Academic and Professional Exchanges. In Copeland, 24–39.

The Foreign Language Competence of Hungarian Scholars: Ten Years Later

Péter Medgyes and Mónika László

1. Introduction

It is a commonplace to state that today English is the lingua franca of the world, and it dominates all walks of life. While there are a host of reasons for its spread in the 20th century – political, economic, social, religious, and literary –, 'linguistic reasons do not rank highly' among them (Crystal 1995: 106). In a similar vein, Kaplan (1993) asserts that the dominance of the English language is not due to its superiority to any other fully-fledged language, but rather due to historical coincidence. As for the future, although there is no guarantee for its eternal hegemony, the chances are that English will reign supreme for several more decades (Graddol 1997).

In a world that has granted English the privilege of *primus inter pares*, native English speakers have a distinct advantage over non-natives and, by implication, non-native speakers of English are handicapped in all areas of competence and discourse (Medgyes 1999). In scientific discourse, specifically, communication takes place in 'cooperative networks formed amongst invisible colleges of scholars' (Baldauf and Jernudd 1987: 98) who normally speak among themselves in the same language – overwhelmingly in English. Unless a scholar is able to use English, he or she will not be admitted to these exclusive clubs by the 'gatekeepers' of the profession, including journal editors, reviewers, and conference organisers. Therefore, a high level of competence in English has become a make-or-break requirement for an increasing number of academics in an increasing number of disciplines.

The primary aim of this study is to supply further evidence to the assumption that the universal language of scientific discourse is English. The country of investigation is Hungary, a non-English speaking country in Central Europe. The participants are Hungarian scholars, who were asked to provide information about various aspects of their foreign language competence in general and their English language competence in particular.

However, this study is designed not only to take a snapshot of the Hungarian academic community in the last year of the second millennium, but also to show the *spread* of English. 'Spread' is a relative term whereby two or more stages of development are measured against each other. In this study, a yardstick for comparison is ensured by the availability of an identical survey which was carried out in 1989. On the basis of the differences exposed in the foreign language competence of the two cohorts of Hungarian scholars over a span of ten years, it becomes possible to rate the speed at which the English language was gaining ground in Hungary, at the expense of both the native language and all the foreign languages other than English.

As is generally known, Hungary underwent dramatic changes in the last decade of the 20th century, which permeated the whole of its political, economic, and social structure. Needless to say, these changes did not leave its scientific life untouched either. Although the percentage of GDP being spent on research and development dwindled from 1.6 in 1990 to 0.7 in 1997 (*Statistical Yearbook 1991*; *1998*), the scientific community of Hungary made great strides in 'going international' during the same period.

Nevertheless, two things have to be borne in mind. Firstly, most Hungarian scholars are awarded academic titles rather late in their career, certainly beyond the age when one would normally engage in comprehensive foreign language learning. Therefore, the participants in this study cannot serve as a typical showcase for the sweeping changes that characterised the foreign language education policies of the 1990s. Secondly, in the

1970s and 1980s Hungary was by far the most liberal country in the communist bloc, jocularly labelled as 'the happiest barracks', and scientists were a particularly privileged lot. By the time communism imploded in 1989, they had already got into the mainstream of scientific discourse. In the 1990s, therefore, their job was to strengthen, rather than establish, ties with international scholarly networks. With regard to foreign language competence, too, the researchers of this study expected to find a steady progress, and not a breakthrough, in the participants' English language proficiency, and a simultaneous slow recession in their command of other foreign languages. This is all the more likely since there must be an overlap between the cohorts of 1989 and 1999.[1]

2. The Hungarian example: the foreign language competence of the population

Hungary is a small landlocked country in Central Europe with a population of slightly more than ten million. The official language, and the mother tongue of nearly 98% of the population, is Hungarian. Hungarian is a unique and isolated language in Central Europe, because it belongs to the Finno-Ugric branch of the Uralic family of languages, while all the surrounding countries use a language of Indo-European origin as their mother tongue. In order to minimise the effects of their linguistic isolation, Hungarians have always attached great importance to the learning of foreign languages.

Prior to World War II, German was the major foreign language taught in Hungarian schools, followed by French, English, and Italian. After the communist takeover in 1949, all foreign languages were virtually banned except Russian, which became the compulsory foreign language in all types of schools. For forty years, every student received eight years of Russian instruction, and those who went on to tertiary study had an additional two years. The very low cost-effectiveness of Russian teaching was apparent from the start, as were also the political motives that kept Russian in the curriculum. From the late 1960s on, in line with the gradual liberalisation of the communist dictatorship in Hungary, other foreign languages slowly crept back into the curriculum, although the scope for learning them remained limited (Medgyes 1993).

In 1989, consonant with the historic changes occurring in Hungary and, in quick succession, over the whole of Central and Eastern Europe, Russian was officially stripped of its privileged status in the school curriculum. Since then, it has been possible for Hungarian school children to learn any foreign language they wish, provided only that the school they attend is able to offer that language – largely a matter of teacher supply. Thus, the distribution of foreign languages at every level of education has changed irrevocably. Recent statistics show that while in primary schools German is the most popular language with English as the runner-up, in secondary and especially in tertiary education this order is reversed (Medgyes and Miklósy in press). Another illuminating piece of data about the German/English divide is that while German is typically spoken by less educated and older people who live in smaller towns and villages, English is favoured by the better educated and the young from metropolitan areas. Nonetheless, the foreign-language competence of Hungarians leaves a lot to be desired. According to a comprehensive survey (Terestyéni 1995), 11.2% of the population claims at least to 'get by' in one foreign language, 3.3% in two languages and 0.8% in three or more languages.[2] It is particularly worrisome that English is spoken by only 5.1% of the population (further details are supplied below).

3. The foreign language competence of Hungarian scholars: the study of 1989

The original questionnaire was circulated among Hungarian scholars and the data were processed in 1989. Since this happened to be the last year of communist rule, this study provided 'a state-of-the-art analysis of a limited segment of contemporary Hungarian society from a socio-linguistic perspective, [but] it may also be regarded as a retrospective, summary document concerning a bygone, historical period' (Medgyes and Kaplan 1992: 69). While it came as no surprise that the foreign language competence of eminent scholars far exceeded that of the Hungarian population, it was assumed that, by virtue of their high standing in the social pyramid, scholars were setting trends for the society at large.

More specifically, the Hungarian scholars participating in the study of 1989 were aware that, as native speakers of a language spoken by few, they were at a distinct disadvantage unless they were able to use English, the dominant language of most of the disciplinary communities. Nevertheless, while the average Hungarian scholar was found to be a polyglot, only half of the participants claimed to possess a high level of English language proficiency. It was further established that knowledge of English was in inverse proportion to age, that is, the older the scholar, the less likely it was that she or he controlled English. At the same time, while scholars from the human/social sciences tended to speak far more languages and far better than their colleagues from the natural sciences, the English language occupied a disproportionately small segment of their language volume. By way of illustration, while 'hard scientists' published more than half of their works in English, 'humanists' published two thirds of their works in Hungarian. (More details about the 1989 survey are provided in the sections below.)

4. The present study: research design

This study strives to replicate the 1989 study as closely as possible, cognizant of the fact that the changes which took place in the 1990s can only be registered if comparable sets of data are available. Therefore, the procedures applied to select the participants, design the questionnaire, as well as collect and process the data are the same in 1999 as ten years ago.

4.1. The participants

4.1.1. Selection and return rate

In the spring of 1999, the researchers of this study solicited the support of the Hungarian Academy of Sciences, the major scientific body in Hungary, for circulating a questionnaire among scholars. With our request granted, we were allowed to use the address list of prospective participants and seek financial help from the Hungarian Ministry of Education.

The questionnaire was mailed to all Members of the Academy,[3] a total of 311, and every fourth Doctor of Science (*Doctor Scientiarum*), a total of 558.[4] Out of the 869 letters that had gone out, 843 reached their targets, and 620 were completed and returned. In other words, 73.5% of the scholars agreed to be involved in the research![5] In our era of 'questionnaire fatigue' and junk mail, the high return rate exceeded all our expectations. In searching for the causes, we like to think that this achievement is mainly ascribable to a keen interest scholars take in foreign language learning in general and in research concerned with language learning in Hungary in particular. This assumption seems to be confirmed by several participants, who noted that the results should also be published in Hungary; some specifically asked to be sent a complimentary copy of the paper.

4.1.2. Age

The data in Table 1 show that while 53.4% of our sample are above 65 years of age, a mere 10.7% are below 51. This confirms our belief that, as a rule, Hungarian scholars cannot hope to be admitted to the highest echelons of the academia until they have reached a fairly advanced age. To be frank, the turbulent years of the 1990s brought little to bear on accelerating academic mobility, and the conclusion drawn in the 1989 study seems to hold true: 'If such titles were merely ceremonial, awarded in appreciation of scholarly achievement, the matter would not be significant, but the fact is that professional prestige is tied to the possession of academic degrees' (Medgyes and Kaplan 1992: 72). In fact, there is more at stake than professional prestige: access to research and development funding is largely dependent on the possession of academic titles.

Table 1: Frequency distribution by age, 1999

	N	*Percentage*
Below 51	65	10.7
51–65	214	35.7
Above 65	341	53.4

Unfortunately, no specific item inquired about the participants' present occupation, and thus it was impossible to separate the active and inactive scholars in the 65+ group. As the data obtained from scholars long retired was not believed to be relevant, it was decided to disregard the senior cohort altogether, even though this implied losing sight of many participants who were at the peak of their professional career. Thus the analysis below is primarily targeted at 279 scholars, 45% of the full sample.

4.1.3. Academic field

The Academy runs eleven scientific sections.[6] Although the demarcation lines between them are often permeable and fuzzy, and hence admittedly arbitrary, this system of distribution is supposed to provide a manageable framework for the fulfilment of research and development functions. For ease of reference in this study, the eleven sections were grouped into two umbrella categories, one which comprised the human and social sciences (Language and Literature, Philosophy and History, and Economics and

Law), and the other which included the natural sciences (the other eight sections). Short of better terms, scholars from the human and social sciences will be called 'humanists', while those from the natural sciences 'scientists'.

On the basis of this categorisation, 88.2% of the scholars below 65 years of age were scientists (Table 2); in the full sample, which also included the 65+ age cohort, scientists took up a smaller proportion (80.9%). If these figures are compared to those of the 1989 survey, in which scientists enjoyed a far less comfortable majority over humanists (71.1%), it may be suggested that the relative position of the natural sciences in the Hungarian Academy is becoming stronger.

Table 2: Frequency distribution by academic field, 1999 (population under 65)

	N	*Percentage*
Human/social sciences	32	11.8
Natural sciences	240	88.2

4.2. The questionnaire

A replica of the 1989 questionnaire,[7] the 1999 questionnaire consists of 22 items. Most of the items were designed for multiple-choice response, including seven which required a certain degree of introspection. With only three items left open-ended, the completion of the questionnaire was estimated to take no more than 30 minutes.

The questionnaire falls into two main parts. The first part inquired about foreign-language competence (who speaks which foreign languages at what level?), language learning attitudes (what factors contribute to the success of language acquisition: aptitude, length of stay in the target-language country, awareness of its importance as a world language?) and various aspects of writing academic papers (language of publications, writing strategies?). The second part of the questionnaire is concerned with issues relating only to English. Specific questions were asked about the participants' learning history (when and where they studied English?), as well as their present use of English (which variety, how often?). A set of items refers to details about their English-language competence (which skills, reading rate, writing ability?). In addition, control variables regarding age, discipline, and publications were introduced. Efforts were made to consistently compare present results with those obtained ten years ago, thereby mapping the linguistic progress Hungarian scholars are supposed to have achieved in the first ten years of democracy.[8]

Before the results are supplied, a final caveat may be appropriate. In an oft-quoted article, Marton (1981) makes a distinction between first-order and second-order research, the former being concerned with what people do, and the latter with what they *perceive* they do. The present study is a typical case of second-order research, insofar as all the results were obtained from self-reports, and no attempt was made at judging the accuracy of the participants' statements by testing their actual foreign language proficiency. The validity of self-assessments is a controversial issue (Ross 1998). For example, Wall et al. (1994) discovered low correlations between students' self-reports and their test scores, which may be due to inadequate self-awareness, insecurity, undue modesty or, conversely, inflated self-esteem. All or any of these factors might also characterise the participants of this study. On the other hand, there is evidence in the literature that self-evaluation is an accurate measure of real language proficiency (Bachman and Clark 1981; Shameem 1998), especially if the population is homogeneous in terms of culture, language, and educational level (LeBlanc and Painchaud 1985). Although Hungarian scholars may be regarded as a fairly homogeneous sample, it is necessary to exercise caution in interpreting the findings of this study.

5. Results

5.1. Language volume

5.1.1. Foreign-language competence

The participants were asked to indicate their perceived foreign language competence on a four-point scale: low, medium, high, and near-native/native. Their aggregate results are expressed numerically in the following way: one point was given for the judgment 'low', two points for 'medium', three points for 'high', and four points for 'near-native/native' proficiency; this matrix was used for each language the participants claimed to speak, and eventually the points were added up. Thus, for example, if a participant scored ten points, he or she spoke two languages at low, two at medium, and one at near-native/native level ($2 \times 1 + 2 \times 2 + 1 \times 4 = 10$ points), or two at high and one at near-native/native level ($2 \times 3 + 1 \times 4 = 10$ points), and so on.

The language competence of the population below 65 years of age ranged between zero and 21 points, the overwhelming majority (86.4%) scoring between 3 and 9 points. The mean value was 6.4, with a standard deviation of 2.9. This implies that average Hungarian scholars, in their judgment, spoke at least two foreign languages quite well, or more than two languages less proficiently. The means for scientists and humanists[9] were respectively 6.0 and 9.1, which shows that the language volume of participants from the human/social sciences was substantially greater (Table 3). On the other hand, the difference in standard deviations (SD = 2.4 versus SD = 4.2) indicates that there was less difference between individuals in the group of scientists than in the group of humanists; in other words, scientists formed a more homogeneous group in terms of their language volume. Furthermore, the comparative results show that the language volume of Hungarian scholars below 65 shrank in the ten-year period (from an average score of 7.3 down to 6.4), and that this tendency of reduction was characteristic of humanists and scientists in equal measure. Nevertheless, even today the average Hungarian scholar need not be ashamed of his or her foreign language competence.

Table 3: Language volume: humanists versus scientists, 1999 (population under 65)

	N	*Mean*	*SD*
Human/social sciences	32	9.06	4.21
Natural sciences	240	6.03	2.42

5.1.2. English-language competence

The next set of data relates to the participants' English language competence (Table 4). Whereas the mean value for the 35–50 age cohort was 2.8, the same values for the 51–65 and 65+ cohorts were respectively 2.6 and 2.4; that is to say, the younger the scholar is, the better his or her English was claimed to be. The same tendency was observed in the 1989 sample, with the corresponding mean values being 2.6, 2.3, and 2.1. These figures also testify to a slow process of improvement in English-language skills over the ten-year-period, possibly owing to a growing awareness of the importance of English in scholarly communication. The enhanced role that English plays in the distribution of foreign languages in 1999 is even more visible if the loss in general language volume is confronted with the gain in English language competence – the relative importance of English rose quite dramatically in the past decade.

Table 4: Comparing language volume and English language competence: 1989 versus 1999

	Language volume Mean (SD)		*English language competence Mean (SD)*	
	1989	1999	1989	1999
Below 50	7.4 (4.3)	6.0 (2.7)	2.6 (0.9)	2.8 (0.6)
51–65	7.3 (3.4)	6.5 (2.8)	2.3 (1.0)	2.6 (0.7)
Above 65	8.8 (4.5)	7.0 (3.1)	2.1 (1.2)	2.4 (0.8)

Next, the data on the English language competence of humanists and scientists for the age cohorts below 65 were contrasted. Table 5 demonstrates that the mean scores for the two groups were 2.9 and 2.7 respectively, which means that humanists judged their English to surpass their peers' from the natural sciences. This is a surprising finding, especially in comparison with the data obtained in 1989, when the respective mean scores were 1.9 and 2.4. Humanists proved to make rapid improvements: whereas in 1989 their English-language competence did not even reach the medium level, in 1999 it came close to a high level of proficiency. The authors of the 1989 article said that 'English has become the universal lingua franca in the natural sciences, but the human/social sciences lag behind', adding that 'the gap may be wider in Hungary than in other global regions' (Medgyes and Kaplan 1992: 74). In contrast, the results of the 1999 survey suggest that the dominance of English has been extended to the human and social sciences, and Hungary has caught up with this universal trend.

Table 5:
English language competence: humanists versus scientists, 1999 (population under 65)

	N	*Mean*	*SD*
Human/social sciences	28	2.85	0.75
Natural sciences	237	2.67	0.66

Table 6:
Frequency distribution of languages at high and near-native level/native, 1999 (population under 65)

Language	*N*
English	171
German	46
Russian	33
French	12

Table 6 displays the relative frequency of foreign languages spoken by the participants below 65; only data on high or near-native/native level proficiency were recorded. Whereas nearly two thirds of the sample claimed to be highly proficient speakers of English (171 participants), the number of highly proficient speakers of German, Russian, and French is 46, 33, and 12, respectively. In an English-German relationship, this means that whereas in 1999 there were nearly four times as many highly proficient English users as German users, ten years earlier the difference was as small as 1.2! The plight of German in scientific discourse was particularly revealing in view of the foreign language competence of the Hungarian population as a whole. As established by a large-scale survey (Terestyéni 1995), 6.1% of Hungarians 'got by' in German, compared to 5.1% in English.[10] In other words, while German is still the most widely spoken language in Hungary, English has become the unrivalled language of science across the disciplines.

As a matter of fact, the participants were asked to name the language they thought to be the lingua franca of their discipline: 96.3% in the below-65 cohorts granted English pride of place, 3% identified English and another language, and merely 0.8% (2 participants) said that a language other than English was the leading language in their academic field.[11] The proportions for the 1989 sample were 82.1, 12.7, and 5.2% – this implies that in the ten-year period English virtually eliminated all other languages in the competition for the title of 'World Language'. Viewed in this light, there is little reason for complacency: there is still a long way to go from recognising the hegemonious role of English to mastering the language itself! (More details on English-language proficiency are given below.)

5.2. Factors contributing to language development

5.2.1. Language aptitude

Participants were asked to assess their language aptitude on a five-point scale. In the sample below 65, the mean value (2.5) fell exactly between 'good' and 'average'. While nearly 10% (27 participants) considered themselves very gifted language learners, only two scholars placed themselves at the bottom end of the aptitude scale. Not surprisingly, a significant positive relationship was found between language aptitude and language volume: those participants who judged themselves to be gifted language learners tended to speak more languages more proficiently than those who perceived themselves as less able learners.

5.2.2. Time spent abroad

The participants were also asked to mark the length of their stays abroad expressed in months, as well as to name the countries of destination. In the age groups below 65, 87.5% of the participants spent a minimum of one month abroad, the mean being 39.7 months, that is, well over three years per person. Interestingly enough, the mean value for the 65+ cohort was not considerably shorter: 37.8 months per person. Although the huge differences in standard deviation between the junior (SD = 39.0) and senior (SD = 69.1) groups beg caution in interpreting the results, it may be assumed that what the older generation lost due to limited opportunities was compensated for by a longer span of academic career. In comparison, the average scholar in the 1989 survey was recorded to have spent 23.9 months abroad, which implies a growth of about 60% by 1999. This growth may be justified by the favourable political and economic climate between Hungary and the leading countries of research and development, which created incomparably better grant and job prospects for academics.

However, it has to be admitted that these figures show a somewhat distorted picture by including the data obtained from participants who, after having spent decades of their professional career abroad, decided to return to Hungary after the 'Iron Curtain' was pulled down. To rectify this methodological shortcoming, the data bank was curtailed by eliminating everyone who had spent more than a straight 36 months in a particular foreign country. As a result, the participants aged below 65 were found to have spent an average of 18.2 months (SD = 14.4) and the 65+ cohort an average of 12.4 months (SD = 13.4) abroad. The striking drop in standard deviation, especially for the senior cohort, seems to legitimise the reviewed procedure.

Next, the lengths of time spent in English-speaking versus non-English-speaking countries were computed, producing 23.5 and 12.2 months as respective mean values for the cohorts below 65. The preponderance of English-speaking countries is even more pronounced for the 65+ cohort: 19.3 and 5.6 months, respectively. In awareness of the leading role that English-speaking countries, especially the US, play in international scientific research, the above proportions cannot be regarded as unexpected.

5.2.3. Other factors

An item in the questionnaire asked the participants to identify, in order of importance, three or more factors which they believed had helped them learn foreign languages. Since this question elicited free responses, evaluation required introspection and subjectivity. After the answers were collated into ten comprehensive groups, a scoring system was applied to produce a rank order. As the method of evaluation was the same as for the 1989 study, the two rank orders could be compared (Table 7).

Table 7: Other factors contributing to foreign language development: 1989 versus 1999

	Rank order	
	1989	1999
Conscious learning	1	1
Reading for professional purposes	2	3
Professional motive	3	2
Diligence	4	6
Family surroundings	5	7
Attending conferences	6	10
Contacts with foreigners	7	4
Reading for entertainment	8	8
Love of target language	9	5
TV/Radio	10	9

Both lists suggest that the participants attached by far the greatest significance to conscious learning, a category which comprised school instruction, private tutorial study as well as self-study. This is no surprise in a monolingual country like Hungary, where the scope for using foreign languages in everyday communication is extremely limited, and where living and studying abroad was, and still is, the privilege of a small proportion of citizens. In looking for major discrepancies between the two lists, the most striking rise may be observed for the item 'love of the target language', which indicates an enhanced level of motivation to study foreign languages. At the same time, the climb of the category 'Contacts with foreigners' and the drop of the category 'Attending conferences' may be ascribable to the fact that encounters with foreigners are no longer restricted to conferences: there is a far wider range of opportunities for maintaining professional contacts today than before 1989 (more about this later).[12]

5.3. *Publications*

5.3.1. Number and language

When asked about the total number of publications, the average Hungarian scholar below 65 years of age claimed to have published between

26–50 works within a period of ten years (1989–1998). More interestingly, the participants reported on the language in which they were publishing, that is, the proportion of their publications in Hungarian, in English and in other languages. According to the figures for the total population, close to half of all publications were written in English (49.6%), leaving 41.9% for Hungarian and 9.6% for other languages.[13] If these figures are contrasted with those for 1989, it turns out that the order of English and Hungarian was reversed, while the scope for other languages became narrower (Table 8). The advantage of English is even more marked for the cohorts below 65, with the percentages for English, Hungarian, and other languages being 62, 32, and 6, respectively – this reveals the increasing pressure on younger Hungarian scholars to write in English. However, when cumulative productivity for the below-65 groups was subsequently examined in relation to humanists versus scientists, some unexpected results were produced. Whereas the proportions for scientists reflected a pattern similar to the one indicated above (67% for English, 27.1% for Hungarian, and 5.9% for other languages), humanists were still found to publish the majority of their works in Hungarian (68.9%), rather than in English (20%) or in other languages (12.2%). This may be due to the peculiarities of certain human/social sciences (for example, a scholar specialising in Hungarian literature, history or art is obliged to write in Hungarian as a matter of course), but a slower 'reaction time' to universal trends may also account for the preferred use of the native language. It has to be reiterated, however, that the representation of humanists in the sample is too small to jump to conclusions.

Table 8: Language of publications: 1989 versus 1999

	1989 Percentage (SD)		*1999* Percentage (SD)	
Hungarian	46.7	(31.4)	41.9	(30.4)
English	39.5	(33.3)	49.6	(33.2)
Other	13.8	(16.2)	9.6	(12.7)

Furthermore, the participants were asked to describe their preferences for alternating between Hungarian- and English-language publications. The main reasons for publishing in English were impact factor and citation indexes, with better-paid jobs, international travel and academic friendships also being mentioned. It was repeatedly pointed out that since only English-language publications could reach an international audience, writing in any other language, especially in Hungarian, a language virtually incomprehensible beyond the national borders,[14] was a waste of time. In Hungary, too, the number of English-language professional journals was on the increase, and, as a participant noted maliciously, those who were still unable to read in English were not worth addressing in the first place.

Others were less antagonistic and assigned Hungarian a multiple role in scholarly communication. The argument most frequently advanced concerned the educational value of publications in Hungarian, including textbooks for university students and professional articles for colleagues with an insufficient command of English. Several participants remarked that their first works, notably their PhD dissertation, had been written in Hungarian. The genres which typically allowed room for Hungarian included monographs, state-of-the-art reports, popular science books and articles, reprints of original papers, and reviews. While some participants admitted that they would turn to Hungarian (and not to refereed international) journals for fear of rejection or out of laziness, others said that 'It's only too fair to accept invitations from Hungarian journals every now and then.' Several participants pointed out, with varying degrees of fervour, that they were writing in Hungarian primarily led by the desire to protect and develop the native tongue as a tool for academic discourse.

5.3.2. Work style

Two questions explored the participants' work style with regard to the language of scholarly papers. According to the data obtained from the age groups below 65, 82.2% of the participants wrote directly in the foreign language; 10.5% were in the habit of translating their Hungarian texts into the foreign language themselves, and a mere 7.2 employed professional translators. The same set of figures for the entire population (65.5%, 18.2%, and 16.3%, respectively) shows that far more younger scholars wrote without recourse to translation, which may be explained by a higher level of writing skills in English. Since 1989, when there was an equal proportion of participants who wrote directly in the foreign language versus those who resorted to translation, the rate of improvement may be deemed remarkable.

The second question consisted of two subquestions. Firstly, the participants were asked whether they would normally have their foreign-language manuscripts vetted before sending them for submission. Slightly less than half (48.4%) of the age cohorts below 65 answered in the affirmative; this proportion was much higher (59.3%) for the total population, indicating that the younger generation was more confident in their language use. Secondly, the participants had to identify the kind of help they would solicit. Not unexpectedly, the majority claimed to have their text revised by a native speaker of the target language (75.2%) and/or an expert in the field (81.8%).

5.4. Specific information about English-language competence

5.4.1. Variety of English

This item supplied three alternatives for the participants to specify the variety of English they perceived to speak: British English, American English, and 'other'. Comparative figures for British and American English (Table 9) show that British English, the prestige variety in Hungary throughout the 20th century, was gradually losing ground between 1989 and 1999 in its use by Hungarian scholars. This tendency becomes even more pronounced if the 1999 sample is broken down into two major cohorts: an increasing number of younger scholars used the American variety, probably due to the pivotal role the US plays in contemporary science. Among the mixed varieties, the British-American hybrid was the most common one, followed by 'Hunglish', 'continental English', 'international English', 'broken English', and 'school English'.[15]

Table 9: Varieties of English spoken: 1989 versus 1999

	1989 Percentage			*1999 Percentage*		
	American	*British*	*Other*	*American*	*British*	*Other*
Below 51	33.3	35.8	30.9	39.7	34.9	25.4
51–65	22.8	54.4	22.8	33.5	47.1	19.4
Above 65	14.6	71.2	14.2	23.0	63.9	13.1
Total	23.0	54.9	22.1	29.2	53.4	17.4

5.4.2. Beginning to learn English

Six categories were provided for the participants to determine the age at which they had begun to learn English. According to figures contrasting the situation in 1989 and in 1999, an earlier comment, namely that scholars had not been considerably affected by the better language learning opportunities of the 1990s, appears to be verified; no significant differences were found in starting age between the cohorts of 1999 either. Table 10 shows that over 40% began to learn English after graduation, probably due to the meagre opportunities available for the study of foreign languages other than Russian at school and university. On the other hand, a significant positive relationship ($p < .005$) was observed between the onset of instruction and the ability to achieve advanced English proficiency, that is, better speakers tended to start learning English at an earlier age.

Table 10: Frequency distribution by start of English study, 1999 (total population)

	Below 6	*6–10*	*10–14*	*14–18*	*18–24*	*Above 24*
Below 65	1.5	6.9	11.3	30.3	9.9	40.1
Above 65	2.0	6.1	14.6	28.6	2.7	45.9
Total	1.9	6.3	12.9	29.4	6.1	43.4

5.4.3. Where English was studied

The participants were also asked to specify where they had studied English, marking more than one option if they wished. In the age cohorts below 65 years of age, adult courses (27%) and self-study (26%) were found to be the preferred ways of learning, which provides additional documentation of the limited school-based opportunity to study English (17%).[16] Incidentally, the data for the total population revealed very similar proportions.

5.4.4. The skills area

The participants were asked to rank their English-language proficiency level in various skills on a seven-point scale (1 being the best). For reasons of reliability, the following method of evaluation was employed: first, those skills for which the participants had assigned 1 or 2 points were identified; then, their frequency was tallied to produce percentages. Thus the percentage for technical reading, for example, means that 82% of the participants in the 1999 cohorts below 65 years of age said they were very proficient in this skill (Table 11).

Table 11: English proficiency level in relation to different skills: 1989 versus 1999

	Rank order in percentages	
	1989	1999
Technical reading	71	82
Formal speaking	49	68
Informal speaking	48	65
Informal writing	43	63
Technical writing	42	71
Literary reading	40	47
Listening	36	51

The rank order for 1999 needs some explanation. Not unexpectedly, the participants marked reading in their own professional area as their best skill. Whereas the next four items on the list followed each other rather closely, listening and literary reading came at the bottom with a wide margin. The deficiencies in listening skills may be due to the long decades of political isolation, as well as to the employment of traditional teaching methodology which paid little heed to listening. The shortcomings of the ability to read literary texts, on the other hand, are likely to be caused by the fact that it was merely a leisure activity for most of the scholars participating in the study.

A comparison of the two rank orders witnesses little change. The only substantial modification concerns technical writing: it not only moved up to the second place by 1999, but displayed an improvement of nearly 30%. This suggests that the tendency of 'publish or perish' has caught up with Hungarian scholars as well: they are obliged to write and publish in English to a much greater extent than ten years ago. The significant positive relationship (at $p < .005$) between the level of proficiency in technical writing and the quantity of publishing in English may not be thought to be a coincidence either. Table 11 also reveals that while technical writing was the most dynamically developing skill, quite rapid progress could be registered for every other skill too. It is reassuring, incidentally, that a high degree of consistency between the responses given to the general level of English-language competence and those given to proficiency in specific skills was found by a primary component analysis.

In search of further information about the main skills, the participants were asked four additional questions. Two concerned the frequency of contact with native English speakers and conference attendance, factors which were supposed to contribute to the development of oral proficiency (cf. Table 7). The other two questions inquired about the participants' reading rate and writing ability in Hungarian and in English. It is the responses to these four questions which are discussed in the final part of the study.

5.4.5. Contact with native English speakers

The results for 1989 showed that the average Hungarian scholar had come into contact with native English speakers between once or twice a month and a few times a year. This is an indication of the scarcity of native English speakers in Hungary, and the proportions for the 1999 population testify to only insignificant progress in this regard (Table 12). However, if the cohorts below 65 years of age are examined separately, it turns out that they had far better opportunities than their senior peers or, for that matter, the 1989 population. In other words, the most professionally active Hungarian scholars engaged in international academic communication with native English speakers, and possibly also with non-native English speakers,[17] on a fairly regular basis, which ranged between once or twice a week and once or twice a month.

Table 12: Contact with native English speakers: 1989 versus 1999

	1989 *Percentage*	*1999* *Percentage*
Every day	2.6	7.5
Once/twice/week	20.6	21.1
Once/twice/month	24.1	25.5
A few times/year	34.4	31.4
Rarely	15.4	10.9
Never	2.9	3.7

To examine the influence of four independent variables (start-up age, length of stays, publications, and frequency of contact) on English-language competence, a regression analysis was carried out (Table 13). The computations produced significant relationships for all the variables (except for one) at the $p < .005$ level of confidence. If the differences between the two main age cohorts are analysed, the following statements may be made: (a) there was no difference between younger and older scholars in terms of the effect that an early start exerts on their language development; (b) in the case of scholars below 65, no significant relationship was discovered between the time spent abroad and their language competence; (c) the influence of publishing activity on language competence was greater for younger scholars; and (d) the frequency of contact with native speakers had a greater impact on the language competence of the junior participants than that of their senior peers.

Table 13: Variables influencing English-language competence, 1999

a) Population under 65 (N = 279)
Adjusted R Square .24276

Variable	*B*	*SE B*	*Beta*	*T*	*Sig T*
Onset of study	−.118084	.026684	−.241565	−4.425	.0000
Length of stays	.105496	.114739	.051370	.919	.3587
Publications	.004861	.001172	.224063	4.149	.0000
Contact	−.193158	.034041	−.313431	−5.674	.0000
(Constant)	3.274108	.292988		11.175	.0000

b) Population over 65 (N=320)
Adjusted R Square .33980

Variable	*B*	*SE B*	*Beta*	*T*	*Sig T*
Onset of study	−.135142	.026492	−.256433	−5.101	.0000
Length of stays	.188202	.093223	.105675	2.019	.0445
Publications	.006645	.001364	.270724	4.870	.0000
Contact	−.183822	.036970	−.277320	−4.972	.0000
(Constant)	3.198200	.268625		11.906	.0000

5.4.6. Conference participation

Restricted to events at which English was the (or one of the) official language(s), this item consisted of two parts: one relating to mere attendance, while the other one elicited information about the number of papers presented. According to the figures, in the past five years the average Hungarian scholar below 65 attended 11.9 events and gave 9.0 papers, which means that she or he was actively involved in three out of four events. This relatively high proportion may be explained by the fact that Hungarian scholars were rarely able, or indeed inclined, to attend unless they were specifically invited to speak. In comparison, the respective means for 1989 were 8.4 events and 6.6 papers, which implies that the scholars participating in this study had a far better chance to attend conferences and present papers than their peers in 1989, while the ratio of attendance to presentations remained virtually the same. Further data revealed significant positive relationships between the frequency of conference attendance and, respectively, English proficiency level and frequency of contact with native speakers; on the other hand, no significant relationship was found between conference participation and publication capacity.

5.4.7. Reading rate

The participants were asked to assess their reading rate in Hungarian and in English on a four-point scale (1 being the best). The results for the below-65 cohorts (Table 14) imply that while 47% of the participants placed their reading rate in Hungarian versus in English at the same level, respectively 46.7% and 6.3% considered their reading rate in English one level or two levels below their reading rate in Hungarian.

Table 14: Reading rate in Hungarian and in English, 1999 (population under 65)

	Hungarian Percentage	*English Percentage*
Very fast	39.8	9.5
Fast	44.5	50.0
Moderate	15.0	36.5
Slow	0.7	4.0

5.4.8. Writing ability

Writing ability in the two languages was evaluated on a three-point scale (Table 15). In comparing the results for Hungarian and English, 44.1% said that there was no difference between their abilities in the two languages, while 46.7% ranked their capacity in English one category and the rest two categories lower than their capacity in Hungarian.

Finally, interesting results were yielded by the self-reports relating to a comparison of reading rate and writing ability in the two languages. With respect to Hungarian, the majority of participants (77.1%) viewed their reading rate better than, and the remainder (22.9%) as good as, their writing ability. As for English, reading was prioritised to a lesser extent; some at least, though only 7.3%, said that their writing was superior.

Table 15: Writing ability in Hungarian and in English, 1999 (population under 65)

	Hungarian Percentage	*English Percentage*
Facile	73.3	27.7
Moderate	24.9	57.3
Laborious	1.8	15.0

6. Conclusion

The researchers of this study set themselves a double goal: they intended to prove that (1) English had become the main language of academic communication for Hungarian scholars, and (2) its significance had increased throughout the 1990s. In order to access relevant data, a survey which entailed 620 top-ranking representatives of Hungarian science was designed and administered in 1999, and its findings were contrasted with those of an identical survey carried out in 1989.

The survey of 1999 demonstrated that while the foreign language volume of the average participant had diminished in the ten-year period, the proportion of English-language competence was perceived to have risen substantially, especially in relation to the cohorts below 65 years of age. The relative weight of English was even more pronounced in terms of publications: during the past decade, foreign languages other than English were relegated to a marginal role in academic discourse, a tendency grounded in the view that scientific achievements had a chance of circulation only with the proviso that they were reported in English.

The enhanced role of the English language and of the countries where English is the native language was displayed by several other factors as well. The participants' overall level of proficiency in English exhibited considerable improvement, with the most distinct progress being registered for the skill of technical writing. In congruence with this finding, the overwhelming majority of the participants reported writing their scholarly papers directly in English, without recourse to first drafts in the native language. Further data revealed that 40% of Hungarian scholars had started to learn English only after graduation, and for more than half of the population adult courses and self-study were the preferred forms of study, an indication of the fact that the majority were unable to avail themselves of the fairly auspicious conditions available for learning English at school after 1989. While the duration of stays abroad was lengthened by 60% between 1989 and 1999, the average Hungarian scholar spent twice as much time in English-speaking countries as in non-English-speaking countries. The favourable social and political situation prevailing in post-communist Hungary enabled its scholars not only to spend more protracted periods of time abroad, but also to attend more conferences and maintain more regular professional contacts with native English speakers than ten years before.

It may be argued that the onslaught of English in the scientific life of Hungary is due to two simultaneous processes: (1) a global trend which has granted the English language a privileged position, and (2) the historical changes occurring in Hungary and the whole of Central and Eastern Europe. With respect to this second process, Hungary differs from other countries of the region in at least one sense, namely that the expansion of English began earlier and therefore has been more gradual in Hungary than elsewhere.

Let us deal with one final issue. Against the irresistible advance of English, what fate may await the Hungarian language? Will it cease to exist in scientific communication? This is a pertinent question raised by several participants in this study. One scholar went as far as to admit: 'These days I write nothing in Hungarian but occasional obituaries.' The outlook of the researchers of this study is not so bleak: although it is true that the share of Hungarian diminished in the ten-year period, and especially in relation to scientists of the younger generation, publications in Hungarian still accounted for 42% of all publications. At the same time, the choice of language became more content-specific; while reports on new scientific achievements tended to be written in English, information for lay audiences was disseminated in Hungarian. On a more general plane, this may be interpreted as an indication that Hungary is on its way to being listed among countries where English is used as a second rather than a foreign language (Kachru 1985), with the implication that in certain areas English becomes the standard vehicle of communication. Undoubtedly, science is such an area. Whether this process is to be condoned or bemoaned is not for the authors of this report to adjudicate.

7. Acknowledgments and notes

The authors would like to express their gratitude to the Hungarian Academy of Sciences and the Hungarian Ministry of Education for their support. They also wish to acknowledge the contribution of Robert B. Kaplan, co-author of the 1989 article, for his encouragement. Our special thanks are due to Éva Gál for her invaluable clerical assistance.

1. The extent of overlap between the two cohorts cannot be ascertained because of the anonymous nature of the survey.
2. The picture becomes even bleaker if these data are compared to those obtained from a similar survey conducted fifteen years earlier (Terestyéni 1981): by 1994 a mere 2% increase could be recorded.
3. Membership in the Academy is the highest honour to which a Hungarian scholar can aspire. Members are ranked in two categories: Ordinary Members and Corresponding Members, the former being the superior category.
4. The reason for including only a quarter of the Doctors of Science was merely a financial one: to save on postage and administrative costs. Selection from the available pool was made according to alphabetical order, except for two other criteria: a proportionate number of scholars were selected according to domicile and gender. As for domicile, Budapest had to be represented in proportion with the central role it plays in the scientific life of the country: 70.5% of all scholars (Members and Doctors together) live in the capital city! As for gender, although the ratio of women rose from 4.7% in 1989 to 8.8% by 1999, it is still too low to justify a distribution by sex. Similarly, the data on Members and Doctors will be examined in the aggregate for sake of convenience.
5. The scholars were sent two consecutive letters: one in May and a reminder letter containing the same questionnaire in August. The distribution of return rate was 65.8% in the first and 34.2% in the second round.
6. The eleven sections are: Language and Literature, Philosophy and History, Mathematics, Agriculture, Medicine, Engineering, Chemistry, Biology, Economics and Law, Earth Sciences, and Physics.
7. Only one item was altered: in Item 4, the time scale was updated and shortened to include only works published in the post-communist era.
8. After the questionnaires had been collected, the data were computerised and then subjected to analysis, using the Statistical Package for the Social Sciences, 6.0 Windows (SPSS+).
9. The low number of humanists (32 scholars) below 65 years of age requires caution in assessing the results.
10. The appropriate figures for Russian and French were respectively 2 and 0.9%.
11. Some participants gave vent to their bitterness over the dominance of English at the expense of other foreign languages, arguing that, for example, 'German or French would have been far more suitable tools for scientific communication.'
12. The wittiest remark upon factors which facilitate language learning was perhaps this one: 'It's impossible to learn languages and love only from books. Both activities have to be practised in real situations as well.'
13. The representation of other languages was too small to deserve much mention: the leading languages were German, Russian, and French – in the same order as in 1989.
14. 'The international academic community still can't read Hungarian,' commented a participant ironically.
15. Further epithets include 'nuclear English', 'idiosyncratic English', 'Esperanto English', 'Hungaro-Californian English', 'English spoken in Rottenbiller Street', etc.

16. Some participants suggested that academics might have preferred self-study even if opportunities for organised study had been more readily available. Others expressed their dissatisfaction with the present state of foreign language instruction at school, recommending ways of improvement.
17. It was pointed out by several participants that encounters with other non-native speakers of English were far more frequent than those with natives – and no less useful!

8. References

Bachman, Lyle F./Clark, J. L. D.
1981 *Self-Assessment of Communicative Competence in English*. University of Utah: Unpubl. mimeo.

Baldauf, Richard/Jernudd, Björn H.
1987 Academic Communication in a Foreign Language: the Example of Scandinavian Psychology. *Review of Applied Linguistics* 41: 220–224.

Crystal, David
1995 *The Cambridge Encyclopedia of the English Language*. Cambridge: Cambridge University Press.

Graddol, David
1997 *The Future of English?* London: British Council.

Kachru, Braj B.
1985 Standards, Codification and Sociolinguistic Realism: The English Language in the Outer Circle. In Quirk, R./Widdowson, Henry G. (eds.) *English in the World*. Cambridge: Cambridge University Press/British Council, 11–30.

Kaplan, Robert B.
1993 The Hegemony of English in Science and Technology. *Journal of Multilingual and Multicultural Development* 14: 151–72.

LeBlanc, T./Painchaud, G.
1985 Self-Assessment as a Second Language Placement Instrument. *TESOL Quarterly* 19.

Marton, Ferenc
1981 Phenomenography: Describing Conceptions of the World Around Us. *Instructional Science* 10: 177–200.

Medgyes, Péter
1993 The National L2 Curriculum in Hungary. *Annual Review of Applied Linguistics* 13: 24–36.
1999 *The Non-Native Teacher*. Ismaning: Hueber Verlag.

–/Kaplan, Robert B.
1992 Discourse in a Foreign Language: The Example of Hungarian Scholars. *International Journal of the Sociology of Language* 98: 67–100.

–/Miklósy, Katalin
In press The language situation in Hungary. Special issue. *The Journal of Multilingual and Multicultural Development* 20.

Ross, Steven
1998 Self-Assessment in Second Language Testing: A Meta-Analysis and Analysis of Experiential Factors. *Language Testing* 15: 1–20.
Shameem, Nikhat
1998 Validating Self-Reported Language Proficiency by Testing Performance in an Immigrant Community: The Wellington Indo-Fijians. *Language Testing* 15: 86–108.
Terestyéni, Tamás
1981 The Knowledge of Foreign Language in Hungary. *Acta Linguistica Academiae Scientiarum* 31: 229–331.
1995 Helyzetkép az idegennyelv-tudásról [State of the art in foreign language knowledge]. *Jel-kép* 1: 47–60.
Wall, Dianne/Clapham, Caroline/Alderson, Charles J.
1994 Evaluating a Placement Test. *Language Testing* 11: 321–343.

Swedish, English, French or German – the Language Situation at Swedish Universities

Britt-Louise Gunnarsson

1. Introduction

1.1. Universities and language choice

For a small country like Sweden with barely 9 million inhabitants to isolate itself academically from the rest of the world is of course impossible. Swedish scholars require access to the knowledge produced in the rest of the world and need to ensure that their own discoveries and insights are made available outside Sweden. Swedish cannot provide an adequate means of communication if the country is to become part of the international academic community, which means that Swedish academics must be able to communicate in speech and writing in foreign languages. Students at Swedish universities have to be trained to use these languages, and researchers encouraged to communicate significant research findings internationally. Most Swedish university teachers would probably agree with these generalised statements, but that does not prevent them from having different opinions about *how far they should go* in adapting to the international, and in fact mainly the Anglo-American, scholarly community.

The prevailing opinion can be illustrated with the help of a language choice triangle as in diagram 1. At one corner of the triangle we find those whose ideal would be a totally English-speaking academic community in Sweden; preparing students from the very start for the major international market in knowledge and accustoming them to understanding and using English. At the other corner can be found those who aspire to a polyglot European scholarly environment. Those who advocate this standpoint claim that we should accept the consequences of our European involvement and focus on the major European languages – Russian, German, French, Spanish and Italian – not only on English. Europe is considered to be an important market for our students and the preservation of its linguistic diversity seems to accord with the concept of the European Union. At the third corner there are those who claim that it is important to retain Swedish for use in Sweden with English and the other languages being used primarily by an elite group likely to have international impact. According to this group Sweden is where the majority of Swedish students will be working and they should be trained to contribute to the Swedish community, which means that they should also be able to use their own language to communicate with the rest of the population.

Attitudes to the extent of internationalisation and how far it should go will obviously vary from discipline to discipline and depend on which research issues departments choose to focus on. It is, however, important to realise that attitudes also reflect underlying principles about the position of the university and of higher education, and that the choice of language will help to define this position – on the international, European and Swedish scene. The choice of language, irrespective of whether this is conscious or unconscious, will undeniably have an impact on the position eventually adopted.

Dangers lurk at all three corners of the language choice triangle, be it Anglo-American dominated international, polyglot European or monolingual Swedish. If all teaching takes place in English, everything read and written in English, students will end up lacking academic language in Swedish. Having learnt their terminology in English, they will lack Swedish terms and expressions to describe what they are doing. It will be difficult for them to discuss their research with their contemporaries and therefore dissemination of their research findings will have to be entrusted to others. The choice of language helps to isolate students and

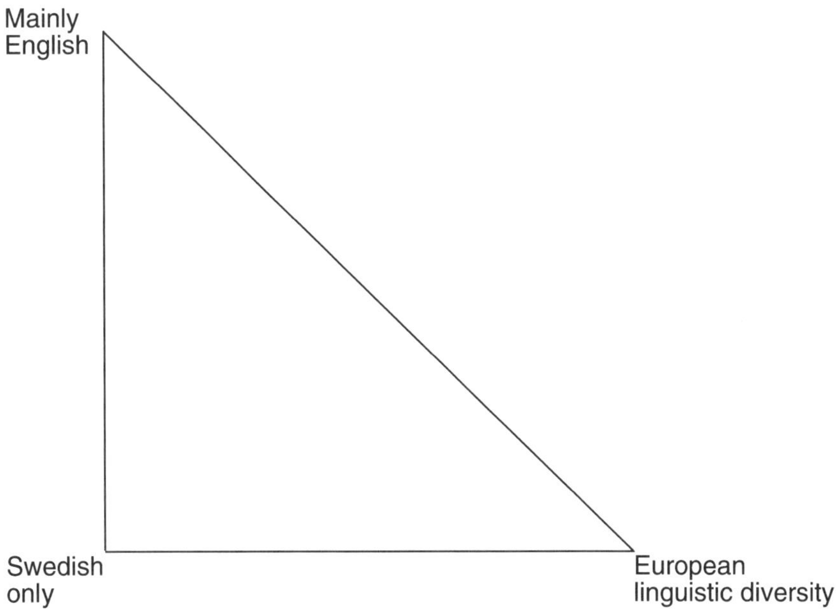

Figure 1: Language choice triangle

researchers from the community they belong to. Democratic objections can also be made to this policy. Through whom and in what way should research best benefit society, how can research be useful, and how can the man in the street, politicians and others gain insight into ongoing research activities?

The dangers can be described from another point of view as well. If the Swedish higher education adopts English completely, in the long run this will naturally have serious consequences for the academic and professional registers of Swedish. If a language is to survive, it has to be used and develop by adapting to new circumstances, otherwise it dies out. If English – or any other foreign language – predominates in the research community, we will end up in a situation that is described in socio-linguistic terms as *diglossia* (in Fishman 1970, for instance). In higher education we would be faced with functional diglossia if English became the language used for academic purposes while Swedish remained the language of ordinary, more informal conversation and for popular science and professional journals.

If, on the other hand, the universities were to focus on the polyglot European alternative, unless there were a noticeable increase in the efforts devoted to language learning by both students and those who allocate resources, we would end up with a highly differentiated academic community, in both linguistic and cultural terms. If, like today, the majority of students were able to master no more than one foreign language well enough to be able to use it as a means of scholarly communication, early in their programmes they would have to concentrate on acquiring this specific language, with some choosing French, others German, Russian, Italian or English. Early in their studies they would be forced to choose the cultural background and the market they intended to devote themselves to. The price of diversity would in this case be the fragmentation of the Swedish university environment. A student with normal linguistic abilities who had focused on French would not be particularly good at German, nor would what they wrote in French be understood by those who had chosen German. However, the European alternative would probably not imply the same risk for academic Swedish. In a polyglot Swedish university with no preference for any one language, Swedish would probably be used as the lingua franca shared by the scholarly community.

Finally, if Swedish were the predominant language for students, teachers and the majority of researchers, they would be screened off from the rest of the research community. Inability to talk or write about their work in any other language than Swedish would restrict them to Swedish settings and contacts. Nor would they understand – except with difficulty – the international discourse of their discipline and its terminology so they would choose material written in Swedish. There is a risk that they would be isolated in their own corner – academically speaking – far from the front line of international research and also from elite groups in Sweden as well. Researchers in Sweden would be divided up into different groups, some facing inwards and others with international contacts, and internal parity would present problems.

It goes without saying that this extreme Swedish stance does not involve the death of academic language in Swedish but on the other hand it could present difficulties for those who seek a wider hearing. They would have to forsake a totally Swedish research environment with its own theories and, to some extent, its own conceptual system for an international research community where other theories and concepts prevail. Often this can involve difficult and sometimes insurmountable problems of comprehension and translation. The difficulty is the transition from one scholarly discourse to another, for instance from Swedish to English.

But this question can also be seen from another point of view. In my opinion it is important to view the choice of language issue in the context of educational issues in general. If we expect students to be able to read text books in English without any problems, or take part in seminars in English, or write papers in English, we must also devote time and resources to further training. The skills they bring with them from the upper-secondary school system will take them some of the way but do not provide them with mastery of the Anglo-American scholarly discourse. And this statement applies to an even greater extent to the other foreign languages. The standard of English in Sweden is high, but not as high as people would like to believe – as has often been pointed out. And when it comes to the presentation of articles and theses, there is often little insight into the differences between the rhetorical devices and textual features of Swedish and those used in English. In order to illustrate this point I would like to describe a study of "abstracts" that I carried out together with a colleague. We analysed a number of abstracts that had been accepted and published in the programme of an international conference in linguistics. The abstracts had all been written in English by scholars from different countries: the USA, Japan, Brazil, Germany, Russia, Ukraine and Sweden. We compared their textual organisation, in other words what was included in the abstract, together with the way in which each individual's research was described. To our surprise we found that the Swedish abstracts were very different from the American ones – in this respect we had expected greater agreement – and also from those from other countries. One could in fact describe it as a Swedish rhetorical style, perceptible even when Swedish researchers were writing in English, which involved diffidence (no theoretical framework) and deprecation of the significance of the research ("my study may be of some interest ...") to such an extent that it probably lacks impact on the international scene (Andersson and Gunnarsson 1995). In this study, researchers from the USA, and also those from Japan, Brazil and Germany, were found to display considerably greater professional awareness and were more distinct and assertive.

The choice of language is linked with thinking. In acquiring a language, one also acquires a cast of thought. Academic thinking is at least to some extent culturally determined. For instance images and metaphors differ from culture to culture and we can undoubtedly find many traces of the Anglo-American metaphorical world in the way we view our disciplines. Different language cultures may also differ with regard to argumentation and rhetorical structure. For instance, fairly substantial differences have been determined between German and Anglo-American textual organisation (Clyne 1987).

The question of choice of language is also intimately linked to the allocation of resources within the universities. The adoption of an international standpoint, i. e. an increase in the role played by English or the other foreign languages in the academic community, should in other words involve particular emphasis on language teaching from undergraduate level and onwards. It is not only the case that a firm command of a language is needed in order to be able to write and speak English, German, French or Spanish, it is also necessary if they are going to be read and understood. If students fail properly to understand what they read, there is a great risk that the knowledge they develop will merely be superficial and fragmentary. During the 60's the term *halvspråkig* (*semilinguistic*) was used within Scandinavian socio-linguistics to refer to the situation in which someone with a specific first language, such as Finnish, grew up in an environment where they were educated in another, like Swedish. Within the individual, the languages were not only used for different functions but also fragmented the acquisition of knowledge to the detriment of the creation of a well-founded conceptual system. Such individuals ran the risk of becoming *halvspråkiga* (*semilingual*) (Hansegård 1968). It is not my intention here to endorse the concept of semilingual, which gave rise to a great deal of discussion, but merely to indicate the learning problems that are linked to the acquisition of dependable knowledge via a language that is not mastered totally. It is not good enough to understand one term here and another there. In order to develop knowledge, students must understand texts in their entirety so well that they can integrate them with what they have assimilated previously. Otherwise there is a risk that their reading will lead to dangerous *semicomprehension*, a state that may inhibit continued learning and their own creative thinking. The decision to internationalise teaching and research should therefore be accompanied by investment in adequate teaching of foreign languages. Obviously such teaching steals resources and time from other courses, but if there is a desire to internationalise the universities while retaining high quality, this is the consequence that must be faced.

1.2. Other periods, other countries

In this context, it can be valuable to glance back a century or two to see what the situation used to be like in Sweden and also to take a look at what goes on in the rest of the world

We often think that our own situation is unique, particularly from a historical point of view. We are sometimes told that never before has the internationalisation of research been such a central issue. And it is hardly possible to compare the degree of internationalisation in different periods. We can, however, determine that the academic community has been international during other periods as well. What has changed is mainly the direction of international contacts and also the quantity of exchange. The expansion of higher education has meant that there are now more researchers who can travel, and this increases the extent of internationalisation. And the easier world-wide travel becomes, the more often scholars participate in international events, which can then be held more often.

Up until the middle of the 18th century, Latin served as the lingua franca of the whole of Europe. If a young Swede sought education, he would most probably apply to one of the famous universities of Europe. He could travel to the university in Prague, Leyden, Amsterdam or Paris. If he wanted to publish a learned work, the language he chose was of course Latin. Eventually the Swedish universities were seen as alternatives but they too used Latin as the language of learning. It was not until the middle of the 18th century, during Sweden's Period of Liberty, that Swedish began to emerge as an academic language. At the same time there was increased emphasis on the sciences; useful research that would help to remedy the deplorable national finances. Students enrolled at Swedish universities in larger numbers than ever before. The ideology of the period also embraced support for the national language and gradually Swedish became the language used for academic purposes, for lectures and for doctoral theses. This transition did not pass without discussion, however. When Sweden's Royal Academy of Science was founded in 1739, there was no agreement about the language in which its quarterly Annals should be published. Those who advocated Latin included Carl von Linné, who wanted Swedish findings to be expressed in a language that could be read outside Sweden lest the nation's scientists become isolated from the world around them. This is one battle he did not win, however, and the Annals were published in Swedish, a development of great significance for the emergence of a Swedish scientific register (Gunnarsson 1988; Gunnarsson and Skolander 1991; Gunnarsson 1997). Scientific journals, books and pamphlets were published in Swedish. While Swedish scholars were obviously able to read and speak foreign languages, the bulk of scientific texts were composed in Swedish. For a long period of time the firmest links were with Germany, as examination of lists of prescribed texts from the beginning of this century demonstrates. In medicine, for instance, there were many works in German, with of course works in Swedish as well.

After the Second World War, however, as we know contacts shifted direction, and we gradually find that more and more of the prescribed texts are in English. What is more interesting, however, is the evolution of English where our scholarly journals are concerned – in 1964 the widely respected Swedish journal *Ekonomisk Tidskrift* started to use English and changed its name to *Swedish Journal of Economics* and subsequently *Scandinavian Journal of Economics*. In other words, Swedish economists choose to publish in a language non-Scandinavians could read, and they were of course not alone in this. It can easily be seen that Swedish doctoral theses are largely written in English, and we know that within many disciplines publication in the international journals is required if articles are to be evaluated when researchers apply for posts. In oral contexts we can also see that English is the language opted for in Swedish settings, one sole non-Swedish speaker can make a large group of people switch to English. Choice of language is not a majority decision, but the rule seems to be that English takes over as soon as any one individual is unable to understand Swedish.

We know therefore that English is pervading the Swedish academic community and in many cases replacing Swedish. The question is of course to what extent this a general problem for higher education. Is the same evolution taking place all over the world or is this an example of the Swedish tendency to yield to foreign influences?

To begin with it has to be pointed out that the Swedish situation is by no means unique internationally. There are many who fear that English will not only become the language spoken abroad but also one that will gradually eliminate other languages in their own settings (Phillipson 1992; Pennycook 1994). Many others have focused on the language situation in Europe (Ammon et al. 1994; World Englishes 1997; Adamzik 1999). In France and Germany more active resistance to linguistic encroachment by English is now being mounted, even though these languages have considerably more native speakers than Swedish. Academics all over the world are concerned about the problem of choice of language, English or the national language, and anxious about the consequences. This does not make it less important to study more closely the language situation in higher education in one specific country. It will certainly add to what we know about the situation in general in the small internationalised nations of Europe and the rest of the world.

We know that English is pervading the Swedish university world, and that in many fields Swedish is being dislodged, but we do not know how far this development has gone nor where it begins – at undergraduate level or in doctoral programmes. And it is here this survey is relevant. Its aim has been to determine the use of English and other foreign languages at a Swedish university. Using interviews and a questionnaire sent to every department at the university of Uppsala, we have tried to form a picture of the language situation in one university. We hope, however, that by describing current circumstances at one university in one country we can initiate discussion not only about the current situation, but also about the future of scholarship in the small nations.

2. Survey of the language situation

In order to determine the extent of internationalisation in Swedish universities today, we devised a questionnaire with 22 questions about the language situation and sent it to every department at Uppsala University. The administrators of each department were asked about the use of English and other foreign languages in the prescribed texts in undergraduate and doctoral programmes, and the language used in lectures, seminars and complete courses. We asked how many of the department's teachers and students did not have Swedish as their first language and whether the department's students participated in international exchanges. We were also interested in the languages in which text – essays, articles, reports and theses – was produced at the department. Finally we attempted to get some idea of the attitude to the use of languages other than Swedish and feelings about the requirements linked to the choice of language. Were people happy about the current language situation or did they want to change it? And if so, did they want more or less use of foreign languages? (All of the questions can be found in the Appendix.)

The number of responses was very high (83%), and these were distributed evenly across all the faculties and disciplines. Our findings derive, therefore, from theology, law, medicine, pharmaceutics, humanities, social sciences, and faculty of science and engineering. Education was also included in the study even though it lacked research programmes at the time – the questionnaire concerned the academic year 1993–94. In presenting the responses from the Uppsala faculty of humanities, I shall distinguish between the language departments which have a special relationship to foreign languages, referred to below as languages, and the other disciplines, referred to as humanities. (For a more detailed presentation of the study see Gunnarsson and Öhman 1997.)

2.1. Oral communication

Some of the questions in the questionnaire dealt with teaching, seminars and guest lectures – with regard to English and the other foreign languages in both undergraduate and doctoral programmes.

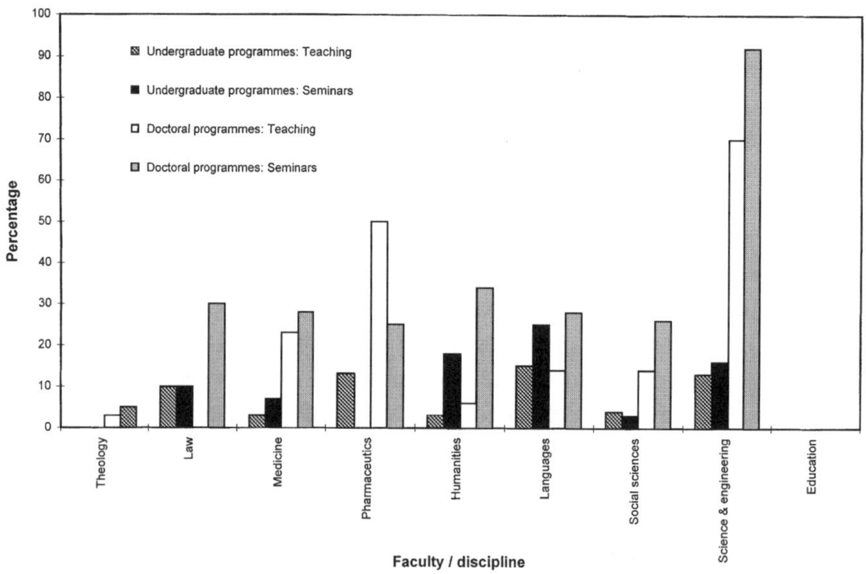

Figure 2: Teaching plus seminars and guest lectures in English in undergraduate and doctoral programmes

As diagram 2 indicates, on the whole English is used considerably less in undergraduate than in doctoral programmes, which is not of course surprising. In the faculties of science and engineering, pharmaceutics, law, and both the humanities and the language departments, between 10–25% of undergraduate teaching and seminars is said to take place in English.

A different picture emerges, however, if we consider English in the doctoral programmes. As far as teaching is concerned, the science and engineering faculty claims that on average 70% of its teaching is in English, in pharmaceutics the figure is 50% and in medicine 23%. The corresponding proportion in social sciences and languages is 14%.

For seminars and guest lectures the science and engineering faculty is in a class of its own. Here, nearly all of the seminars and guest lectures, 92%, are in English. The amount of English at the other faculties is considerably less, although not insignificant. In law, medicine, pharmaceutics, the humanities and social sciences between 25–35% of the seminars and guest lectures are in English. Theology is the odd man out, however, with only a minor proportion of its doctoral seminars, 5%, conducted in English.

Finally, where the other foreign languages are concerned, the questionnaire clearly shows what an insignificant role they play in oral communication. It is basically only the language departments that use other foreign languages for teaching, seminars and guest lectures.

2.2. *Written communication*

In the study we were also interested in the language situation in written communication and included questions about the language of the prescribed texts used in undergraduate and doctoral programmes and of the written work produced – essays, theses, articles and conference papers.

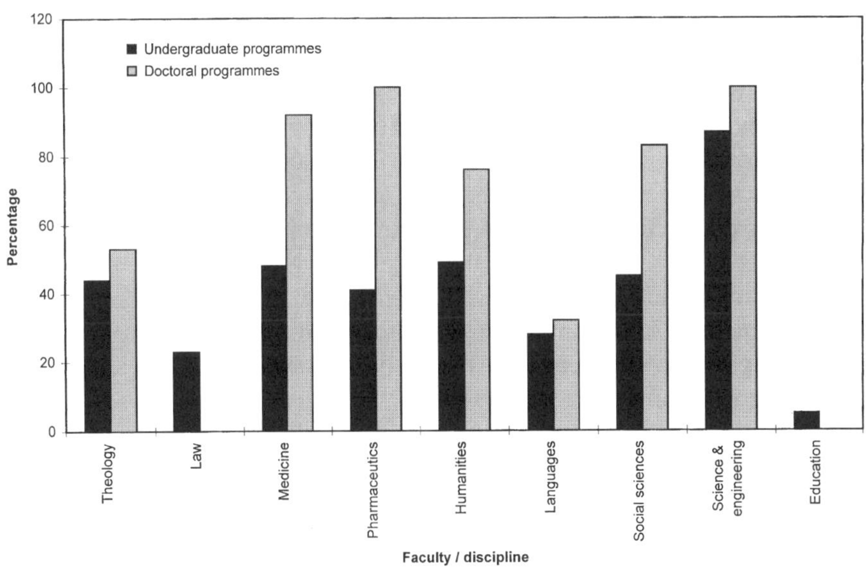

Figure 3: Text books in English in undergraduate and doctoral programmes

Diagram 3 shows what proportion of the prescribed texts are in English at each of the faculties. As the diagram reveals, the science and engineering faculty stands out by virtue of the large element of English in its undergraduate programmes, with as much as 87% of its prescribed texts in English. In theology, medicine, pharmaceutics, the humanities and the social sciences a relatively large proportion of the texts, between 41–49%, are in English. The corresponding proportion in law and the language departments is lower, however, at between 20–30%. It could be considered remarkable that in teacher training almost none of the texts are in English – only 5%.

In doctoral programmes the proportion of prescribed texts in English is even higher. The science and engineering faculty and pharmaceutics claim that all the texts used in doctoral programmes are in English, and nearly all in the faculty of medicine, 92%. In the social sciences and the humanities the majority, or between 75–85%, are in English. It is perhaps hardly surprising that in the language departments not such a large proportion of the texts used are in English, 32%, in view of the fact that here studies deal with various foreign languages and not just English. For theology the proportion of texts in English is comparatively small, with only half in English. If we take other foreign languages into account, however, we find that as much as 26% of the prescribed texts used in doctoral programmes in theology are in one of these languages.

We were also interested in the departments' own written production, and to form some opinion of how much was in English we asked what language was used for third and fourth term papers in undergraduate programmes and also the language of various publications connected with doctoral programmes and research – theses at different levels, articles and conference papers. Diagram 4 contains a graphic representation of written production at the different faculties.

The first thing to note is that the papers produced within undergraduate programmes are largely in Swedish, although in medicine, science and engineering, and in the language departments about half (from 47–57%) are in English. In the humanities as well, a noticeable and not insignificant proportion are in English (14%).

When it comes to text production that is linked to research and doctoral programmes, English predominates in several faculties. This is true, of course, of the science and engineering faculty, where all the theses and conference papers and nearly all articles (96%) are written in English. In pharmaceutics, too, English is the language used in writing about research: all the theses and articles are in English. The same applies to

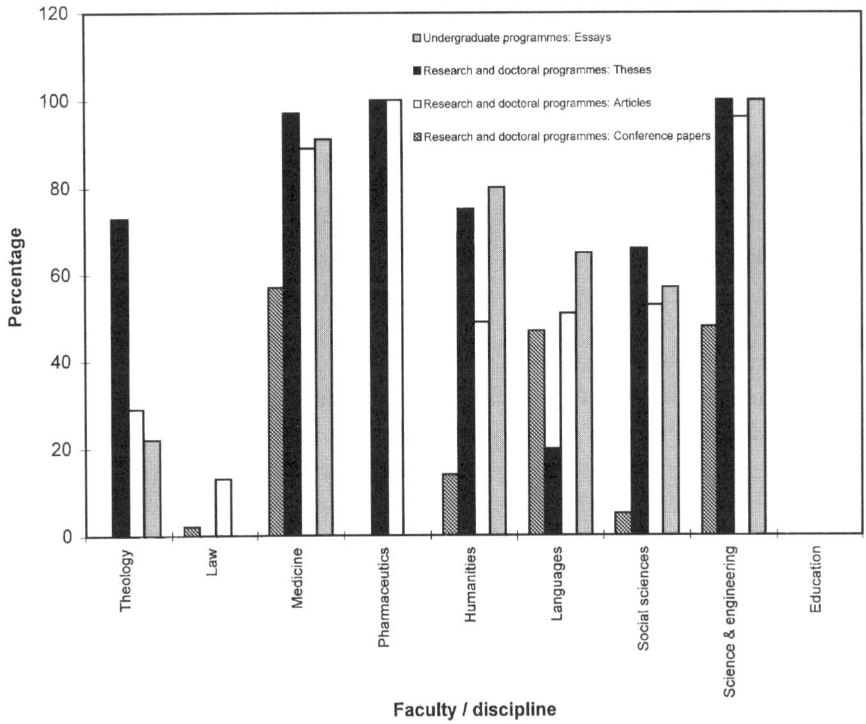

Figure 4: Written production in English in undergraduate programmes, research and doctoral programmes

medicine where nearly all the theses (97%), conference papers (91%) and articles (89%) are in English.

English is also the dominant language within research and doctoral programmes in the social sciences and in the humanities. In theology most theses are in English, 73%. However, where articles and conference papers are concerned Swedish and other foreign languages were equally as common. In the language departments the picture was more varied as well, with both Swedish and the other foreign languages being used.

If science and engineering, medicine and pharmaceutics are characterised by the large amount of texts produced in English, law represents a contrast. No theses in law and only 13% of articles during the period concerned were written in English.

As far as the departments' own production of text in the other foreign languages is concerned, it was only at the faculty of theology and the language departments that other foreign languages were used to any considerable extent. It should, however, be pointed out that articles and conference papers in other foreign languages than English were also found in social sciences and in the humanities.

2.3. Attitudes to language choice

Our questionnaire was intended not only to determine how English and the other foreign languages were actually used but also attitudes to their use as well. Were departments pleased with the language situation or did they want to change it, and if so did they want more or less English and other foreign languages?

In order to find out the attitude in the departments to the relationship between Swedish and English in teaching we asked them to tick whether they considered English should be used "much more", "more", "as it is now", "less" or "much less" in both undergraduate and doctoral programmes. On the basis of the responses we then calculated not only the proportion for each alternative from each faculty or discipline but also the mean with regard to attitudes to English as a teaching language. In doing so we placed each alternative on a scale of 5–1: much more = 5, more = 4, as it is now = 3, less = 2 and much less = 1.

Diagram 5 presents the responses to our questions about attitudes to the use of English and other foreign languages in undergraduate and doctoral programmes and in research. As can be seen from the diagram, it is the faculty of theology that is least content with the role played by English in both teaching and research. There much more English is desired, particularly in doctoral programmes. It is important to remember that this faculty stands out in a number of respects because of the remarkably small proportion of English prescribed texts and the use of English in teaching and written production. On the other hand, the theologians are happy with the use of the other foreign languages, and we have already shown that it is only here, together with the language departments, that the other foreign languages play any significant role.

The law faculty also claims that it would like "more" English in its teaching, at both undergraduate and doctoral level, and as has already been observed, the law faculty is characterised by its comparatively low usage of English in both programmes. When it comes to the languages used in research, the lawyers are satisfied with the role played by English but would on the other hand like increased use of the other foreign languages.

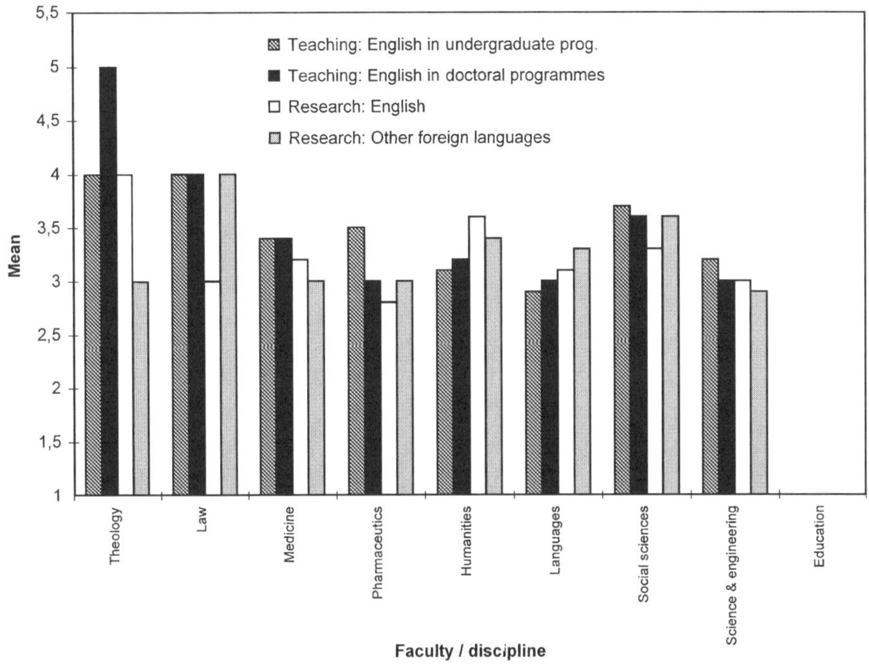

Figure 5: Attitudes to use of English and other foreign languages

Other faculties are more satisfied with the current situation, but at several there is a tendency to want "more" English. In the social sciences and medicine "more" English is desired primarily in teaching and in research as well, while the humanities mainly want "more" English in their research. In pharmaceutics "more" English is desired in undergraduate programmes but the situation is fine "as it is now" in doctoral programmes. Where research is concerned, it is even considered that the use of English could be reduced.

Finally, the departments in the science and engineering faculty are content with the current use of English in teaching and research, as are the language departments. It has, after all, already been made clear that this faculty is the one that uses by far the most English at different levels and for different purposes. Indeed, no efforts seem to be made to change the balance between the languages.

In considering the attitudes to the use of the other foreign languages, it is interesting to note that more of these languages are desired in law, the social sciences and in the humanities (including the language departments). We have already observed several times that these languages play a rather subordinate role as means of communication within higher education. Judging from the responses, there does, however, seem to be some desire to change this state of affairs.

3. Discussion

3.1. The current language situation

As the analysis of the responses to the questionnaire has shown, there are major differences between the faculties both in the way language is actually used and attitudes. There is also a clear difference between undergraduate programmes on the one hand and doctoral programmes and research on the other. In an attempt to summarise the current situation with regard to the use of different languages in either undergraduate programmes or in doctoral programmes and research, the different faculties and sections have been placed on the language choice triangle described above in which "Mainly English" constitutes one corner, "European linguistic diversity" the second, and "Swedish only" the third.

Diagram 6 represents undergraduate programmes. Arrows are used to indicate the direction of any changes the faculties seem to want to make – judging from the responses to the questions on attitude. It goes without saying that this representation of the language situation at the university is highly simplified and sketchy. Positions have been selected on the sides of the triangle on the basis of the average proportion for (1) teaching+seminars, (2) prescribed texts, (3) texts produced within the faculty, i. e. third and fourth term papers. The higher up on the axis "Swedish only – Mainly English" the closer we approach a situation in which only English is used. And correspondingly the further to the right on the axis "Swedish only – European linguistic diversity" the larger the proportion of communication taking place in different languages. Positions close to the 0 indicate that the predominant language is Swedish.

Diagram 6 shows that where undergraduate programmes are concerned, the science and engineering faculty can be found in the middle of the "Swedish only – Mainly English" axis. Medicine is a little lower. The language departments can be found in third place but they are also moving towards "European linguistic diversity". Theology, which is closer to "Swedish only", is also moving in the same direction. We find teacher training programmes closest to the left-hand corner, "Swedish only".

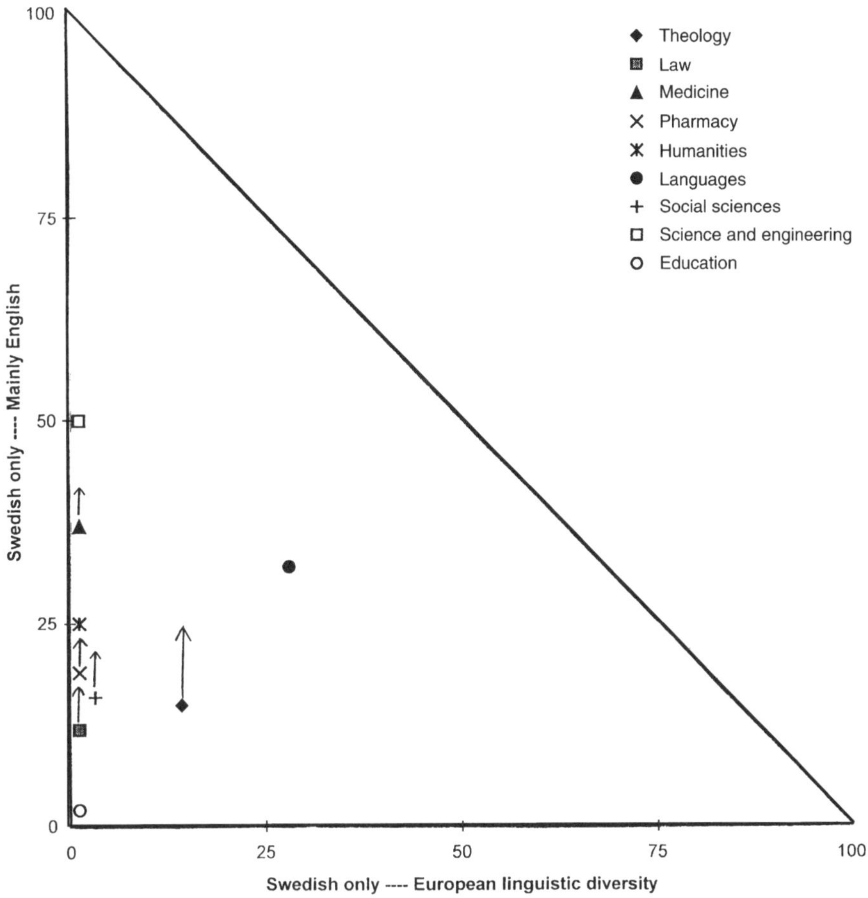

Figure 6: Language choice situation in undergraduate programmes

If we then examine the tendency of changes, we find that several faculties want to increase the proportion of English – medicine, social sciences and pharmaceutics to some extent, theology and law more. The development we can therefore expect where choice of language is concerned is a movement towards more English in undergraduate programmes.

Diagram 7 represents doctoral programmes and research. Here too the position of the faculties on the axes is based on responses concerning the use of language for (1) seminars, lectures and teaching, (2) prescribed texts, (3) written production within the departments such as theses, arti-

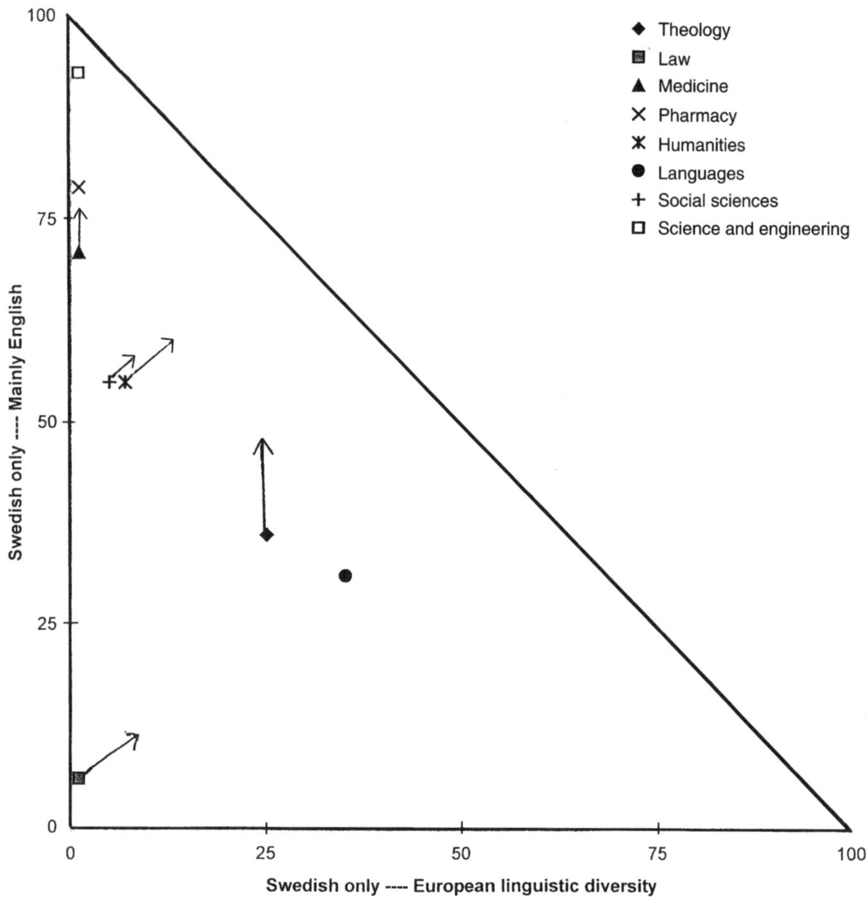

Figure 7: Language choice situation in doctoral programmes and research

cles and conference papers. Arrows indicate the change of direction in use of language that the faculties seem to be trying to achieve.

Where doctoral programmes are concerned, diagram 7 reveals that most of the faculties end up on the upper half of the "Swedish only – Mainly English" axis. At the top we find the science and engineering faculty, followed by pharmaceutics and medicine. The social sciences and the humanities are also found in the upper half, in other words the language used in doctoral programmes for teaching, seminars, prescribed texts and written production is somewhat more often English than Swedish. In the faculty of theology and the language departments the proportion of English is smaller, but in return the proportion of other languages noticeable. Theology and the language departments tend therefore to some extent to be moving towards "European linguistic diversity". Finally, the law faculty can be found relatively close to the left-hand corner, "Swedish only".

We also find that several faculties would like to increase the proportion of English in doctoral programmes. This is true above all of theology, but also applies to law, the social sciences, medicine and the humanities. It is also interesting to note that in three cases an increase in the other languages is desired as well – primarily in law, but also in the social sciences and the humanities. We can therefore expect any development where choice of language is concerned to be towards even more English in doctoral programmes and research and also towards greater European linguistic diversity to some extent within the social sciences and humanities.

From this we can therefore conclude that, as a whole, higher education in Sweden has come quite a long way in adapting to the Anglo-Americanised scholarly community. This is obvious most of all in doctoral programmes, and then mainly in the experimental faculties, science and engineering, medicine and pharmaceutics. But the social sciences and the humanities are well on the road towards internationalisation and judging from attitudes in these faculties are prepared to go further. The role played by other foreign languages is, on the whole, marginal, and virtually confined to the language departments and the faculty of theology. Nor can any major desire to alter this state of affairs be discerned, or at least not in the experimental faculties.

On the other hand internationalisation has not had such a major impact on undergraduate programmes. There is, however, a significant proportion of English, mainly in science and engineering but also in medicine and the humanities. Above all, a large proportion of the prescribed texts are in English in many faculties, and here too, the trend is to wanting to include more English in undergraduate programmes as well.

3.2. The consequences of allowing developments to take their course

It is possible, therefore, to establish that English is the dominant language in doctoral programmes and research. This applies principally to the three experimental faculties, science and engineering, medicine and pharmaceutics, but English is also relatively dominant in the social sciences and humanities, where there also exists the desire to increase its importance. Linguistic adaptation – and with it certainly cultural adaptation as well – to the Anglo-American world of scholarship must be described as advanced. There is virtually no use of Swedish as an academic language in much of higher education in Sweden, nor are German, French, Spanish or Russian used in the faculties concerned either. English has become the dominant language for both oral and written scientific communication. This inevitably faces us with the question of what consequences a one-sided choice of language like this may have for the position and role of Swedish research – in Sweden itself, in Europe and in the world at large.

Today it is scarcely an exaggeration to assert that we are exposed to Anglo-American "imperialism" of the kind that has been dreaded in many parts of Europe and the rest of the world (Phillipson 1992; Pennycook 1994). It is perhaps no less important to consider what consequences this may have in terms of knowledge. What in the long run will be the impact on our disciplines when our thinking is so intensively – and so one-sidedly – influenced by American and English scholarship? What will be the effect of being deprived of texts that are neither written in nor translated into English – of many of the texts produced in German or French? What will our relationship be to the European scholarly community, to Germany or France, for instance?

If we turn our attention to Sweden, it is of course the case that higher education is threatened with *diglossia*, the allocation of different functions to different languages discussed at the beginning of this article. I am not claiming that there is a threat of Swedish society as a whole succumbing to English, but there is good reason to believe that in certain sectors, in areas of commerce and higher education, for instance, the use of different languages for different functions is beginning to develop. If a language is not used for certain functions, it can hardly develop, in other words we may end up lacking words and terms to refer to new phenomena and concepts in the sectors where Swedish is no longer used. In the long run this will of course give rise to the impoverishment of Swedish, with the language failing to provide an effective means of academic communication in fields such as engineering, the natural sciences and medicine.

The state of diglossia can also result in the isolation of higher education and research from the rest of society. It is obvious that this poses problems for democracy. There will be little contact between researchers, whose world is, in reality, an international English-speaking one, and the general public with no access to the international journals in which research findings are published and discussed, and which they would anyway be unable to read and understand. This problem is of course linked to whether researchers want and are able to "describe" their research in Swedish. If not, there is a risk that other, less knowledgeable individuals will undertake the popularisation of research results and the dissemination of knowledge. This will probably not be particularly successful. Another democratic problem linked to the choice of language is the community's right to insight into research activities.

One question that must also be asked in relationship to the large amount of text production in English within research and doctoral programmes in Sweden is whether the linguistic and stylistic level of these texts can maintain the same high quality as texts written in a first language. Are Swedish doctoral students and researchers as capable of presenting their intellectual approach, arguing for their opinions, and presenting their findings as they would be if they wrote in their own mother tongue. I contend that there is good reason to believe that this is one aspect of the problem of quality that we should not underrate.

In systematic linguistic comparisons of scientific texts – articles, abstracts, theses, grant applications – written in English by British and American researchers with similar texts written in English by researchers from other countries it has been possible to demonstrate considerable differences not only in linguistic variation and accuracy but also in the organisation, argumentation, presentation of the writer's own contribution, theoretical framework etc. (for an overview, see for instance Connor 1996). Comparisons have also been made of the way in which editors assess articles written by British and American researchers and those written by scholars from other cultures, and this has revealed that the deviations by non-native speakers from the Anglo-American norm are interpreted as lack of scientific rigour, vagueness, lack of stringency etc. (Clyne 1987; Mauranen 1993). When talking of the internationalisation of Swedish research that is taking place, we must not ignore the way in which linguistic competence and language training can affect the impact of presentations of research projects. In many areas, people have become increasingly aware of the vital role played by academic argumentation and rhetoric for the way in which a communication will be understood by its readers, and how cultural norms govern the way we write and the way we read. A researcher's discoveries in the laboratory or research cubicle only becomes an accepted research finding when it has been formulated in text, and not even a Newton, Edison or Einstein would have got very far if they had not also possessed mastery of the correct scientific rhetoric and presentation techniques (Latour and Woolgar 1968; Mc Closkey 1986; Bazerman 1988).

The problems linked to the production of their own text apply, of course, mainly to active researchers and doctoral students, or in other words those who chiefly use English for their own scholarly writing. For undergraduates the problems are different, and here it is reading rather than writing that is the main issue. In my opinion more should have been done to find out how much students in their first and second terms understand of the English works they are forced to read. To the best of my knowledge no systematic study of the level of comprehension has been carried out, but I believe that there is reason to assume that the ability of undergraduate students to benefit from what they read is not infrequently inadequate. In many cases the result is *semi-comprehension*, a kind of understanding that can in the long run give rise to serious deficiencies in the way they approach learning. These students come directly from upper-secondary schools, where they have been trained to read and understand normal English prose. When at university they are suddenly confronted not only with a new type of text, in the specialised books using the new and difficult language of their discipline, but also English terminology and expressions, this is not infrequently too much for them. Both as teacher and parent, many of us have heard students groaning about the difficult English text books. This almost certainly has an impact on quality, and is a problem that has not received the attention it warrants.

3.3 Proposed measures

Internationalisation of higher education in Sweden has therefore come a long way, at some faculties a very long way indeed. Obviously one could maintain, in view of the consequences of internationalisation outlined above, that attempts should be made to restrain developments that are increasingly leading to more and more use of English at universities. The first question is, however, whether this is at all possible, and secondly if it is desirable. Turning the clock back is hardly ever appropriate, and if attempts were made to isolate the country from the world around it would certainly have devastating consequences for the Swedish academic community and society as a whole. In what follows, I will instead assume that internationalisation is given and that we are gradually going to encounter more of it rather than less. The measures proposed below therefore focus on what has to be done to prevent a deterioration in the quality of teaching and research.

The texts prescribed for undergraduate courses

In undergraduate programmes there is a problem related to the students' understanding or rather lack of understanding of the English books they are required to read. As the questionnaire shows, undergraduate reading lists often contain titles in English, so this is not merely a problem for a few faculties. One can also assume that the English works included in these lists are central to the courses and the subjects – otherwise Swedish works would probably have been listed instead – and that it is therefore especially important for students to understand their contents completely.

My arguments are based on my doubt as to whether students always understand what they are reading. I claim that there are good grounds for believing that the learning load – new knowledge, difficult text, foreign language – can be too large for undergraduates to succeed in assimilating what they read. Comprehension requires more than grasping individual terms and the textual context must be understood as well.

My first proposal therefore involves training undergraduate students to read. The development of a reading comprehension programme must of course be based on a systematic study of the comprehension of specialised texts in foreign languages by various categories of student. How much do they understand? At what level does their comprehension function? What causes them problems?

A reading comprehension programme can then be developed on the basis of the results that can be aimed at students in different faculties and disciplines and those with different kinds of difficulties. Admittedly, not all students will need help of this kind, but what could be provided is some form of remedial training for those who do not find it that simple to crack the linguistic code of English. Many students almost certainly need to develop the vocabulary that is specific to their discipline, and also the more general vocabulary of scientific prose. Reading comprehension training should be offered during the first and second terms of undergraduate study and in many cases doctoral students with the necessary linguistic competence could act as instructors.

Writing in doctoral courses and research

As the responses to the questionnaire reveal, a great deal of the text written by doctoral students and researchers is in English. Within large areas of higher education English is the language of theses, and we have seen that this also applies to articles, reports and conference papers. What underlies my proposal here is hesitation about the linguistic and stylistic level of the texts written in English by Swedes. Various comparative analyses of texts have shown clearly that writing in a foreign language is not without its complications. It is not just a question of vocabulary but of the overall organisation of the text, the arrangement of various sections, argumentation etc. In the questionnaire we also asked departments whether it was customary to have texts written in English checked for use of language. What is surprising in this context is that recourse is so rarely made to such professional assistance.

If the quality of the research conducted at a university and then disseminated through various international channels is to maintain high and consistent quality, initially measures must be introduced intended to increase the ability of doctoral students and researchers to write texts of various kinds in English themselves. Secondly, universities should make it possible to have texts checked or translated by professional translators without too much expense or delay.

My second proposal therefore concerns courses in academic writing for doctoral students and researchers. Courses in academic writing in English should be arranged at faculty or departmental level. These should be based on the texts that participants have themselves produced and should obviously aim to improve writing skills, but should also be intended to make writers aware of the differences between various cultures and the textual traditions of different languages.

It is also important for doctoral students and researchers to have access to effective assistance with translation and language checking. Universities should draw up contracts with a number of qualified translators so that employees could be provided with these services at a reasonable cost and without delay. In addition, internal language service units should be set up, staffed by advanced students of the various languages, who could work part-time and offer advice and check or translate the texts produced by other undergraduate and doctoral students.

The role of German and French in the university

One finding of the study that must be described as striking is the marginal role that the other foreign languages play in all the faculties, apart from theology and the language departments, of course. We are far from any kind of balance between the various foreign languages, between English, French, German, Italian, Spanish and Russian. The current state of affairs is quite simply that English is the paramount foreign language. Linguistically speaking, we can talk about total Anglo-American domination. What it is important to bear in mind in this context is that this domination is not only a matter of language as an isolated phenomenon but is also linked to the cultures and the attitudes of mind linked to the language. Higher education in Sweden has reached the situation where undergraduate and doctoral students are exposed to considerably more influence from Britain and America than from France or Germany. They are introduced to an Anglo-American scholarly network and risk in consequence exclusion from the European scholarly community.

Because of the range of languages now offered on an "equal" basis in the Swedish school system, allowing pupils to choose between a number of second and third languages, heterogeneous cohorts of students are now arriving at the universities. Previously university teachers knew what skills they could expect from students who had studied English, German and French before enrolment and could therefore prescribe texts in these different languages. This is no longer the case. Because some students have French as their third language and others Spanish, Italian, German or Russian, the only foreign language they have in common is English. The variety of languages offered in the school system has therefore reinforced the monopolisation of higher education by English. If this domination is to be challenged, a system has to be reintroduced in which certain foreign languages are given priority over others. In my opinion it would be reasonable to focus on German and French.

My third proposal therefore concerns firm action to strengthen the role of German and French in higher education. This would mean that universities would ensure the creation of a range of courses offering practical skills in German and French, primarily intended for those who are not studying foreign languages. It should also be possible for students in the various faculties to take these courses and have them credited towards their degrees. The next stage would of course be to introduce more texts in German and French and to have seminars, lectures and teaching in these languages.

Swedish and the research community

I have saved what I consider to be the most serious problem until last. This is the fact that because of its dominance in higher education today, English constitutes a serious threat to the continued existence of a scientific register in Swedish. I am not painting too pessimistic a picture when I claim that in the natural sciences, engineering, medicine and pharmaceutics, at least, the current situation is not dissimilar from what sociolinguists refer to as *diglossia*, i. e. a division of functions between two languages, with one of them used as an official formal language while the other is an everyday vernacular. In the experimental faculties, and also as we have seen increasingly in the other faculties, English is the language used for formal purposes, i. e. in academic contexts, while Swedish is used as a vernacular, or in other words for popularisation and teaching at lower levels. In many areas in higher education Swedish is definitely under threat: young students are not being taught to use Swedish for professional purposes.

There are of course several reasons for the prevailing division of functions, but one that is often mentioned is the way in which qualifications are evaluated. Doctoral students and researchers have to publish in the international journals if their articles are going to be given any weight. This obviously has impact on their writing in Swedish and fewer and fewer researchers in these fields themselves describe their research in Swedish. They leave it to journalists and other professional popularisers to disseminate their findings to the public. I can see a danger for democracy in this state of affairs.

Here I should like to refer to a study that I carried out a few years ago together with a couple of colleagues. We analysed and compared the textual systems of some articles on medicine published in either *Läkartidningen* in Sweden, *The British Medical Journal* and *Deutsche Medizinische Wochenzeitschrift* in Germany, during 1994. What mainly struck us was the difference between the Swedish article on the one hand and the British and German articles on the other. In simple terms one could say that the Swedish article was considerably more journalistic. It is interesting that the Swedish article not only differed from the German and British articles but also from the article about the same research written by the same authors in English for international publication (Gunnarsson, Bäcklund and Andersson 1995). It is also interesting that the textual features found in the article in *Läkartidningen* in 1994, and in other Swedish articles of the same year, also differed from those that had earlier applied in the same journal. Analysis of textual features of articles in *Läkartidningen* from the first decade of the 20th century and the 1940s and 1980s have been able to show how these successively assumed more and more of the characteristics found in the articles in German and English in 1994 but not in those in Swedish (Gunnarsson 1998). A change had in other words taken place since the 1980s. What we found in 1994 was a new Swedish medical article style, one which according to the editor was now the norm and which he explained as being the result of deliberate adaptation on the part of the journal to prevailing circumstances in medicine. Because of the way in which qualifications were now being evaluated, Swedish researchers chose to publish their original articles in international journals, which had obliged *Läkartidningen* to redefine its role in the medical profession. Now, according to its editor, its prime function was to publish popular presentations and surveys (Gunnarsson, Bäcklund and Andersson 1995).

This is nothing more than an example of the actual adaptation to a prevailing state of diglossia in which English is the scientific language of medicine in Sweden whereas Swedish remains the language used for popular presentations. In this case the two researchers had written both the articles in English and in Swedish. In many other cases, however, others – journalists or other professional popularisers – undertake the Swedish version. The question is, of course, what this kind of presentation by writers not involved in the research will be like. The quality issue does not concern the description of the research but the way its findings are disseminated.

My third proposal therefore involves widespread support for Swedish as a scientific language. This kind of support is, in my opinion, needed if we are to avoid a definite state of diglossia in the universities.

To begin with, the way in which qualifications are evaluated needs reviewing. Publication in Swedish should be ranked as a qualification. It may well be that in certain quarters there exists exaggerated confidence in the value of inclusion in international journals and international series of publications.

Secondly efforts are needed to make it easier for researchers to publish their work in both Swedish and English or some other foreign language. There should be more generous grants available for translating works published originally in Swedish.

Thirdly there should be more emphasis on training in Swedish and Swedish as a professional language at the faculties where Swedish is under threat. Training of this kind could, for instance, take the form of specialised courses in oral and written communication in which students could practise speaking and writing about their fields in Swedish.

4. Conclusion

This survey has provided insight into the language situation at one specific university in one specific country. It is, however, my belief that the description of the current language choice situation and of the different problems that can then ensue has a more general application, at least for the smaller nations in Europe. I also believe that the threat to Swedish reflects a more general threat to the lesser used languages in the world today. The proposed measures that I have outlined in my concluding remarks could probably also be applied in many other areas. I would like to emphasise, however, that they should be viewed as nothing more than proposals. What is important is that we should all begin to consider where the language choices made in our universities will lead and what can be done to prevent deterioration of quality. Active linguistic planning is not only possible but also necessary if we are to continue to be able to feel pride in our students and researchers.

5. References

Ammon, Ulrich/Mattheier, Klaus J./Nelde, Peter H. (eds.)
1994 English Only in Europe? *Sociolinguistica* 8.

Adamzik, Kirsten
1999 Ist die Linguistik eine "anglophon geprägte" Disziplin? Eine Analyse am Beispiel der Fachsprachenforschung. Paper presented at the 12th Symposium on *Language for Special Purposes, LSP '99, Bolzano/Bozen, 30th August – 3rd September 1999.*

Andersson, Bo/Gunnarsson, Britt-Louise
1995 A Contrastive Study of Text Patterns in Conference Abstracts. In Wårvik, B./Tanskanen, S.-K./Hiltunen, R. (eds.) *Organization in Discourse. Proceedings from the Turku Conference* (Anglicana Turkuensia 14). Åbo, 139–148.

Bazerman, Charles
1988 *Shaping Written Knowledge. The Genre and Activity of the Experimental Article in Science.* Madison, WI: University of Wisconsin Press.

Clyne, Michael
1987 Cultural Differences in the Organization of Academic Texts. English and German. *Journal of Pragmatics* 11: 211–247.

Connor, Ulla
1996 *Contrastive Rhetoric. Cross-Cultural Aspects of Second-Language Writing.* Cambridge: Cambridge University Press.

Deneire, Marc G./Goethals, Michael (eds.)
1997 *Special Issue on English in Europe* (World Englishes 16 (1)).

Fishman, Joshua A.
1970 *Sociolinguistics. A Brief Introduction.* Rowley, MA: Newbury House.

Gunnarsson, Britt-Louise
1988 Medicinsk facktext i 1700-talets Sverige. *Ingemar Olsson 25 augusti 1988 (MINS 28.) Inst. för nordiska språk.* Stockholm, 125–139.

1997 On the Sociohistorical Construction of Scientific Discourse. In Gunnarsson, B. L./Linell, P./Nordberg, B. (eds.) *The Construction of Professional Discourse*. London/New York: Longman, 99–126.

1998 Academic Discourse in Changing Context Frames: The Construction and Development of a Genre. In Allori, P. E. (ed.) *Academic Discourse in Europe. Thought Processes and Linguistic Realisations*. Rome: Bulzoni, 19–42.

–/Bäcklund, Ingegerd/Andersson, Bo
1995 Texts in European Writing Communities. In Gunnarsson, B.-L./Bäcklund, I. (eds.) *Writing in Academic Contexts*. Uppsala: Uppsala Universitet, 30–53.

–/Skolander, Björn
1991 *Fackspråkens framväxt: terminologi och ordförråd i facktexter från tre sekler 1. Projektpresentation och materialbeskrivning*. (FUMS rapport nr 154). Uppsala universitet.

–/Öhman, Katharina
1997 *Det internationaliserade universitetet. En studie av bruket av engelska och andra främmande språk vid Uppsala universitet*. (TeFa nr 16). Uppsala: Uppsala universitet.

Hansegård, Nils E.
1968 *Tvåspråkighet eller halvspråkighet?* Stockholm: Bokförlaget Aldus/Bonniers.

Latour, Bruno/Woolgar, Steve
[1986] 1979 *Laboratory Life. The Construction of Scientific Facts*. Princeton, NJ.

Mauranen, A.
1993 *Cultural Differences in Academic Rhetoric. A Textlinguistic Study*. Frankfurt a. M. etc.: Lang.

McCloskey, Donald N.
1986 *The Rhetoric of Economics*. Brighton: Wheatsheaf.

Pennycook, Alastair
1994 *The Cultural Politics of English as an International Language*. London/New York: Longman.

Phillipson, Robert
1992 *Linguistic Imperialism*. Oxford: Oxford University Press.

World Englishes.
1997 Special Issue on English in Europe. Guest editors, Marc G. Deneire and Michael Goethals. Volume 16, Number 1.

6. Appendix: questionnaire

The questionnaire contained the following 22 questions:

1. Number of full time student places.
2. Number of doctoral students.
3. Number of teaching and research posts.
4. Number of students travelling abroad to study.
5. The countries they travelled to.

6. Number of exchange/scholarship students received by the department.
7. The countries they came from.
8. How many of the foreign teachers taught in English or any other foreign language.
9. How many Swedish teachers taught in English.
10. What proportion of teaching took place in English.
11. What proportion of teaching took place in other foreign languages.
12. What proportion of seminars/guest lectures took place in English
13. What proportion of seminars/guest lectures took place in other foreign languages.
14. What proportion of prescribed texts were in English.
15. The number of papers and theses written in English.
16. The number of other kinds of publication written in English.
17. Whether the department used language checkers/translators for papers, theses and publications.
18. Who carried out language checking/translation.
19. The relationship between Swedish and English in teaching.
20. The relationship between Swedish and English with regard to research.
21. The use of other foreign languages with regard to research.
22. The degree of agreement to statements about demands and attitudes to languages used.

IV. Countries with an Own International Language of Science

The Languages of Science in France: Public Debate and Language Policies

Claude Truchot

1. English and French in conflict (1981)

The questions of the use of English in scientific research and of the future of French as a language of science came to be publicly debated in France in the early 1980s. The discussion was opened by two conferences, which took place in 1981, and was developed in the daily press. The first conference, entitled *Le français chassé des sciences*, gathered in Paris people from different fields who were concerned with the future of French in the sciences. It dealt mostly with the situation in France (CIREEL 1981).The second one, *L'avenir du français dans les publications et communications scientifiques et techniques*, organised on a much larger scale, brought together in Montreal the same year top-level scientists and members of governmental institutions from Canada, France and Belgium (Conseil de la langue française, 1981). The press coverage of these events was important. The articles in the papers were fed by a large corpus of facts and figures presented at these conferences. The analysis of the situation revealed that English was used as a language of publication by a large proportion of French-speaking scientific communities, and was the main language used in international conferences held in the three countries. The indicators showed a certain amount of variability according to fields, and to countries. The use of English was more important in Canada (Quebec) and Belgium than in France. It was more important in chemistry and biology than in mathematics or engineering.

Scientists knew this situation though most of them did not have a global approach towards it. However the general public was not aware of it. This explains the importance given to the situation by the media. Besides proper articles, several declarations from scientists and policy deciders were published. Most of the problems raised were exposed and debated. The situation was largely perceived as a form of language conflict, and the discussion largely focussed on what was described as evidence of hegemonism. Thus a focal point was the influence of the 4000 journals which in the world, out of 50 000 scientific publications, published the results of advanced research. It was stressed that more than 80% of them were owned by a handful of powerful publishing houses, mostly American, British and Dutch, and used exclusively English as a language of publication. These journals were able to publish just a small part of the world scientific production, but the articles they published were those which were used everywhere as references. They were indexed in the data banks compiled to store scientific information. Furthermore the largest number of data banks were located in the USA. They were the most widely used, and like the *Science Citation Index (SCI)*, the most influential. Some 90% of the information recorded in the US data banks came from contributions written in English, the 10% left being shared between Russian, German, Japanese and French. Scientists in order to be recognised had to publish in these journals. Articles published elsewhere in other languages had little chance of being recognised. Their situation was summed up by this phrase: "Publish (in English) or perish".

This public debate encouraged governmental authorities to take measures. In France in September 1981, the Minister for Science and Technology issued a document recommending the use of French in French scientific journals and at conferences held in France and francophone countries. Scientific committees were strongly advised to take more into account contributions written in French when assessing researchers. The creation of a fund to help translation and interpretation was announced and at the same time it was stated that governmental financial support would not be given to international conferences from which the use of French would be excluded. The announcement of these measures was followed by a very heated debate, and also by a broad discussion among scientists, which revealed the feeling of uneasiness most of them had towards this question. They found a way out of this uneasiness by trying to see more clearly the fields of language use in which English was indispensable from those in which French could be retained. This appeared clearly in a study conducted in 1984.

2. The 1984 Strasbourg study

This study was carried out in Strasbourg, which is an important pole of research with many international connections, among a large sample of full-time researchers (in C. Truchot 1990; 1996). The study dealt with the following points: language proficiency of scientists; language uses for publication, quest of information, participation at conferences, daily work in research centres; attitudes towards the use of English. The analysis of language proficiency showed that all the members of the sample had learned English but disclosed important variations. Some 55% also declared they had an operative command of German. However this knowledge of a major European language was hardly of any use as only 2% of the information they treated was in this language, against 85% in English and 12,5% in French. 95% of them answered they delivered their paper in English at conferences that took place in foreign countries. In international conferences taking place in France, 76% said they spoke in English and 24% in French. On average about 70% of their scientific production was in English and 30% in French, though it varied considerably according to disciplines and to individuals. In fact the higher they ranked in the scientific profession, the more English they used.

But this study also revealed that most researchers felt it important to maintain the use of French in some areas. The first area is that of communication inside the places where science is produced, laboratories and research centres. Whereas in many countries English is used when researchers from several language origins work in a laboratory, this practice was not found in the sample except in one laboratory. Researchers coming for more than short-term visits were reported to adopt the use of French. The social environment in and outside the laboratory encouraged them to do so. The second area is that of postgraduate studies . They considered it was important that doctoral research, as an activity taking place in France, should be in French, and theses written in this language.

The study of attitudes revealed that 70% considered that the use of English was a necessity. But 52% considered it was a handicap relative to English-speaking scientists and to scientists who had a better knowledge of English than they had. For 8% it was even a form of alienation. On the other hand 16% found it was an advantage, mostly people who felt they were proficient in English. A large proportion were in favour of policies to support the use of French in the sciences, but only 20% were convinced that the measures decided by the Minister for Science and Technology, as listed above, could change the situation.

3. An attempt to plan language pluralism in the sciences (1990)

After the early eighties, the debate came back to the public scene just on occasions. This was the case in 1989 when the *Institut Pasteur* decided to give up the French title of the renowned review *Les Annales,* and gave it a title and content in English. This caused uproar in France and francophone countries, particularly in Quebec where the press was very critical. The *Institut Pasteur* argued that this decision had to be taken because the results published in French were not reviewed and indexed by data-banks, and because the French title did not attract contributors and readers. This event and some others revealed that, in spite of the measures taken some years before, the situation had not changed. This prompted decision makers to make a new assessment and to have a debate around it.

The decision was taken by the Ministry for Francophonie and the Ministry for Science and Technology. A leading role to organise a debate was given to *Le Monde diplomatique*, a political monthly paper well read in influential circles and able to bring out a broader approach to the problems. The title given to the Forum organised in 1989 and to the book that followed, *Quelles langues pour les sciences?* (Cassen 1990), expresses a position based on linguistic pluralism, much different from the one prevalent in the early eighties based on a conflict between English and French. Starting from a thorough analysis of the situation, measures were discussed with the aim of finding a balance between English, French and other languages. As the Minister for Science and Technology, Hubert Curien, put it, a balance had to be found between two objectives: making good science and keeping it close to citizens. In this perspective, resorting to English at conferences is not seen as the main problem, nor publishing in English. But promoting high standard journals written in French and recognised by the international scientific community can be considered as a desirable objective. So is finding the means to help foreign researchers coming to France to acquire competence in French. As 75% of foreign researchers in France are doctoral students, it means helping them to be sufficiently proficient to write their thesis in French. To keep science close to citizens the use of French should be related to its popularisation outside the restricted field of scientists, towards education, industry and technological development, and scientific information.

4. The language of the Republic and the sciences (1994)

A more radical attitude was adopted a few years later. In 1992 when Parliament voted a Constitutional Law to ratify the Maastricht Treaty transforming the European Economic Community into the European Union, a language amendment was added. It modified the Constitution by stating explicitly that "The language of the Republic is French (La langue de la République est le français)". The debate around that amendment was limited, as there was consent between Right and Left (then in power) on the relevance of the decision. Only the circles defending regional and minority languages pointed out the problems raised in front of the efforts made to have these languages recognised. Their reluctance proved right a few years later when the French government, which wanted to sign and ratify the European Charter for regional and minority languages, realised that the Constitution forbade them to do so.

This article in the Constitution provided a basis for the *Law on the use of the French Language* which was adopted by Parliament in 1994. Its purpose was to ensure the use of French in France in domains where it tended to be in competition with other languages, in fact almost only with English. It includes status planning and corpus planning measures. The public debate on the law was highly controversial but concentrated nearly exclusively on terminology planning: words coined to name in French new objects and new techniques originally named in another language. For most standardised languages the decision to adapt or not such words is usually taken by language users. But in France it was decided in the early seventies that government institutions should also contribute. And the 1994 law planned to make obligatory the use of the words they coined. This obligation, which was adopted with the law by a majority in Parliament, was soon after suppressed by the Constitutional Council as infringing on public and individual liberties. It was maintained for state institutions (education, administration). As a result of this controversy, status planning measures went unnoticed. This was the case of two series of provisions having bearings on the sciences.

The first one deals with the use of French at conferences and in scientific journals. It provides some room for French, by making it a right for a contribution to be read in French at a scientific conference taking place in France, by making it an obligation to have a French version for the documents of presentation, and for the other documents to include an abstract in French. Scientific journals should also publish abstracts in French as soon as they are received public funds. Compared with the large use made of English in international conferences, these measures are rather symbolic. A more compelling measure was included in the version of the law adopted by Parliament, stating that public funds could be given to scientific research only if the results were published in French.

But this measure was also declared anticonstitutional and therefore suppressed.

The second one is less symbolic. It declares French to be the language of education, particularly of examinations and theses (to the exception of the teaching of regional and foreign languages). In fact up to now its main impact has been on the writing of theses. The Ministry of Education insists on the fact that in France vivas are public and should therefore take place in the language of the public sphere. Most universities have inserted this obligation within their own regulations. Though there has never been any assessment, there seems to be consent on its application. In fact most professors, when they direct research, prefer using a language they can master, even if it is at the expense of some time spent in correcting the texts of their foreign doctoral students. Special provisions are made for doctoral theses that have a co-director from another country. Co-directions are favoured by academic authorities, and valued by being given a European or an international label. In those cases (except when everyone speaks French) the student and his/her directors choose the main language of the thesis, but a substantial abstract should be given in another language.

In fact writing a thesis in French raises problems for the students, whatever their mother tongue, who in some fields deal only with documents in English. They have to transfer information from one language to another, an exercise they are not trained for. They can rely on terminological databases that are devised and kept up to date in most areas by scientific institutions, but this does not solve all the problems. On the other hand writing in English for non-native speakers raises other difficulties.

5. Language shift

There has not been any new assessment since the ones made in 1981, 1984 and 1989, giving a reliable description of the language situation in the sciences in France nor on the position of French as a language of science at the beginning of the 21st century. But we have some figures provided by scientific institutions and evidence given by scientists, which help to determine areas of language shift but also areas of language maintenance.

According to OST (*Observatoire des sciences et des techniques* 1996), in PASCAL, the database of CNRS (National Centre for Scientific Research) for exact sciences, the quotations in English have risen from 75% in 1978, to 80% in 1988 and 85% in 1994. The quotations in French were 12% in 1978, 8% in 1988 before rising to 11% in 1994. The situation for French is somewhat better in FRANCIS, the database for human sciences and mathematics: in 1994 quotations in English were 52%, those in French, 38%. We do not have an analysis of the national origins of the scientists having written the articles in French.

In 1996, OST made an analysis of the languages of publication of French scientists based on information taken in the *SCI*. It reckons that 17,4% of their publications were written in French and 82,4% in English. In 1982 31,6% were in French and 68,1% in English. When they publish in French journals, they tend to publish in French (73%). But they publish in these journals only 20% of their scientific productions against 32,4% in American ones. And French journals tend to turn to English: 26,5% in 1993. Lists of publications provided by applicants to posts of professors give revealing evidence. They contain almost entirely articles from American, British and European English-written journals, whereas some 20 years ago such lists would be shared between English and French titles.

A survey commissioned by the Délégation Générale à la Langue Française (DGLF 1998), revealed that out of 102 international scientific conferences which took place in France in 1997, 53% had English as their only official language, 32% English together with other languages. However the use of French together with English for external communication (announcements, reports) is now of general practice by the organisers of conferences. This was initiated by the 1994 Language Law, but according to scientists, it seems to be more a habit taken than a real constraint.

6. Language maintenance

If the shift to English is important and regular, there are different fields where French has remained in common use. Thus whereas English made its way in primary scientific information, that is to say in the journals disclosing the results of research, French is very much present in secondary information, giving the state of the art in each discipline. There are journals devoted mostly to secondary information, written entirely or partly in French and which are well read. The most flourishing ones try to attract French-speaking audiences and are organised on a francophone basis.Thus *Médecine-Sciences*, recently founded, stems from a co-operation between Québec and France. It publishes both primary and secondary information and has acquired sufficient reputation to be indexed in the *SCI*.

In research centres, French is reported to be the main language of everyday use. Foreign long-term visitors, as pointed out earlier, tend to make their way in French. Quite a few come from countries where sciences are taught in French in higher education. Courses in French as a foreign language are also provided, particularly for students who are on scholarships. But people tend to be practical, and adapt to visitors, particularly short-term ones, when they have language problems. In fact sociolinguistic analysis would probably reveal that most situations are bilingual, with English being largely used in the quest for information and in email exchanges. But efforts are also made to network French-speaking researchers.

As we said previously French is the usual language in which theses are written in France. Most postgraduate courses are in French, a part of the written and computerised information being in English. Undergraduate courses are also in French. But for some subjects in exact sciences it is difficult to find good textbooks. There are also areas in higher education, training for professions where the use of languages is important, like management, commerce, information and communication technologies, in which subjects are partly taught in foreign languages, mostly English. But the purpose is language proficiency, not using English as a lingua franca, though the difference is sometimes difficult to make. Most courses given to students in exchange programmes, like the Socrates (Erasmus) programmes of the European Union, are in French. This does not prevent universities from attracting students. In 1995–1996, out of 18 countries, about 17,5% of the students taking part in Socrates (Erasmus) programmes of the European Union went to France, just after the United Kingdom (20,5%) and before Germany (13%), Spain (11%) and Italy (7,5%).

If we consider that in Europe, internationalisation of universities is the main factor which makes them use English as a lingua franca, French and French-speaking universities still rely on French. This maintenance clearly depends on the international standing of French. At present it is a fact more than a choice. To convert it into a strategic choice would mean an assessment of the present and prospective international standing of this language, an assessment which francophone institutions have not made up to now. With the process of globalisation becoming more intensive, it is an assessment which will have to be made.

Scientific magazines published in French popularise the results of research outside circles of scientists. Some aim at audiences with a high scientific culture, like *La Recherche*. Some aim at a larger public (*Sciences et Vie, Sciences et Avenir…*). All of them are thriving and regularly increasing their circulation.The question of the use of French for the quest of scientific information on the Internet is for the time being an open one, as French, like most languages other than English, is in the process of being inserted on this medium.

7. Which debate for which problems?

What is striking is that the question of the languages of science is in 2000 hardly debated in France, much less than in the 1980s and even later, both in scientific circles and in the general public. It may be that every one is satisfied with a situation in which English and French appear to share different functions and to be complementary. It is also true that the ways in which problems were debated some years ago are not considered as being politically correct nowadays. But it does not mean that problems have ceased to exist. Such questions as 'how can science be kept close to citizens?' and 'in what languages can they have access to scientific knowledge?' are still raised. The problems are less acute in France than in many countries including the ones that have languages of wide communication (Conseil de la langue française du Québec 1996). But this means that questions should be raised on a much larger scale than that of a single country and not that they should be eluded.

8. References

Cassen, Bernard (ed.)
1990 *Quelles langues pour la science?* Paris: La découverte.
Cireel (ed.)
1981 *Le français chassé des sciences.* Paris: CIREEL.
Conseil de la langue française
1981 *L'avenir du français dans les publications et les communications scientifiques et techniques.* Vol. I, II, III. Québec: Gouvernement du Québec.
1996 *Le français et les langues scientifiques de demain.* Québec: Gouvernement du Québec.
D. G. L F. (Délégation générale à la langue française)
1998 *Rapport d'activité 1998.* http://www.dglf.culture.fr.

Deneire, Marc/Goethals, Michael (eds.)
1997 *Special Issue on English in Europe, World Englishes* 16. Oxford/Boston: Blackwell.

Observatoire des sciences et techniques
1996 *Rapport 1996*. Paris: Ministère de la Recherche.

Truchot, Claude
1990 *L'anglais dans le monde contemporain*. Ch. 4: L'anglais dans les sciences. Essai d'interprétation de la diglossie des milieux scientifiques. Paris: Le Robert, 95–181.
1997 The Spread of English: From France to a More General Perspective. In Deneire, M./Goethals, M., 65–76.

English in Linguistic Research in Belgium[1]

Roland Willemyns

1. Prolegomena

The direct occasion to write this paper was a presentation I was invited to deliver at the 14th World Congress of Sociology in Montréal (30. July 1998) on the dominance of English in linguistic research in Belgium. Since it is obvious that a large portion of the effects of this dominance are related to the dominance of English in scientific research in general (Ammon 1998), I shall try to focus here on other, more specific aspects. Yet, one should be aware that even so, the fact just mentioned cannot possibly be ignored, since it is indeed the paramount factor outweighing all others.

I have no knowledge of any studies or essays previously published on this particular topic. Therefore, I largely draw on my personal experience and on my familiarity with the scene of Belgian linguistics. I am perfectly aware, though, that in so doing, I may have under- or overestimated some items, maybe neglected or forgotten others and, in general, given a too personal impression of a situation others may possibly judge differently. This also partly explains why the situation in Flanders (Dutch-speaking Belgium) is treated more extensively than the situation in Wallonia (French-speaking Belgium). Another important reason, though, is that the latter situation has been far less surveyed and that, consequently, information is simply not available.

2. Introduction

Belgium is a small country with some 10 million inhabitants, divided into various language territories: Flanders with some 5.8 million Dutch-speakers, Wallonia with some 3.2 million French-speakers, Brussels with approximately 1 million inhabitants speaking Dutch or French or both and a small German-speaking territory with some 65.000 inhabitants (Willemyns,1988).

There is a rather close contact between Flemish linguists and their colleagues from Holland, not only because of the institutional cooperation between several of their universities, but also because all matters pertaining to language and literature in both Holland and Flanders are managed by the *Nederlandse Taalunie*, an intergovernmental, supranational body, to which both the Dutch and the Flemish governments have delegated their prerogatives to decide autonomously on linguistic and cultural affairs (Willemyns, 1984)[2]. Also, a special fund, created and financed by the research foundations of Holland and Flanders (NWO and FWO) succeeded in generating a considerable number of joint research projects in the fields of Dutch language and literature. This cooperation between Flemish and Dutch linguists on the basis of their common mother tongue proved to be a much more important factor in generating joint research projects than a theoretically possible cooperation of Dutch- and French-speaking Belgian linguists on the basis of citizenship. Also, the fact that there exists no institutional cooperation between the Dutch language and the French language science foundations in Belgium, makes it not easy to initiate joint research projects between Flemish and Walloon linguists.

The effects of English on linguistic research in Belgium may theoretically surface in various domains. I'll discuss what I feel to be the most important or interesting ones:

- the publication of the results of research both internationally and domestically
- the popularization and diffusing of similar results to a larger audience
- the teaching of linguistics in Belgium at the undergraduate, graduate and postgraduate levels
- the publication of text books and other popularizing or teaching materials based on linguistic research
- the internal, i. e. domestic communication among linguists in a bi- or multilingual country.

– the consequences of the Anglo-Saxon dominance on the methodology and the concepts used in the field of linguistics as such.

In all these domains English as a matter of fact is competing with the vernacular languages and although its influence varies according to the domain under consideration, my overall impression is that "dominance" may not be the appropriate term to designate the influence of English on linguistic research in Belgium. Research as such is mostly conducted in one of the national languages, which are also used for the teaching of courses on linguistics at all levels in all universities.

3. Publication infrastructure

In Belgium's seven universities (4 Dutch speaking and 3 French speaking ones) most of the linguistic research is conducted within the so-called "language departments", the largest of which are the Dutch departments in Flanders and the French departments in Wallonia. Yet, the importance of English immediately surfaces as soon as we look at the publication of the results and conclusions of linguistic research.

As far as books are concerned, there is usually no problem to publish monographs on linguistic issues in Dutch or French respectively, albeit it that through a lack of domestic publishing companies, many of them are published in Holland or France.

As far as articles are concerned, there appears to be an acute shortage of Belgian journals and reviews on linguistics. In this field, though, the same applies to the Netherlands since the list of linguistic journals in the Low Countries at large is very short indeed. The main ones are *Taal en Tongval* and *Nederlandse Taalkunde*, the former one specializes in dialectology and sociolinguistics, the latter one, successor to *De nieuwe Taalgids*, on linguistics in general, though with a special focus on formalized linguistics. The journal of the Royal Academy of Dutch Language and Literature in Belgium (*Verslagen en Mededelingen van de Koninklijke Academie voor Nederlandse Taal- en Letterkunde*) is one of the few remaining outlets for linguistic papers, yet publishes on literature as well. A wide variety of linguistic items is generally to be found in *Leuvense Bijdragen*, a journal which, despite its name, only occasionally publishes articles in Dutch. Apart from occasional publications (Proceedings of conferences, year books, readers, university preprints, Festschrifte etc., and a few smaller ones as *Gramma* or *Glot*) this sums up the very restricted possibilities of domestic publishing on linguistics in the Low Countries. Also, it should be mentioned that, to my knowledge, there do not exist any journals abroad in which Dutch is the only or one of the possible languages of publication. Although specialized journals on Dutch linguistics and/or the culture of the Low Countries do exist abroad, their language of publication is not Dutch. Consequently, as far as journals are concerned, the Flemish linguist lacks the outlet he usually has at his disposal for the publication of monographs. French speaking Belgian linguists or Dutch speaking ones in the field of Romance languages are in a much more advantageous position since they have access to the whole publication infrastructure of the Romance languages territories.

Summarizing, the real pressure on the linguist to publish in English has to do with two well-known factors:

1. the country is too small to offer adequate opportunities for publication (in journals that is)
2. in order to reach the international linguistic community there is no alternative but to use the only language understood by all, viz. English.

The latter factor, of course, the Belgian linguist has in common with his colleagues all over the world, who are not native speakers of English. Using another language of greater circulation (e. g. German or French) is no real help, unless one specifically wants to address that part of the linguistic world. As a consequence, a considerable part of the work of Flemish linguists, anxious to be more widely known and recognized, is published outside the country and in English (there are but very few English language publication opportunities inside Belgium).

It is rather easy to list a number of considerable disadvantages in this situation. Some of them are quite specific, others are experienced all over the non-English speaking world:

(1) some specific topics appear to be of only limited interest to most journals abroad (e. g. some aspects of Middle-Dutch or of local dialects, studies in toponymy etc.)

(2) publication in journals or readers abroad is usually subject to special limitations as far as the number of pages is concerned and mostly the interest of the editors (though not necessarily of the readers) is limited to the mere outlines of the research results. The research methodology and a complete overview of the results usually have to be cut down. Consequently, they have to be published domestically as well, if an appropriate

means of publication can be found. Survey articles, on the other hand, are very much appreciated, since their specific aim is to inform linguists abroad on facts or situations they're not familiar with, yet are keen to know more about. Consequently, and according to my own experience both as a contributor and as a reader, we are faced with the rather odd situation that the real scholarly articles are hard to publish abroad in extenso, whereas it is relatively easy to find a publication opportunity for the more surveying ones, including in very prestigious international journals. This is no less than a paradox, since all domestic academic and research instances and institutions value publications in foreign prestige journals very highly, very often to the detriment of those published domestically and in the indigenous language.
(3) to write in a foreign language is always difficult and demanding and puts the non- native linguist in a disadvantaged position as compared to his native English speaking colleague. Also, the English lingua franca as is often used orally in many international conferences is not suitable for use in writing and is, consequently, rejected by the editors of most English language publications, both inside and outside the English speaking world.

On the other hand, the number of publications on linguistics in Dutch or French (in monographs, textbooks and other teaching materials, vulgarizing journals etc.) continues to be considerable and, consequently, there is no danger (neither immediate nor in the long run) for the continuation of either of them as a language of science in linguistics, as we will discuss in a moment.

The next item I mentioned is the possible usefulness of English as a means of internal communication among linguists in a bi- or multilingual country. In this respect, it is interesting to observe that English is hardly ever used in order to avoid to have to choose between one of Belgium's domestic languages, the more so since, as I stated earlier, there appears to be very little contact between Flemish and Walloon linguists in the first place. The *Belgian Journal of Linguistics* (publisher: John Benjamins) which is the Yearbook of the Belgische Vereniging voor Linguïstiek ('Belgian Association of Linguistics') is published in English. The decision to do so was taken some years ago, following a consultation of its members in writing. Although there is no way of knowing for sure, I don't suppose the majority who decided to ban Belgium's official languages (Dutch and French) did so in order to avoid domestic language problems. Rather, their main preoccupation may have been with reaching a more international audience. Whether this has been successful I don't know, but given its rather limited impact in Belgium itself, serious doubts are permitted.

The Belgian Association for Applied Linguistics (*Belgische Vereniging voor Toegepaste Linguïstiek*) does publish its internal communications in Dutch and in French. It occasionally also publishes short monographs or editions on specific topics, for which either English, or Dutch or French are used.

4. The linguistic publication profile

A survey carried out in 1997 by the *Vlaamse Interuniversitaire Raad*, a government body representing all Flemish universities, among Flemish scholars in linguistics revealed some interesting details about their language use or preferences as far as publications are concerned[3]. The data relate to publications published from 1992 through 1996 and the response rate to the survey was 44.4%. The figures quoted here were published in a "draft version" of the report, dated April 26., 1999 (Moed et.al., 1999).

An analysis of the publication lists of the participants reveals that, as far as all publications put together are concerned, the following languages were used to the following extent:

Dutch:	32%
French:	25%
German:	7%
English:	33%
Other:	3%

Since the scholars were asked to differentiate between "real scholarly" publications and more popular ones for a wider audience (vulgarization), it was possible to detect a most interesting difference between these two kinds of publications.

As far as the latter category is concerned (vulgarization), the language use is as follows:

Dutch:	64%
French:	17%
German:	6%
English:	10%
Other:	3%

Here we not only observe the predominance of the mother tongue of the authors and their intended readers, but also the fact that the use of

English is not really impressive. Consequently, it may be put forward that the audience one has in mind is the paramount criterion guiding the scholars in their language choice.

The scholars qualified 77.5% of their production as "real scholarly" ones and 22.5% as publications for a wider public (vulgarization). Since they were asked to make this distinction themselves there is no way of knowing on the basis of which criteria the selection was performed nor, obviously, whether all were using more or less similar criteria.

Anyway, the difference between the two categories of scholarly publications is striking. Here, the language use is as follows:

Dutch:	22%
French:	27%
German:	7%
English:	41%
Other:	3%

The use of Dutch is reduced by two thirds, the use of English, not surprisingly, has quadrupled and it emerges by far as the most important language in which four out of every ten scholarly items by Flemish linguists are published. This, of course, must include the so-called survey articles.

The only real surprise is the rather large number of publications in French and it is, therefore, interesting to have a closer look at the relationship between sub-disciplines and topics and language usage. First of all it appears that publications in English are rather unevenly distributed among the various sub-disciplines used in the survey. The "object language English" (i. e. a publication on some aspect of English)reveals the highest percentage of publications in English, followed by the sub-disciplines *Computational and Mathematical Linguistics*, *Psycho- and Neuro-Linguistics* and *Semiotics*. Dutch as a publication language is dominant in the "modules" *Morphology, Toponymy* and *Lexicology* and in the "aspects" *Applied Linguistics, Sociolinguistics* and *Dialectology* (Moed et.al. 1999, 155) . As far as French as a publication language is concerned, its surprisingly large usage may, to a considerable extent, be explained by the fact that 20% of the linguists who participated in the survey are working in the field of Romance linguistics. Most of the international publications of these colleagues are directed toward French language journals in the Romance language community abroad ("The publication language French is mostly used for publications on the object languages *French* and *Other Romance Languages*" Moed et.al. 1999, 155). Anyway, it is clear that, as far as Flanders is concerned, English is the main but not the only international language of publication.

Some data seem to indicate that the publication language chosen (or felt necessary to use) may also influence the way the research results are presented or even the way in which the research is conducted. If we take the examples of publications on more than one object language or with more than one author, it appears that:

– 73% of the publications that have both Dutch and German as object languages are written in German; 90% of those with Dutch and English as object languages are written in English;

– Co-authorships, less frequent in the human than in the natural sciences, do not occur very often, as was to be expected. Yet, the sub-disciplines in which they reach the highest frequency are exactly those in which we observed the highest amount of English usage, viz. *Computational and Mathematical Linguistics*, *Psycho- and Neuro-Linguistics*. Also, as Moed et.al (1999, 161) observe: "Sub-disciplines in which English is the dominant publication language, tend to show a higher degree of collaboration as reflected in co-authorships than sub-disciplines in which English is a less prominent publication language".

5. The language of linguistic instruction

As far as the use of English as a language of instruction in Belgian universities is concerned, we have to take into account the effects of Belgian legislation on the use of the languages in education. In all school types, from kindergarten to university the law decrees that no official degrees can be delivered unless the instruction has been carried out completely in Dutch in Flanders or French in Wallonia. Consequently, only post-graduate courses and studies not leading to an official degree could possibly be carried out in English. It has to be noted, though, that this legislation has nothing to do with English and surely was never meant to constitute an obstacle for the possible use of that language. Rather, it is a consequence of the way Dutch and French speakers in Belgium decided to secure the homogeneity of their respective linguistic territories, after decades of language problems in the country. Since the group of German speakers in Belgium is too small a German language university does not exist. When, after the federalization of the country, education passed from the federal to the regional governments, the same type of legislation remained in effect. Using the example of Flanders, (the English translation of) art. 61 of the relevant decree says: "The language of teaching and administration in the universities is Dutch". The very few exceptions to this general rule are to be found in the field of postgraduate courses and in the case of visiting scholars. Yet, even in that case, students have a right to request to pass their exams in Dutch even on courses which, due to the exceptions just mentioned, should have been taught in a foreign language. Also, the number of postgraduate programs in English is very restricted. Taking the example of my own university (the *Vrije Universiteit* in Brussels) it appears that 33 out of the 47 postgraduate programs are conducted in Dutch (70%). Of the 14 programs offered in English, 11 (78%) are to be found in the fields of natural sciences (6 out of 7) and economics (5 out of 8). None of the postgraduates in language and literature is in English, and even in medicine and pharmacy (1 out of 9) and engineering (1 out of 6) the number of English language programs is extremely low. Also, in Law School, only 1 program out of 6 is in English. A quick survey revealed that a similar pattern is to be discerned in other universities as well. 75% of the University of Ghent's postgraduates, e. g., are in Dutch, 25% in English. Here too, most of the latter are in the same field, viz. 11 out of 19 (= 57.8%) in "Agriculture and Applied Biology" and none of the postgraduates in language and literature is in English. The overall conclusion, therefore, is that even in postgraduate education the Flemish universities overwhelmingly stick to Dutch as the language of instruction. As directly regards the scope of this paper: none of the postgraduate programs in linguistics is offered in any language but Dutch.

This decree, which came into effect on 1 October 1992 (the complete text of the decree and the implementing orders are to be found in the Internet on the following address: <http: //www.ufsia.ac.be /~ingelaer/>), is mirrored by a similar one valid for Belgium's French speaking community and both are the continuation of legal provisions having existed previously at the Belgian national level.

Foreign language departments, of course, are exempt from this rule, as far as their courses on those languages are concerned. There exists a very long tradition indeed that, from their very first university course onward, students of Germanic and Romance languages are taught all their courses and seminars on both literature and linguistics in the foreign language they're studying. This does, of course, not apply to the same degree for languages which haven't previously been taught in high school (e. g. Slavic or African languages).As far as linguistics is concerned, at least a reading competence in English by *all* students is taken for granted.

Since a considerable part of the professional literature is in English and even suitable textbooks may not always be available in Dutch, it is compulsory to (be able to) read books and articles in English, even on topics specific to Dutch linguistics. Both scholars and students, therefore, are forced to acquire the necessary proficiency in English. Usually, this is not really a problem nor does it meet with protests from the part of the students. In many other departments English reading comprehension courses form an integral part of undergraduate training, based on the belief that students should be able to consult original sources for personal work. All of this is, at least, the case in the Dutch language universities.

Consequently, the role of English seems to be limited to a language of science since it plays hardly any role at all as a language of instruction or as a language of popular publications aimed at diffusing the insights won by the scientists to a larger audience. Although this may certainly be different in other fields, it is obvious that at least in the field of linguistics English is playing a complementary role, rather than usurping functions formally performed by Dutch. Until a few decades ago, Flemish linguists, eager to publish abroad or attract an international audience, did so in German or in French. What English has done in the recent past is taking over the role of these other foreign languages, not so much taking over functions formally held by the vernacular.

On the other hand, the exploding expansion of English as a language of science certainly is the co-instigator of discussions in the Dutch speaking community in Holland and Belgium (and probably elsewhere as well) on whether English is to be considered a "threat" to the use and even the development of the mother tongue. Although, as I have demonstrated, the basis for this perception is an impression rather than a fact, it may lead to attitudes which are to be considered as harmful. Consequently, it is important to make the population at large understand that the use of English as a language of science does not constitute a threat to the status of the vernaculars nor to their continuing functioning.

6. On the use of linguistic terminology

As far as traditional grammatical terminology is concerned Dutch has a certain tradition of using its own terminology for many concepts which, in many other languages, are usually only referred to by their Latin derived counterparts. Consequently *zelfstandig naamwoord* 'noun' is used alongside with *substantief*, *bijvoeglijk naamwoord* 'adjective' with *adjec-*

tief, *spraakkunst* 'grammar' with *grammatica* etc. With terms like *werkwoord* 'verb', *bijwoord* 'adverb', *voorzetsel* 'preposition', *meervoud* 'plural' or *lijdend voorwerp* 'direct object' the Latin derived counterparts are hardly ever used. As far as more recent linguistic terminology is concerned, the situation is a little more complex. Terms as *dieptestructuur* and *oppervlaktestructuur* are very common for *deep* and *surface structure.* Also, the bulk of phonological terminology is Dutch (*distinctieve kenmerken* 'distinctive features', *gespannen* 'tense', *foneemdriehoek* 'phoneme triangle') or "international" (e. g. *foneem* 'phoneme', *dentaal* 'dental', *correlatie* 'correlation', etc.). *Phonological space* or *incidence* are some of the few exceptions. On the other hand originally German terms as *Umlaut*, *Ablaut*, *Inlaut*, *Auslaut* still are in frequent use, as is the case in many languages. In sociolinguistics, though, a lot of terminology is simply borrowed from English, as, e. g. *change in progress, code switching, communicative competence, language maintenance, language shift, linguistic insecurity, matched guise, power and solidarity, upward social mobility,* and more of the like. The use of Dutch counterparts or translations is not frequent. Interestingly, this phenomenon does not apply in the field of dialectology, probably thus stressing the Anglo-Saxon origins of sociolinguistics. Although German dialectology used to have a considerable influence on the work of Flemish dialectologists, most of the relevant terminology is in Dutch, with the possible exception of a few very specific terms as *das geographische Nebeneinander des historischen Nacheinanders* or *Vordringen mit punktuellen Zielen.*

7. Effects on the scientific evolution of linguistics

Although it is not specific for the situation in Belgium, one more aspect of the dominance of English cannot be left out here, viz. the potential perverse effects of English on the scientific evolution of linguistics.

The vastness of the English language scientific output means its impact on scientific thinking, its clout, is all-pervading. Good science published in other, smaller languages, leads to fragmentation and smaller diffusion. Let me quote a few examples: Ferguson is credited in most studies as the inventor of the concept *diglossia* (1959) when in fact the concept and its implications were known since ca 1890–1930 in French Hellenistic studies. Ferguson may be the scholar who anchored it in sociolinguistics, through publishing in English, but he was not the originator of the concept as so often stated. Similarly, Joshua Fishman is credited as the founder of the Sociology of Language, yet G. Cohen published *Pour une sociologie du langage* a thick volume in 1964, i. e. one year before Fishman began his impressive series of publications in English. These examples do not take away any of the credit from Ferguson or Fishman but do show how other scientists tend to perceive things discovered through English as the pioneering work.

Most of the linguistic research originally produced in and on English has great impact in training future non-English linguists, merely because of numbers. This then leads to assumptions that many of the findings published in English language studies are universal and it takes more time and effort to disseminate corrective theories, produced in and on other languages, once the English based or produced publication has become widespread. This even holds true for some of the most common terms in our profession as, e. g., *dialect*, *continuum*, *diglossia, social class, language community* etc., all concepts which are very specifically different in the Anglo-Saxon world and are too often simply transferred, both by Anglo-Saxon and other linguists, to situations in other language territories where the concepts in question may refer to a very different reality. On the other hand, for concepts which apparently don't exist in the Anglo-Saxon world, we are often lacking an English counterpart or equivalent. *Umgangssprache* may be a good example of this.

8. Conclusions

By way of conclusion, I would like to formulate some hypotheses and make some recommendations.

(1) An impressionistic comparison between the situation in Flanders and in Holland leads to the conclusion that in the latter country English may be used more extensively in the academic world than is the case in Flanders, mainly as a consequence of two structural differences. The habit in Holland, that PhD dissertations have to be printed before they are defended entails that more dissertations in linguistics are published in English than is the case in Flanders, since very often the so-called "defense copies" are also used for the commercial edition. Also, the use of English in academic teaching seems to be more widespread than in Flanders, where the legislation mentioned before is a strong means of preventing the expansion of English in that particular domain. Yet, it seems not unreasonable to believe that in the future the use of English in academia in Belgium will rather increase

than decrease. Also, it is worth mentioning that in other fields the use of English appears to be more widespread than in linguistics, also as far as (postgraduate) teaching is concerned.

(2) Although having to publish in English certainly presupposes a certain proficiency and particular skills, there seems to be no direct correlation between the amount of language proficiency and the extent of using English in research.

(3) The audience one has in mind is the paramount criterion guiding the scholars in their language choice or preference. Also, publications dealing with a particular object language tend to be written in that language, whereas publications on more general linguistic topics tend to be written in English.

(4) The use of English as thé international lingua franca in academics does not seriously harm the use, development and functioning of the vernaculars, at least in the so-called developed world. Consequently, English has to be considered as a complementary means of expression and communication, specifically meant to enable and establish contacts with the international scientific community.

(5) In many people there seems to be a confusion between the role of English as the habitual tongue in the Anglo-Saxon world and as the lingua franca of the international scientific community. Therefore, it has to be kept in mind that English is as useful for contacts with the non-English speaking world as it is for contacts with the Anglo-Saxon world.

(6) Many negative attitudes against English as the international language of science seem to be inspired by:
   - a misjudgement of the role of English as a complementary means of communication in domains where many vernaculars could not possibly function anyway;
   - negative connotations in countries where domestic language conflicts in which English plays a part happen to exist in other domains than the scientific one alone;
   - the fact that the international use of English gives an undeniable and often enviable advantage to native speakers of that language, as opposed to all others. Yet, since the reintroduction of Latin is not a serious alternative, this is something we have to live with.

(7) The advance of English cannot possibly be stopped. Therefore, measures to enforce its position are as unnecessary as countermeasures are doomed to fail.

(8) The effects of the dominance of English on the concepts used in linguistics may be harmful to the scientific research of the linguistic

situation in non-English speaking countries and, consequently, to the development of linguistics in general.

(9) The function of English should remain what it is right now: a complementary means of communication in a specific domain. Domestically, therefore, it should not be used in any other functions or domains.

9. Notes

1. I am grateful to my colleagues Hugo Baetens Beardsmore and Sera de Vriendt for their comments on an earlier version of this paper.
2. A similar Francophone institution does not exist.
3. Similar surveys have never been carried out in French-speaking Belgium.

10. References

Ammon, Ulrich

1998 *Ist Deutsch noch internationale Wissenschaftssprache? English auch für die Lehre an den deutschsprachigen Hochschulen.* Berlin/New York: de Gruyter.

Moed, H. F./Luwel, M./Nederhof, A. J./De Samblanx, V./Verbrugghen, K./van der Wurff, B.

1999 *Towards Indicators of Scholarly Performance in the Humanities: An Exploratory Study in the Fields of Law and Linguistics at Flemish Universities.* Brussel: Vlaamse Interuniversitaire Raad (Draft Version).

Willemyns, Roland

1984 A Common Legal Framework for Language Unity in the Dutch Language Area : The Treaty of Linguistic Union. *Multilingua* 3/4: 215–223.

1988 Belgium. In Ammon, U./Dittmar, N./Mattheier, K. (eds.) *Sociolinguistics.* Berlin/New York : de Gruyter, 1254–1258.

English as a Future Language of Teaching at German Universities? A Question of Difficult Consequences, Posed by the Decline of German as a Language of Science

Ulrich Ammon

It is common knowledge among reasonably well-informed sociolinguists that the dominance of English as a world language of science has developed in the course of the twentieth century. At the beginning of the century and during its first two decades there were actually three roughly equal languages of science, namely English, French and German, to list them alphabetically. It is hard to say which of them was altogether of more importance, since this can be measured in various ways for which adequate data are not always readily available.

One valid way of measuring the respective importance of a language is via its share in the overall publications of science. Tsunoda (1983) has provided data for the major languages in a study which covers the centenary from 1880 to 1980. He drew on the most comprehensive national bibliographies available of France, Germany, Russia and the United States, counting all the publications registered according to language of publication for four natural sciences (biology, chemistry, physics, medicine) and mathematics. His findings are illustrated in Figure 1 which results from a two-step calculation done separately for each language, first of the arithmetic means of the various sciences for each national

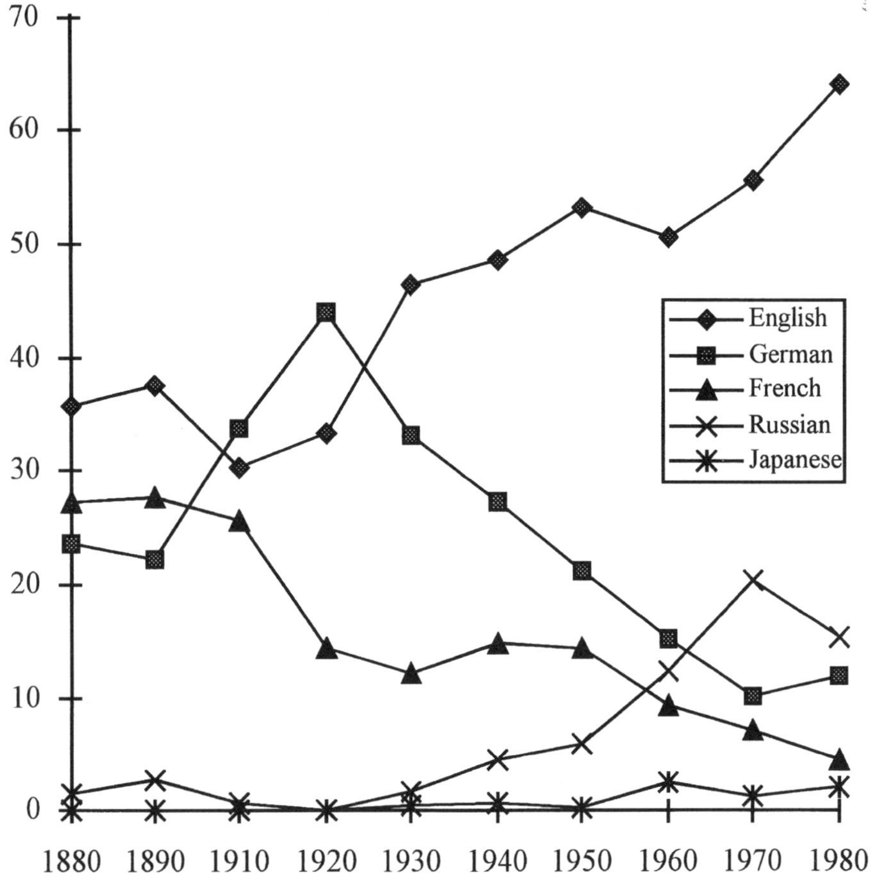

Figure 1: Shares of various languages in scientific publications in the course of one centenary (following Tsunoda 1983)

bibliography and then the arithmetic means of the various national bibliographies.

Though more studies of this sort are desirable to refine insights, the general course of development seems to be rather well established even now, since Tsunoda's findings are basically confirmed by numerous additional fragmentary data. An example is Shenton's study of language use in international conferences in the 1920's for which he finds: 'French, English, and German, in the order named, are the three major languages of cosmopolitan conferences.' (Shenton 1933: 223)

Table 1: Number of international organizations regularly using the languages for their conferences in the 1920's (following Shenton 1933: 255, 261, 264, 268)

Status at Conference Language	Official	Translation	Permitted
French	220 (31)	18 (4)	22 (23)
English	160 (21)	26 (4)	34 (25)
German	132 (15)	29 (4)	31 (25)

The difference between the three languages is noticeable but not overwhelming, as table 1 shows. One has to take into consideration that, after WW I, German was banned from all international scientific conferences for years and could never regain its previous status (cf. Schroeder-Gudehus 1966; 1973). Though most of the organisations examined by Shenton were not, or at least not primarily, concerned with science or the humanities, many of their international conferences were to some degree. In addition, those organizations explicitly labeled 'ART-SCIENCE' in Shenton show about the same proportion; their figures are added in brackets in table 1.

We know that in international conferences English nowadays is no less dominant than in scientific publications (cf. Ammon 1991: 242–251). The reasons for this dominance can partially be deduced from figure 2 which shows the development of share in Nobel prizes in science of the English-speaking and the German-speaking countries in the course of the 20th century but also includes once more, for a comparison, the development of share in scientific publications of the English and the German language. Obviously the English-speaking countries' share in Nobel prizes has dramatically increased, while the German-speaking countries' share has dwindled. If the share in Nobel prizes is taken as a rough indicator of scientific achievement, which seems valid, then the English-speaking countries can be said to have outstripped the German-speaking countries in scientific achievement. The French-speaking countries would, by the way, show a course of development similar to that of the German-speaking countries.

Figure 2 also shows the rather parallel development of (share in) language of science and (prizes for) scientific achievement. It seems plausible to take the latter as a major reason of the former, i. e. to assume that the lead the English-speaking countries have acquired in science has been a reason, perhaps even the main reason, why English has became the domi-

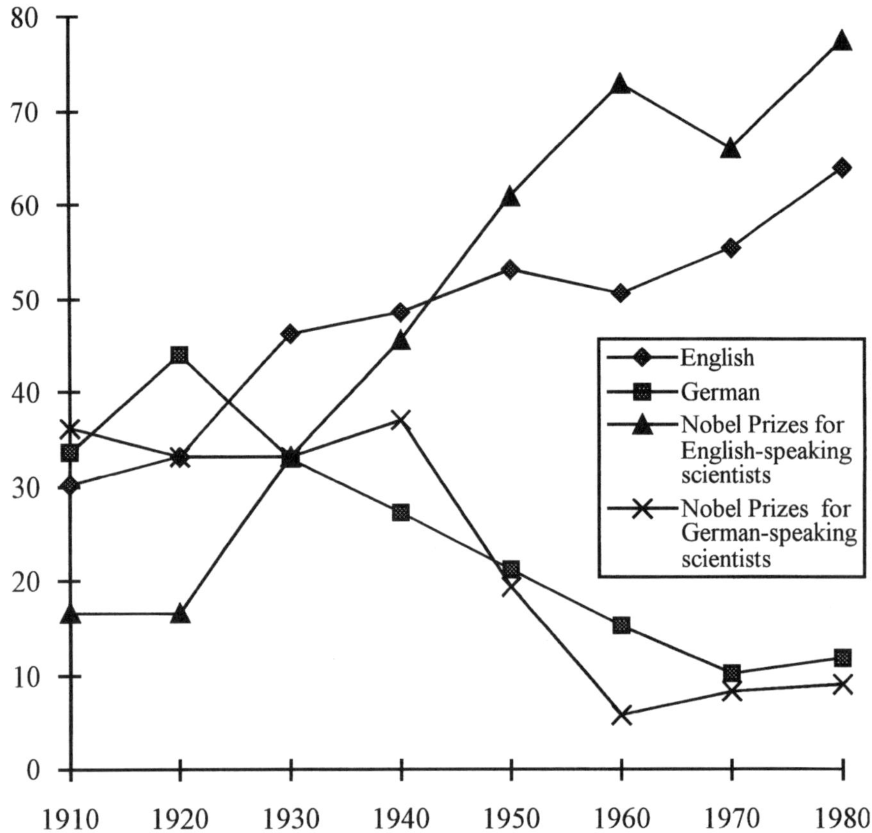

Figure 2: Shares of English-speaking and German-speaking scientists in Nobel Prizes compared to their languages' shares in scientific publications of the 20th century

nant language of science. Actually, a feed-back model of explanation seems more realistic than such a one-sided view, but at the moment the simplified one-sided view suffices to put forth an important argument, namely that French and German declined as languages of science because their countries did not keep up the same level of scientific achievement as the English-speaking countries. Some of the reasons can be found in the political history of this century during which the major French-speaking and German-speaking countries were twice thoroughly destroyed by war, while major English-speaking countries, particularly the United States, remained unscathed; in addition, a substantial part of German scientists, particularly Jews, were expelled or murdered by German Nazism and those who escaped found refuge mainly in the United States. By 1936, 1617 scientists and scholars from universities had been forced to leave, of whom 1160 went to English-speaking countries (825 to USA, 318 to UK, 17 to Australia), numerous others followed later (Kröner 1983: 13).

Though these events certainly had considerable impact on the development, it seems likely that the English-speaking countries and, accordingly the English language, would, even without them, have gained the lead over their German and French rivals, though more slowly and to a lesser extent. This may be assumed because of their greater numerical and, particularly, their greater economic strength, which is not only due to the course of history within this century but also to previous successful British colonialism. The data in table 2 show the recent proportions, though figures may diverge to some degree depending on methods of measuring (cf. for similar data Graddol 1997). (The numerical strength is given in form of a span derived from various figures.)

Table 2: Numerical and economic strength of languages in the 1980's (numerical strength according to different sources) (following Ammon 1991: 41 f., 49)

Numerical strength (native speakers in Million)			Economic strength (GNP in US $)		
English	French	German	English	French	German
300–415	55–95	92–120	4271	669	1090

Since the basic social and economic structure of the major English-, French- and German-speaking countries is similar, size may be assumed to be the crucial reason for divergent recources for scientific research and, consequently, scientific achievements (cf. de Solla Price 1986: 142; 191–194). The figures presented in table 2 may also be taken as a warning against wishful thinking that the English-speaking countries could lose their scientific lead in the near future; they probably will not, at least not to the French-speaking or the German-speaking countries. That the English language could lose its dominant position as the world language of science in the near future, seems even less likely. Its dominance has even increased after 1980 (cf. for detailed figures until 1996 Ammon 1998: Chapter D). Also, once achieved, such a function tends to have considerable stability because world-wide linguistic readjustment would be very costly (cf. the "law of inertia" of linguistic regimes; de Swaan 1993: 222).

It seems therefore more realistic for the other language communities to plan on how to deal with the dominance of English rather than how to abolish it.

2. Effects on the declining international languages of science and their speakers

The effects of the dominance of English as a language of science is presumably very different on various functional types of language and also on various types of society. As to functional language type, the following distinctions seem to be particularly important in our context:

(1) languages which have until recently been international languages of science on an equal footing with English (French, German),
(2) languages which aspire for the function of international language of science, or for a greater share in this function (e. g. Japanese, Chinese, Spanish),
(3) modernized or elaborated (*ausgebaute*) languages which, however, have always been limited to the national level as a language of science and do not show international aspirations (e. g. Danish, Dutch, Swedish),
(4) not yet modernized languages which, consequently, are only used to a very limited degree at best, as a language of science, even on the national level (e. g. Haus(s)a, Quechua, Hiri Motu).

All these functional types of language are somehow hampered in their development if another language dominates international scientific communication; the actual impact and the social consequences are, however, presumably different for each type. One should, by the way, not confuse the effect on the functional development of a language with the effect on its native speakers: Negative effects on the development of a language may for instance be compatible with overall personal advantages for its speakers, and vice versa. I shall in the following concentrate on some effects of the dominance of English on functional language type (1) above and its speakers and illustrate them mainly with examples from German.

2.1. Domain limitations of the national language within the country

German scientists have in recent years had experiences unheard of in the first half of the 20th century. One of them has been that their own language was not accepted even within their country, e. g. for international conferences which took place in Germany or for journals published there, but that the use of English was required. Such a development amounts to domain limitations of the German language in the realm of science within the own country. German is now probably used less than English, even within the German-speaking countries, for international communication; from some international conferences and journals of the natural sciences it has even come to be totally excluded.

In addition, there are quantitative limitations to the use of German, even within the German-speaking countries, in various domains. They are generally greater

(i) in the natural sciences as compared to the humanities,
(ii) in the 'pure' or theoretical sciences as compared to the applied sciences, and
(iii) in research as compared to teaching – in each case particularly in written, but also to some degree in oral communication (cf. Skudlik 1990; Ammon 1998).

Though the use of German is still the rule in all these domains, English has made noticable inroads.

Another experience German scientists have had in recent years and about which there has been much complaint is the neglect of publications in German by the international scientific community, especially from the side of the Anglophone scientists. Many of the latter show a strong tendency not to take notice of publications in any other language but English. This is true even of linguists, of whom some try to find linguistic universals through scientific literature only in English, or even – what seems more problematic – the analysis only of the English language. Thus, to mention just one example, Levinson confesses awareness of the possibility that '[t]he reader may (...) be disappointed to find little reference to languages other than English,' which he, however, justifies by maintaining that 'other languages (...) have simply not been subjected to the same kind of analysis.' (Levinson [1983] 1992: xi) This statement is highly questionable, particularly for the time of the edition from which this is quoted. Brown and Levinson [1978] (1987) on the other hand, a study with the aspiration of universal validity, presents examples from numerous languages but does not refer to any publication in a language other than English except a single one in French. Such observations are of course not intended as an argument against the content of such books but only as an illustration of the neglect of research published in languages other than English by Anglophone scientists and scholars. There have been various reactions from the side of German scientists or scholars to observations that their Anglophone colleagues ignore publications in German; gradually however, the opinion seems to become prevalent that 'if you can't beat them, join them' (e. g. Markl 1985).

2.2. Slow-down of modernization of the national language

The English-speaking nations have at present a greater research capacity than has any other language community in the world. Therefore, they have also the greatest capacity for modernizing (or elaborating) their language. This capacity is even enlarged by the fact that not only the English-speaking countries' own scientists and scholars are engaged in this modernization process but also researchers from numerous other countries for whom English serves as the preferred medium of international communication. It is therefore hardly astonishing, that all other languages trail English in degree of modernization. In this situation, some language communities try to coin as much new terminology for their own language as possible by way of translating from English, with the French probably leading this course of action. Some terminology is naturally also created from within these languages, of which a small portion again finds its way into English.

Accordingly, researchers who work in the front lines of their field, or even teachers who try to deal with the newest developments in their subject, find themselves regularly in a situation where they have to either use English terminology or to coin new terms ad-hoc in their own language. As a consequence, English keeps making ever new inroads into these languages, actually into any language used in advanced science or technology. In fact, the languages which are relatively far advanced in terms of modernization, are more affected by this development than the so-called underdeveloped languages which are not used for scientific research or teaching.

There is still another, but rather peripheral consequence of the dominance of English as an international language of science for the other fairly modernized languages which has been pointed out lately by some German scientists. They observed that, in some cases, originally German terms have been anglicized, or that their English translations were re-translated into German, thus establishing new German terms which then have been substituted for the original ones. An example is the term *Bekräftigung* of the Russian psychologist Pavlov, who gained his international reputation through German. It was translated to English *reinforce-*

ment. From there it was retranslated to German *Verstärkung* which has become the preferred term in the German-speaking countries – though the new term, as has been pointed out, captures less of the concept than the old one (Wickler 1985: 30). Such observations, together with occasional exaggerations of the English-speaking countries' contribution to scientific development in new histories of science, have been used as a criticism of the linguistic and cultural 'imperialism' of the Anglophone world.

2.3. Scientists' unpreparedness and their absence from the international scene

German scientists have observed, not without envy, that their colleagues from the smaller Northern European language communities often have a better command of the new international language of science than they themselves. As a slight consolation it has been pointed out occasionally that others might still be doing worse, for instance the French. Two factors seem to be involved here, among others:

(a) the degree of loyalty to the own language (L), and
(b) the degree of linguistic distance (D) of one's own language from English.

The higher the loyalty to the own language (L+) and the greater the linguistic distance of one's own language from English (D+), the poorer on average the skills in English, and vice versa. We arrive thus, in very rough terms on the basis of crudely simplified binary concepts, at the following hypothetical ranking from the best to the worst skills in English, on average:

−L, −D (good skills): e. g. the Netherlands, the Germanophone Scandinavian countries,
−L, +D: e. g. Finland or +L, −D: e. g. Germany,
+L, +D (poor skills): e. g. France, Spain.

The scientists of the smaller language communities have never had the chance to develop a strong loyalty to their own language, at least not in recent times, since they have always been dependent on some other language for their international communication. They only had to shift from one foreign language (French, German) to the other (English). Naturally, the loyalty conflicts involved in such a shift were smaller, if not negligable, than they must be for the native speakers of French or German. The latter have to overcome greater barriers of emotional attachment to and national pride in their own language to become adjusted to the new realities. Some even seem to believe that these realities are not yet firmly established and can perhaps be reversed. Therefore, they are loathe to give in, or adjust, to the dominance of English. In a survey I did among a sample of German scientists (n = 69), about half of them from the universities and the other half from industry, considerable reservations against English-language dominance became noticeable.

25% confessed insufficient reading skills in English,
38% insufficient skills in oral comprehension,
57% insufficient writing skills – always according to their own standards.
19% would not participate in conferences if the use of English was required,
24% would rather not have contacts with colleagues under such circumstances, and
33% would not publish, if it could only be done in English. (Ammon 1990)

It has not, to my knowledge, been investigated so far, whether German scientists and scholars are underrepresented in international conferences, and in international research and publication projects, but I assume – on the basis of bits and pieces of information – that this might be true. It seems also possible with respect to French, or perhaps even Spanish, researchers. It might be worth while probing into this hypothesis and its various ramifications.

However, being absent from the international scene is certainly not what is expected from the German scientists or scholars. The former president of the *Deutsche Forschungsgemeinschaft,* the main funding source of research in Germany, has for instance raised expectations indistinguishable from those that would be expected from the Danish or the Dutch scientists by declaring publicly that German scientists should not hesitate to use English for international communication. At the same time he showed awareness of the different situation in which the German scientists find themselves as compared to their Anglophone colleagues. With respect to this difference he pointed out the additional expectation that German scientists not only ought to continuously translate from German, still their preferred language of research, to English for international communication, but also back from English to German for internal communication with their own society, to which they are responsible (Markl 1985: 24 f.). Such expectations highlight the struggle with language which all those scientists face for whom the international language of science is foreign. It seems difficult to convince German scientists, for instance, that becoming bilingual should in the end give them an overall advantage, as some linguists believe, over their monolingual English-speaking colleagues.

2.4. Publishers' difficulties

The shift towards English as the dominant international language of science has also been felt by the German publishers. One effect, among others, was that sales of German-language journals dramatically declined. Some of them were stopped altogether for economic reasons, others have carried on with the help of subsidies, mostly from German research foundations. A considerable number managed to survive but had to change their language of publication. Some turned plurilingual, with most of their contributions in English however. Others became monolingually English, i. e. do not accept contributions in German any more, which seems to have been the best choice from an economic viewpoint. An economically very successful example is *Applied Chemistry* which was formerly *Angewandte Chemie.* The change in language of contributions has regularly been reflected in the title of the journals. Some first chose a Latin title indicating plurilinguality, while others right away shifted to an English title usually indicating therewith that no other language would be accepted for contributions. Examples of these two main types of title changes are:

Archiv für Verdauungskrankheiten > Gastroenterologia > Digestion
Radiologische Rundschau > Radiologia Clinica > Diagnostic Imaging
Psychologische Forschung > Psychological Research
Zeitschrift für Tierpsychologie > Ethology

The shift from German to English has, however, not been as easy as might be imagined. German publishers keep complaining about considerable additional costs as compared to their competitors in the English-speaking countries, due to the fact that for the personnel, as well as for many of the authors, English remains a foreign language. Often, additional native speakers of English have to be co-opted to existing boards of editors, and of course also to be payed. Texts have to be checked by native speakers, or sometimes even translated, before publication, and printers need more time with the foreign language.

Even then, most of these texts produced at extra costs are linguistically of a lower quality than those produced in the English-speaking countries, according to present language norms. This seems unavoidable in a situation where the translators or the proofreaders, i. e. the individuals who produce the final text version, are often neither experts of the field nor professional writers.

It seems unlikely that the readers particularly appreciate these often less than perfect texts. An indication are the comments found in reviews, even from the side of sociolinguists, who are probably linguistically more tolerant than others, and who know, how difficult it is to produce a good text in a foreign language. Let me quote from some reviews by Anglophone sociolinguists on texts in English published by Germans. (They relate, in fact, to two books of which I myself was an editor or, respectively, co-editor.):

> "A decidedly German substratum peeks through in many of the papers written in English. I counted several instances of *und* for *and,* as well as German-style syntax." "In the first 26 pages, this reviewer counted thirty-four errors of proofing, style, and grammar. In some instances, the grammatical mistakes are so severe that near unintelligibility results (...)" (DiPietro 1990) "It also has to be said that some of the English written by non-native speakers is so bad, particularly when combined with the not infrequent misprints, as to be almost incomprehensible. Obviously, the editors did not have native speakers vet all such contributions." (Trudgill 1990: 194 f.)

The native speakers' evaluation of texts in English produced by non-native speakers is a sociolinguistic topic in its own right (cf. Coulmas 1987: 106 f., Ammon 1991: 266–268, 237). Fragmentary observation seems to indicate that norm expectations tend to be rather rigorous – in contrast to wide-spread assumptions of the considerable linguistic tolerance of the English-speaking world. If this proved true it would underline the difficulty for publishers from non-English-speaking countries to compete with publishers of the English-speaking world, not to speak of the difficulty for authors.

This may be the right moment to return to the feed-back-model of an explanation for the dominance of English as a language of science mentioned at the beginning, which could be hypothesized along the following lines. The scientists for whom English is a foreign language are disadvantaged in presenting their research, or also their teaching materials, on an international level. Part of this disadvantage is the linguistic unattractiveness of some of their contributions. This also adds to the unwillingness of other scientists to study these scientists' or scholars' languages, or for that matter any languages other than English, since not much of interest can be expected there as it seems. Instead, the language which already generates a greater quantity of publications also promises more attractive texts. It contains more comprehensive information on average anyway, as a result of the greater quantity of texts, and it of course also provides better channels for the dissemination of the own research results. Therefore, a language which has once gained a clear lead as a language of science naturally develops a tendency to even increase its dominance, irrespective of the scientific achievements of its native speakers.

2.5. *Costly but perhaps counterproductive reactions and an utopian proposal*

There have been various attempts and proposals at reducing those disadvantages of the non-English-speaking countries. One attempt which is costly and perhaps counterproductive, is language-spread politics. France seems to rank first in promoting its own language, or to counteract its retreat as an international language of science but also of other functions. Germany, however, is similarly engaged spending now annually close to 3 Billion Marks (roughly 1.5 Billion US $) for promoting its language. Though only a small portion of it is used directly for supporting German as a language of science, for example through subsidizing journals, a good deal is so used indirectly. Examples for such indirect support are:

- financing German language courses to help prepare students for studying, or researchers and professors for working, at German universities,
- making available teaching materials in German, or materials for German as a foreign language,
- sending German language teachers abroad,
- maintaining cultural institutes, mainly the *Goethe-Institutes,* or
- contributing to German-language schools and media abroad (cf. Ammon 1989).

Promoting the German language is only one of the various objectives of these efforts, but one which ranks high. This has become obvious in recent years when, after the breakdown of the Soviet Union, an additional amount of 50 Million Marks has been made available annually to support the teaching of German in Eastern Europe, or when practically all sectors of the Federal budget were cut to pay for the costs of German unification, but the funds for teaching German as a foreign language were increased.

So far, no serious attempt has ever been made to evaluate the effect of this expensive language policy. To my knowledge, no evaluative study of language-spread politics has been done elsewhere either, though a considerable number of countries are engaged in it (cf. Ammon & Kleineidam 1992; Phillipson 1992; Ammon 1994). It is quite uncertain, what actually can be achieved with respect to spreading a language, or averting its retreat, by such a politics, not to speak of the political or moral objections which could be raised against it. Notably, the German and the French language once spread around the world without such a policy. One could argue, in face of history, that spending more money on research, or on teaching at the university, rather than on language-spread politics would be the better investment, even for the sake of language spread. It might in the end contribute more to the language's appeal than does making it more affordable to study. In any case, so far none of the French- or the German-speaking countries' efforts have had the desired effect of giving a noticable boost to French or German as international languages of science.

A somewhat utopian but in the end perhaps more realistic possibility of reducing the disadvantages of the non-English-speaking countries may be *International English* – a term to be taken as a proper name. This would be a variety, or a set of varieties, of English for which not only the English-speaking countries themselves would define the norms. It can be conceived in analogy with the plurinational (pluricentric) languages (cf. Clyne 1992). In their case, each national center (country) is responsible for its own language norms, i. e. those of its own national variety, and not any other. Thus, Britain is responsible for, and only for, the norms of British English, the USA for those of US-English, etc. Attempts at norm control beyond these boundaries, which do in fact occur, should be refuted by the other countries. Accordingly, none of the English-speaking countries individually, nor all of them together, should be allowed to dictate the norms of International English. This proposal would be in line with the fact that the non-native speakers of English have come to be much more numerous than the native speakers (cf. Graddol 1999; also cf. for a related proposal which calls for the non-native speaker's right to linguistic peculiarities, Ammon 2000).

The consequences of this view are complicated. One would be, for instance, that as long as there is no agreement on the norms for International English, national peculiarities from non-English-speaking countries cannot really be evaluated as incorrect. If they cause difficulty of comprehension here and there, even for native speakers of some national variety of English, it should be these individuals' responsibility to remedy it by additional language learning. This relates also to national discourse peculiarities of scientific texts which do not always coincide with Anglosaxon norms (cf. Clyne 1987).

Another consequence of accepting International English as a variety in its own right would be the need of making plans for establishing its norms. This opens up a host of language planning questions, for instance with respect to the possibility of creating a supranational norm-setting institution, and so forth. The norms of International English would not have to be uniform but could – to some degree – vary for different language communities. As there is British English, US-English etc. today, there could be for example Chinese, Indian, French, German and other International English in the future. Related to these divergent, but of course not too deeply divergent, norms would be the question of how to link the various national languages with International English, for instance by regularly coining the terminology necessary for translating back and forth. This and numerous other ramifications of the idea of establishing International English as a variety, or set of varieties, in its own right, beyond any kind of national English, need of course much more elaboration and in my view deserve it (cf., also for more literature, Ammon 2000). A practicable solution along these lines might in the end, to some degree, reconcile the divergent linguistic interests of the English-speaking and the non-English-speaking world.

3. English even for teaching in the non-English-speaking countries?

For a good deal of German scientists, particularly in the natural sciences, English has come to be a more important language of research, or research-related communication, than is German. Recently, English has even been introduced for teaching at university level in Germany, in addition to German. The main reason given for its introduction is the necessity to make German universities more accessible to foreign students, which is considered important for the country's economic and political future. German universities drail American and British universities in number of foreign students, even though they ask for no tuition. Among various possible reasons, the requirement of German language skills is considered crucial; it is believed that many more foreign students would flock to German universities, if they could study in the language they already know, English. The same argument has been raised earlier for various smaller language communities in Europe and has led to the introduction of English as an additional language of university teaching (e. g. The Netherlands, Denmark, or Norway).

In the winter semester 1997/98, so-called "international programs" with English as the language of teaching have been started, mainly in the natural sciences and technologies and in economics, at undergraduate and graduate levels, with financial support from the central ministry of science and education. They have spread from initally 18 to 42 universities, roughly a third of the total number, in the winter semester 1999/2000. They are open to foreign as well as to German students, and they are taught by German professors with a sufficient command of English as well as guest professors from various English-speaking countries.

Evaluation of these programs will have to deal with the possibilities of insufficient English-language skills of the German professors, of difficulties of Anglo-Saxon professors in adjusting to students for whom English is a foreign language, of the availability of adequate teaching materials, or the question of whether high-quality foreign students are at all attracted who might want to acquire native English alongside knowledge in their subject. Beyond these more practical questions, it seems reasonable to foresee possible consequences for the German-speaking countries, or the German-language community, which deserve closer scrutiny. They can, for the moment, be distinguished into desirable and undesireable consequences. It goes without saying that the question of desirability *for whom* precisely, would have to follow. It would, however, carry us beyond the frame of the present article (cf. for details Ammon 1998: chapter E).

Desirable Consequences:

(1) Germany, and perhaps the other German-speaking countries, might develop a more refined English-language culture, with facets like scientists' more comprehensive English-language skills (at least scientists involved in the programs), scientists' increased confidence in using English (perhaps in spite of peculiarities of German English), less linguistic isolation of German scientists, and better chances for German publishers to find authors with a solid command of English.

(2) More foreigners might be attracted to study in Germany. Most of them would at least acquire some sort of Survival German. That way, studying German might even get boosted.

Undesirable consequences:

(1) Studying German abroad might decline if no knowledge of German would be required for studying in Germany.

(2) It may prove impossible to keep up full-scale teaching programs in German alongside English-language teaching. In this case, modernization of the German language might fall behind even further.

(3) The universities might become even more isolated from the rest of the society with a foreign language as the medium of teaching. Communication between academics and non-academics might become even more difficult, or at least appear so.

(4) The German nation is still largely seen as based on the common language (language nation). Thus, German citizenship is, as a rule, tied to a command of the German language, as the courts have confirmed vis-à-vis recent immigrants from Russia. If this language should be squeezed out of the nation's highest educational institutions, would it not appear like dissolving the nation, beginning from its head?

(5) Though the symbolic dissolution of the traditional German nation might be in line with European integration (which German politicians have spearheaded), supporting English over other European languages (which its introduction into German university teaching amounts to) seems hardly compatible with the idea of a multilingual Europe. Rather, it fosters the fear, raised so often particularly by France (not without its own linguistic interests), of a Europe dominated by the English language.

It will be an interesting sociolinguistic task to observe whether these potential conflicts will develop, and it will be a political challenge to resolve them, hopefully with the help of language planning.

4. References

Ammon, Ulrich

1989 Zur Geschichte der Sprachverbreitungspolitik der Bundesrepublik Deutschland von den Anfängen bis 1985: Kommentierte Dokumentation der Diskussion im Bundestag mit anschließendem Ausblick. *Deutsche Sprache* 3: 229–236.

1990 German or English? The Problems of Language Choice Experienced by German-Speaking Scientists. In Nelde, P. H. (ed.) *Language Conflict and Minorities/Sprachkonflikte und Minderheiten.* Bonn: Dümmler, 33–51.

1991 *Die internationale Stellung der deutschen Sprache.* Berlin/New York: de Gruyter.

1994 *Language[-]Spread Policy Volume 2: Languages of Former Colonial Powers and Former Colonies* (International Journal of the Sociology 107). Berlin/New York: Mouton de Gruyter.

1998 *Ist Deutsch noch internationale Wissenschaftssprache? English auch für die Lehre an den deutschsprachigen Hochschulen.* Berlin/New York: de Gruyter.

2000 Towards more Fairness in International English: Linguistic Rights of Non-Native Speakers? In Phillipson, R. (ed.) *Equity, Power and Education.* Mahwah, NJ: Lawrence Erlbaum Associates, 111–116.

–/Kleineidam, Hartmut (eds.)

1992 *Language-Spread Policy Volume 1: Languages of Former Colonial Powers* (International Journal of the Sociology 95). Berlin/New York: Mouton de Gruyter.

Brown, Penelope/Levinson, Stephen C.

[1978] 1987 *Some Universals in Language Usage.* Cambridge etc.: Cambridge University Press.

Clyne, Michael

1987 Cultural Differences in the Organization of Academic Texts. *Journal of Pragmatics* 11: 211–247.

1992 *Pluricentric Languages. Differing Norms in Different Nations.* Berlin/New York: Mouton de Gruyter.

Coulmas, Florian

1987 Why Speak English? In Knapp, K./Enninger, W./Knapp-Potthoff, A. (eds.) *Analyzing Intercultural Communication.* Berlin/New York: Mouton de Gruyter, 93–107.

DiPietro, Robert J.

1990 Review of Ulrich Ammon (ed.) “Status and Function of Languages and Language Varieties (…).” *Language Problems and Language Planning* 14: 298–301.

Graddol, David

1997 *The Future of English? A Guide to Forecasting the Popularity of the English Language in the 21st Century.* London: British Council.

1999 The Decline of the Native Speaker. *AILA Review* 13: 57–68.

Kröner, Peter

1983 *Vor fünfzig Jahren. Die Emigration deutschsprachiger Wissenschaftler 1933–1939.* Münster: Gesellschaft für Wissenschaftsgeschichte.

Levinson, Stephen C.

[1983] 1992 *Pragmatics.* Cambridge: Cambridge University Press.

Markl, Hubert

1985 Die Spitzenforschung spricht englisch. In Kalverkämper, H./Weinrich, H. (eds.) *Deutsch als Wissenschaftssprache.* Tübingen: Narr, 20–25.

Phillipson, Robert

1992 *Linguistic Imperialism.* Oxford/New York: Oxford University Press.

Price, Derek J. de Solla

1986 *Little Science, Big Science … and Beyond.* New York: Columbia University Press.

Schroeder-Gudehus, Brigitte

1966 *Deutsche Wissenschaft und internationale Zusammenarbeit, 1914–1928.* Diss. phil. Genf.

1973 Challenge to Transnational Loyalties: International Scientific Organizations after the First World War. *Science Studies* 3: 93–118.

Shenton, Herbert N.
1933 *Cosmopolitan Conversation. The Language Problems of International Conferences.* New York: Columbia University Press.

Skudlik, Sabine
1990 *Sprachen in den Wissenschaften. Deutsch und Englisch in der internationalen Kommunikation.* Tübingen: Narr.

Swaan, Abram de
1993 The Emergent World Language System. *International Political Science Review* 14: 219–226.

Trudgill, Peter
1990 Review of Ammon, Ulrich/Dittmar, Norbert/Mattheier, Klaus J., eds. 1987/1988 "Sociolinguistics/Sociolinguistic (...)" *Sociolinguistica* 4: 191–195.

Tsunoda, Minoru
1983 Les langues internationales dans les publications scientifiques et techniques. *Sophia Linguistica:* 144–155.

Wickler, Wolfgang
1985 Englisch als deutsche Wissenschaftssprache. In Kalverkämper, H./Weinrich, H. (eds.) *Deutsch als Wissenschaftssprache.* Tübingen: Narr, 26–31.

English as a Language of Instruction at Austrian Universities

Rudolf de Cillia and Teresa Schweiger

1. Introduction

Today English is without doubt the international language of science. The following contribution tries to rate the importance of English as a scientific language at the universities of the German-speaking country Austria, in particular based on the role English plays as a language of instruction at university level. The authors decided to restrict this essay to the dimension of teaching because there are no previous empirical surveys about this question in Austria and there is minimal public interest in this subject. An Internet search in the APA (= Austrian Press Agency) database[1] for "*Wissenschaftssprache*" ("scientific language") and "*Unter-*

richtssprache" ("language of instruction") in the context of English in newspaper articles published between 1 November 1990 and 7 January 1999 did not yield even half a dozen entries, of which only one text was explicitly about English as a language of instruction. Given this background, a comprehensive treatment of the subject English as a language of science in Austria, i. e. its importance as a language of research, scientific publications and teaching, has been impossible. The authors therefore confined their research to the use of English in scientific instruction. This can also be justified by the fact that there is a strong reciprocal influence between scientific instruction and research. Although in Austria scientific instruction and research are no longer considered an integrated whole as defined by Humboldt, they are still closely related to one another. One of the guiding principles of the 1993 University Organisation Act (*Universitätsorganisationsgesetz* [UOG 1993]) is the connection of research with teaching (Article 1, § 2 UOG 1993). Therefore, a chasm between research (in English) and scientific instruction (in German) may neither be assumed nor does it seem desirable. At any rate the description below of the current state of teaching in Austria, will give an insight into the language use at Austrian universities.

2. Legal framework of language policies

2.1. Austrian language policies

On the one hand, the political situation of languages in Austria can be characterised by the fact that Austria defined itself as a German-speaking country in Article 8 of the Federal Constitution[2]. In addition, there are two general provisions on policies concerning the German language. The language of instruction at Austrian schools[3] is in general German and foreigners who want to become Austrian citizens have to prove their knowledge of German as well.[4] However, this provision does not apply to the legally recognised language minorities, i. e. Croats, Slovenes, Hungarians, Czechs, Slovaks, Roma and Sinti (see de Cillia et al. 1998, de Cillia 1998). German also dominates the media. Only five percent of the Austrian newspapers and journals are published in languages other than German and the public radio and television corporation (ORF) broadcasts almost exclusively in German (see Landsiedler et al. 1998).

Another essential postulate of Austrian language policies was formulated by Michael Clyne as follows: "German (Austria) is not a language with strong planning agencies." (Clyne 1995: 199). Austrian language policies can be characterised as laissez-faire policies and this does in fact apply not only to the German language in Austria. There is no targeted and prospective planning of any of the languages used on the Austrian territory and there is no national institution responsible for language planning and policy co-ordination. As a rule, language-related measures are taken in response to political trends. The recurring conflicts with language minorities triggered for instance new legislation for schools (see de Cillia et al. 1998, de Cillia 1998). Yet another example is the inclusion of a provision on Austrian German in Austria's accession treaty to the European Union (Protocol No. 10) (see de Cillia 1996, 1997). Austria's EU membership, phenomena of globalisation, the opening of Eastern Europe and, connected with that, the increased cultural and language activities in Central and Eastern Europe have resulted in a loosening of the legal provision stipulating that only German can be used as a language of instruction in Austrian schools. The amended School Organisation Act (*Schulunterrichtsgesetz* [SchUG]) of 1997 allows the school board to institute a foreign language as language of instruction (working language) "if this appears useful in view of the number of foreigners speaking foreign languages and living in Austria" (see Article 3 § 16 of the SchUG as laid down in the Federal Gazette 1996/767). This provision deals with the languages of "migration minorities" (in Austria these are above all the Bosnian, Croat, Serbian and Turkish minorities) and in practice this is especially true for English as a working language in Austrian schools.

2.2. Legal framework of language policies at universities

The loss of importance of the German language as an international scientific language has hardly stirred public interest and is neither discussed publicly in research nor in the media or politics. A reason for this could be the laissez-faire character of Austrian language policies. As far as our research has shown, there are no relevant scientific publications about English as a language of scientific instruction and research at Austrian universities (regarding teaching in different subjects at schools see de Cillia 1998).

The legal framework at Austrian universities can be characterised as follows: On 1 August 1999 a new federal law, the University Studies Act (*Universitätsstudiengesetz* [UniStG]) entered into force which is binding on all Austrian universities and will override all former laws (e. g. General University Studies Act [AHStG] or Technology Studies Act [TechStG]) as of 30 September 2002. Previous laws and the UniStG will co-exist until then.[5] The UniStG enumerates the students' rights and duties. The rights and duties of the teachers are additionally laid down in the Civil Servants Act (*Beamtendienstrechtsgesetz*).

The provisions of the UniStG concerning language policies may be summarised as follows: the German language is regarded as a sine qua non for studies at Austrian universities (§ 37), even for PhD students. It is recognised on the other hand that German alone is not sufficient as a scientific language for scientists (§ 5). Admission to diploma or PhD programmes requires knowledge of the German language. People whose mother tongue is not German have to prove their knowledge of German by obtaining a so-called "*Reifezeugnis*" (the equivalent of A-levels or a high school diploma) at a German-speaking school. Another option is a supplementary exam, the "*Hochschulsprachprüfung in Deutsch*" (German Proficiency Examination) that serves as proof that a person has an adequate knowledge of both written and spoken German to study successfully and that the person is able to use the German language to the extent required *for the understanding of scientifically relevant texts* (§ 48 supplementary examination, emphasis added by the authors). § 5 (a constitutional provision) stipulates that the use of foreign languages in teaching and examinations, in scientific papers, in academic degrees and in documents may be regulated by a federal law.

There are hardly any explicit provisions concerning the knowledge of foreign languages as a requirement for admission to an Austrian university (see below). As a so-called "*Reifeprüfungszeugnis*" (A-levels, high-school diploma) is an admission requirement, knowledge of at least one foreign language is guaranteed. People who have completed higher secondary education in Austria in most cases also know a second foreign language and/or Latin. In fact no comments may be made on their degree of language proficiency. Theoretically it is possible that none of these foreign languages is English although in reality this hardly ever happens.

For some study courses it is a requirement to take additional exams in a modern foreign language before obtaining the first degree, e. g. business administration (*Betriebswirtschaftslehre* [BWL]), commerce (*Handelswissenschaft* [HW]), international business administration (English is explicitly required), social sciences, sociology, statistics, economics, business applications of computer science, economic and business pedagogics (*Wirtschaftspädagogik* [WiPäd]), translation and interpretation, as well as comparative literature studies. (BMWuV/BMUK 1999: 28) For all other studies the knowledge of foreign languages based on the A-levels examination or high school diploma is considered sufficient.

The technical universities are an exception: for the time being a section of the Technology Studies Act (*Technikstudiengesetz* [TechStG]) regulates the use of English for science teaching as follows: "Teaching in foreign languages § 16. It is the duty of the university boards to ensure that at least *8 teaching units a week per semester of the entire course of studies have to be offered in English on a regular basis*, with 2 of these 8 units in the first part of studies. Students shall take exams on these 8 units in English." (emphasis added)

There are more detailed legal provisions applying to Latin and Ancient Greek: a certain knowledge of Latin is required for about 40 scientific disciplines, e. g. modern languages, divinity, history, medicine, pharmacy and law. Ancient Greek is required for a dozen subjects, e. g. classical languages, archaeology and divinity. While there are hardly any regulations concerning the obligatory knowledge of the dominant scientific language, namely English, attention is focused on the key languages (*Erschließungsprachen*). In addition, universities do not offer an adequate range of foreign language and technical terminology courses. What is more, the courses offered are very often held under conditions that may be considered problematic from a didactic perspective.

Austria's biggest university, the University of Vienna (about 34–35% of all Austrian students), is not equipped with the appropriate infrastructure for foreign language training, e. g. a language centre. Language policies tacitly assume that the language skills required for studying must be either acquired before starting to attend university or in private language courses.

Regulations relevant for university teachers are contained not only in the UniStg but also in one section of the Civil Servants Act (*Beamtendienstrechtgesetz*) (Article IV.7, on the recognition of qualifications at the tertiary level). In accordance with this article, a university teacher has to produce evidence that his/her knowledge of the language of instruction or any other defined language/s meets the requirements. Admission of a teacher to a certain academic institution may depend on this evidence. The passage refers to "languages" (N. B. in the plural) although criteria for choosing one or several language/s are not defined anywhere.

This first assessment of current language policies in Austria reflects a conflicting picture: The legal provisions are rather vague in terms of language use at universities, with the central status of German, however, being retained. The role of English as the dominant scientific language is hardly taken into account in the relevant legislation. Despite this strong emphasis on German as the (only) "state language"[6], there is no indication that measures will be considered to counter a possible loss of status of the German language. However, English is about to replace German as the most important scientific language, which could bring about changes for the status of German in general and might lead to a situation where German becomes less attractive as a foreign language (see Ammon 1991, Ammon 1998).

Our highly limited sample showed that, in contrast to Germany[7], this question is neither addressed by politics nor the media and that it is more or less ignored by the Austrian public. There are no scientific studies on the implications of English as a scientific language for teaching at Austrian universities, either. This study is to provide some remedy for this unsatisfactory state of research.

3. An empirical investigation of the importance of the use of English in Austrian science

3.1. Difficulties in data gathering

As the present survey could not build on former research and studies we had to limit ourselves to "the role of English as language of scientific instruction at Austrian universities". First of all, research was restricted to universities, i. e. we did not include extra-university research centres and scientific institutions like the Austrian Academy of Science or the Ludwig Boltzmann Institute. Secondly, within the universities we confined the study to the use of English as a language of instruction.

A restriction to teaching at universities seemed sensible insofar as in Austria teaching in the strict sense of the word (e. g. classes open to all students) exists almost exclusively at universities while extra-university research institutions concentrate on specific research areas. The fact that we focussed our paper on instruction at universities may be explained by several reasons. The importance of German within the scientific communities may be similar in all German-speaking countries (e. g. Germany, Austria, and the German-speaking parts of Switzerland) (see Ammon 1998) but possible implications for scientific instruction depend much more on the political situation of the countries and especially on the specific situation of science. A survey of the role of English as a language of instruction at Austrian universities is more relevant for the Austrian universities themselves, particularly as there are no previous studies.

There are some publications[8] on the role of "English as a language used for publishing research findings at Austrian universities". Besides, numerous publications on English as a language of research for German-speaking scientists compensate for the above-mentioned lack of research data to a certain degree (see the bibliography in Ammon 1998, especially Skudlik 1990). In fact, this serves to explain why we have opted for addressing the subject "English as a language of instruction" instead of "English as a language of research" but by no means do we want to imply that there was no need for studies investigating the specific situation in Austria.

3.2. *Formulation of the questions and conception of the survey*

We restricted our survey to "language(s) of instruction" but this term is synonymous with "language used for teaching at universities". The following questions were raised: Is English a language of instruction at Austrian universities, i. e. are classes offered in English? If yes, are these subject-related classes or language classes, are they optional or compulsory? Is there a demand for classes in English at all?

According to Skudlik 1991, we can distinguish three different types of sciences: a) (pure) natural sciences, b) social sciences and c) arts and humanities. The influence of English as the dominant scientific language varies depending on the type of science. Science type a) can thus be called anglophone, type b) mainly anglophone and type c) dominated by the national language or polyglot.[9] This classification was by and large taken over by Ammon 1998 and his data on research and instruction in Germany confirm this classification.

Based on our data on university instruction, we should be able to formulate hypotheses about language use within these three types of sciences. In a second stage the attitudes of teachers towards risks and opportunities of using English for teaching at university level can be discussed. The random sample consisted of two parts: first a small random sample based on information on classes offered at all Austrian universities was taken and then a questionnaire was sent to university teachers to explore their feelings and attitudes towards teaching in English. Teachers were chosen from one representative subject of natural sciences (mathematics), social sciences (sociology) and humanities (history) respectively.

4. Classes in English at Austrian universities

4.1. *Scope of the survey*

The questionnaire should give an overview of the general situation in different disciplines offered by Austrian universities. The following data were of particular interest[10]: number of classes offered in a language other than German, the number of classes in English and the number of English classes. The period of investigation was four semesters, from 1996–1998, i. e. the academic years 1996/97 and 1997/98. Questionnaires were sent to all Austrian universities and university institutes.

Altogether there are 19 universities in Austria, more than half of this number are situated in Vienna, the capital. Graz, the second biggest Austrian town hosts 3 universities, as does Linz. Salzburg follows with two universities. There is one university each in Innsbruck, Leoben and Klagenfurt. The universities can be grouped into four categories:[11]

1) Universities which at least comprise one faculty of humanities, natural sciences, and law, in some cases also a faculty of divinity, medicine and economics. The following universities fit into this category: Karl-Franzens University of Graz, Leopold-Franzens University of Innsbruck, University für Bildungswissenschaften Klagenfurt, Johannes-Kepler University of Linz, Paris-Lodron University of Salzburg und Vienna University. As there is only one University of Economics in Austria but several faculties or disciplines of economics, we put Vienna University of Economics and Business Administration into this category as well. The same applies to the Catholic University of Linz.
2) Universities of Technology: Vienna University of Technology and Technical University Graz
3) Universities and Academies of Arts: Graz University of Music and Drama, Linz University of Art and Industrial Design, Salzburg University of Music and Drama Arts "Mozarteum", Vienna Academy of Fine Arts, Vienna University of Applied Arts and the Vienna University of Music and Performing Arts.
4) Engineering Universities and Universities of Veterinary Medicine: Leoben University of Mining and Metallurgy, Vienna University of Agricultural Sciences, Vienna University of Veterinary Medicine

All 19 universities were contacted (= 100%). Ten of the answers were positive[12] (= 52.6%), three negative[13] (= 15.8%) and in six cases no answer was returned (= 31.6%). The overall response rate was 68.4% but distribution was very uneven among the different types of universities.

The analysis of data[14] made it possible to answer the above questions concerning the significance of English as a language of scientific instruction and to limit the field of research even further. At any rate it should be stated that the individual questions raised would need in-depth investigation, preferably in a separate study. The following outline should give a first impression of the language situation at Austrian universities.

4.2. Analysis and interpretation of the data

4.2.1. Classes at type 1 universities:

No completely positive answer was received: two partly positive answers (Innsbruck, Vienna University of Economics, 25%), four negative answers (Vienna, Linz, Graz, Salzburg 50%) and in two cases there was no reply at all (Klagenfurt, Catholic University Linz, 25%). Due to this inadequate data situation we conducted an independent survey on a small scale . At least these sporadic returns indicate that the universities do not seem to keep records on English speaking classes themselves. For the survey conducted by us, one discipline per faculty at Vienna University was studied. Vienna University was chosen for several reasons: It is the biggest type 1 university in Austria[15] and it consists of all faculties mentioned in the definition. These are the following: catholic and protestant divinity, law, social sciences and economics, medicine, humanities, formal and natural sciences.[16]

The findings classified by faculties are as follows: at the catholic-theological faculty 218 classes were offered within the academic year 1996/97 and 214 classes during 1997/98. No classes were held in English. At the faculty of law 1.7% of all classes were in English (12 out of 700) during the academic year 1996/97. In the following year the number of classes offered increased by about 100 (700 to 800) but the percentage of classes in English remained constant (approx. 1.5%). Sociology was chosen to represent the faculty of social and economic sciences:[17] classes in English made up about 1.3–1.4% (1996/97 2 out of 141, 1997/98 2 out of 153 classes offered). Less than one percent of classes were offered in English at the department of philosophy at the Faculty of Humanities: 1996/97 2 out of 237 (approx. 0.6%) and during 1997/98 no classes were offered in English. The number of classes in English at the Faculty of Medicine was also negligible[18]: 2 in 1996/97 (approx. 0.05%) and 3 in 1997/98 (approx. 0.08%) out of about 3000–4000 classes. However, the Faculty of Medicine additionally offers postgraduate studies for foreign doctors in German and English ("*Hochschullehrgang für postpromotionelle Ausbildung ausländischer Ärzte in deutscher und englischer Sprache*"). Chemistry was chosen as representative discipline of the Faculty of Formal Natural Sciences, for which we obtained the following data: while 1996/97 8 out of 863 classes (approx. 0.9%) were held in English, the number decreased in 1997/98 (6 out of 935, approx. 0.6%). A two-hour class per semester was also on offer: "Reading technical journals in English language for scientists and students of medicine".

4.2.2. Classes at universities of technology (type 2)

The response rate was 75% (3 out of 4). Universities of this type have more precise regulations (section of TechStg 1991) concerning the use of English as a language of scientific instruction. One can assume that English is already a language of instruction and therefore classes in English must be offered at any rate. Our results corroborate this fact: In 1996/97 a total of 323 classes in English were offered at Vienna University of Technology (approx. 2.5% out of approx. 12,800), in 1997/98 the number amounted to 344 (approx. 2.6% out of approx. 12,900). At the Graz Technical University the figures are as follows: In 1996/97 144 out of approx. 4,100 (approx. 3.4%) and 1997/98 163 out of approx. 4330 (approx. 3.7%).

There were also classes on English as a technical language at both technology universities, e. g. at the TU Vienna "Technical English I+II (in English)", at the TU Graz "English for Engineers Practising Level I" and for faculty members a tutorial on writing publications in English.

4.2.3. Classes at universities of arts (type 3)

The response rate was approx. 83.3% (5 out of 6). At all universities of this type courses are offered in German only, with the exception of the Salzburg Mozarteum where "Advanced Studies in Music and Dance Education" are offered in English. However, this homogeneous picture of German does not hold true for two reasons: On the one hand, some theses for obtaining MA and PhD degrees, albeit only a few, are written in English, e. g. one MA and one PhD thesis at the Vienna University of Applied Arts. On the other hand, not every class scheduled in German is actually held in German, which can be deduced from the following respondents' answers: "[...] At our university there are only classes in German as the official language of instruction is German. (From experience, though, I know, that some teachers also hold classes in English) [...]" (e-mail from Michaela Heuberger, Graz University of Music and Drama, 16 Dec 1998); "[...] Basically, all classes at the University Mozarteum Salzburg are held in German. Due to the high percentage of foreign students, more than 50 percent, however, many teachers give individual tutorials, e. g. violin classes, in a foreign language, which in most cases is English." (e-mail from Robert Schiller, University Mozarteum Salzburg, 11 Nov 1998)

4.2.4 Engineering and veterinary universities (type 4)

The response rate was 66.6% (2 out of 3). The Leoben University of Mining and Metallurgy is a special case as it offers an academic discipline entirely in English (Studies in Petroleum Engineering) for the following reason: "Several years ago, the graduate studies in petroleum engineering at the University of Leoben, Austria (ULA) became aware of these new challenges their graduates have to accept in their active lives. [...] it was taken into consideration that the majority of the graduates begin their careers in English-speaking companies and countries. Consequently, English has been introduced as the official language of instruction and examination of courses in petroleum engineering."[19] In all other disciplines no class in English was offered in the academic years 1996/97 and 1997/98. Among the recommended facultative classes one could choose several classes for technical English, e. g. "English for Engineers", "English for Plastic Engineers or Ceramic Abstracts" (literature seminar for technical English + workshop).

At the University of Veterinary Medicine 2 classes were held in English within the given period of time (0.02% out of approx. 1000 per academic year). It was not possible to retrieve data from the University of Agricultural Science.

4.3. Conclusion

Without doubt English is used as a language of instruction at Austrian universities, albeit on a very small scale. In most cases adequate language policies are practically non-existent. If there is an explicit legal basis, there are automatically more classes in English, which applies to the universities of technology, for instance. There classes in English are compulsory and thus many more classes are offered in English so that students can actually take more than the required eight weekly hours.

The big type 1 universities with their wide range of faculties show a quite heterogeneous picture. It is remarkable that there are hardly any classes on English for specific purposes and the number of classes offered in English does not correlate with the anglophone profile of the sciences. The lowest profile of English is to be found at universities of type 3, i. e. the universities of arts. Officially there are no classes taught in English and there are hardly any language classes. The influence of English as the dominant scientific language seems to have the least effect upon universities of type 3. In general universities of type 4 also show a very low anglophone profile. The University of Leoben offers however an entire discipline in English.

In any case the issue of English as a language of instruction at Austrian universities would call for a focussed discussion, based on which more explicit language policies could be formulated. Possible structural changes of the instruction system of certain disciplines (introduction of both bilingual and English-only classes) and the debate on the status of German as a scientific language should not be left to the incidental results of laissez-faire policies. The following survey should give an impression of what teachers of different disciplines think about this issue.

5. Which language(s) of instruction? – Teaching in English from the teachers' point of view

5.1. Formulation of the questions and scope of the survey

The above-mentioned results as well as the data presented in secondary literature (see bibliography in Ammon 1998) suggest that the different use of English correlates with the differences between the individual sciences. (cf. Skudlik's tripartite system). A questionnaire was distributed to university teachers to examine these interrelations more closely and to complete the picture of university teaching. Also languages other than English were included in the survey. The questionnaire addressed the following subjects:

According to the teachers, are there differences regarding the use of English between the natural sciences, social sciences and humanities?
Does the English proficiency of teachers differ? Which other foreign language skills do teachers have and use in research and teaching?
To what extent are German and English used as languages of instruction?
How do they assess the students' language skills ?
How do they judge the "supply and demand" of classes, i. e. is there a balance between the number of classes offered and attended?
What attitudes do teachers have towards German / English / bilingual teaching at Austrian universities?

The questionnaires were sent to all heads of the department, who were asked to distribute them to their colleagues.[20] The survey was focussed on one representative discipline of natural sciences, social sciences and humanities. We chose this classification as it reflects empirical reality. Nevertheless it should be pointed out that the classification is not generally recognised.[21] The disciplines finally chosen were history, mathematics and sociology.

The following factors proved decisive for selecting these disciplines: If possible, the discipline was to be offered at all Austrian universities. Secondly, it was also important that curricula were not too different. Moreover, the objective of the discipline was to provide instruction in "pure science", i. e. the training of high-school teachers and applied branches of the science like technical mathematics or science of industrial management were not studied.

Five questionnaires (i. e. in total 100 questionnaires) were sent to twenty departments at the universities of Vienna, Graz, Innsbruck, Klagenfurt, Salzburg and Linz. The individual departments belonged to a university of type 1. The response rates for the individual disciplines varied greatly as is shown in table 1; a total of 46 questionnaires out of 100 were returned. The below percentages have to be interpreted bearing this overall response rate (less than 50%) in mind.

Table 1: Response rates of the individual disciplines

History (H)	23 (30)	76.7%
Mathematics (M)	16 (40)	40%
Sociology (S)	7 (40)	17.5%

Table 1 shows that the interest in "English as a language of instruction at universities" seems to be highest in the two disciplines labelled anglophone and polyglot, whereas it is clearly lower in sociology. Before embarking on an evaluation of the data, the curricula of the individual disciplines analysed should be examined. Compulsory classes, as defined in the curricula, give further information about the situation: In history there are neither compulsory classes in English nor English language classes. The situation is similar in mathematics. The curricula of sociology differ according to university and faculty:

At the Department of Sociology at the University of Graz one class in English is compulsory, i. e. "Introduction to sociology". Sociology at Vienna University (at the so-called *Grund- und Integrativwissenschaftliche Fakultät* [faculty of basic and integrated sciences]) offers one facultative class: "One foreign language (especially English, French, Spanish of an advanced level). If a foreign language is chosen: It is not possible to choose language classes but these must be classes on subject-relevant contents held in English." Furthermore, to complete the first part of the study course successfully, students have to demonstrate their skills in one language "relevant to the studies of sociology" other than German (either by means of A-levels/high-school diplomas or an additional examination). Another optional class exists for sociology students at Vienna University (at the Faculty of Social and Economic Sciences): an optional class in a modern foreign language can be chosen for the first diploma exam. The curriculum does not contain any provisions on the languages to be chosen. As regards sociology at the University of Salzburg, an optional class may be attended: "A modern foreign language with the educational aim of acquiring advanced skills in a foreign modern language in order to be able to participate in the international sociological discussion. English, French, Italian, Spanish, Rumanian, Portuguese (Brazilian), Russian, Polish, Bulgarian, Serbo-Croatian, can be chosen provided the University of Salzburg will be able to offer courses."

5.2. Language use of teachers

Information given by university teachers about their English language skills (table 2) support our hypothesis concerning the diverging anglophone profile of sciences. All numbers of the following tables should be read as percentages of the overall response rate (46 = 100%). The numbers are expressed in round figures.

Table 2: English language skills of university teachers[22]

	Understanding spoken English			Reading			Writing			Speaking English		
	H	M	S	H	M	S	H	M	S	H	M	S
Perfect	6(26)	4(25)	1(14)	7(31)	6(38)	1(14)	3(13)	4(25)	1(14)	3(13)	3(19)	0(0)
Excellent	7(31)	4(25)	4(58)	9(39)	8(50)	5(72)	7(31)	3(19)	2(28)	7(31)	4(25)	3(44)
Good	7(31)	7(44)	1(14)	6(26)	2(12)	0(0)	8(35)	7(44)	3(44)	8(35)	7(44)	2(28)
Moderate	2(8)	1(6)	1(14)	1(4)	0(0)	1(14)	4(17)	2(12)	1(14)	3(13)	2(12)	2(28)
Rudimentary	1(4)	0(0)	0(0)	0(0)	0(0)	0(0)	1(4)	0(0)	0(0)	2(8)	0(0)	0(0)

H = history, M = mathematics, S = sociology

According to their self-assessment, all teachers have some knowledge of English, albeit sometimes rudimentary skills but more than two thirds claim to fall within a range between "perfect" and "good". Analysing the skills listed above, we find that the highest percentage of perfect skills is attained by teachers of mathematics. Only history teachers claimed in three of four cases to have rudimentary skills. English language skills vary greatly much in sociology: According to the information provided by themselves, a remarkable number of teachers (72%) have very good reading skills but few have a perfect knowledge.

The results on the use of English and German for reading and publishing of scientific literature, scientific correspondence and lecturing support our hypothesis that degrees of anglophone profile vary in the different sciences. 13 teachers of mathematics claim to publish exclusively or frequently in English, none of the history teachers publishes exclusively in English. Two historians never publish in English, and two never give lectures in English. Three mathematicians and one sociologist read exclusively English scientific literature; no historian stated to read exclusively English scientific literature.

5.3. Language(s) of instruction

Data on the skills and the use of foreign languages other than English indicate that history departments are more polyglot and linguistically more diverse as compared to the departments of mathematics and sociology. 16 historians mentioned French, 13 Italian and 8 Spanish (cf. with no Spanish in mathematics and sociology). In addition, they mentioned Dutch, Latin, Slovenian, Bulgarian, Hungarian, Serbian-Croatian-Bosnian, Portuguese and Danish. The case of Russian is of particular interest. In all three disciplines 13 persons stated to know Russian; this may indicate the relative importance of Russian as a scientific language (see Ammon 1998).

To investigate the use of language(s) in teaching, we included questions about the use of German and English for lecturing at domestic universities and at the universities of German-speaking and non-German-speaking countries.[23] Multiple answers were possible. As expected, German dominates the disciplines and lectures in Austria: 14 historians, 10 mathematicians and 5 sociologists (i. e. 70% of all interviewees) claimed to use only German.

As regards lectures in non-German-speaking countries, English is the dominant language in all three disciplines; only 3 mathematicians (2%) never use German. 1 teacher of history never uses German, and 3 historians never use English. This can either mean that they do not give lectures in non-German-speaking foreign countries or that another language is used if it is not an English-speaking country. For example, 4 historians (9%) often/sometimes teach in Italian or 3 (6%) often/sometimes teach in French. This is also true for mathematicians (2 [12%] sometimes teach in French) but does not apply to university teachers of sociology.

Technical terminology is an important indicator of the influence of one scientific language on others. It is assumed that practically "German with a cluster of English terminology" (Ammon 1998: 245) is used and that teachers do not use terminology in a uniform way. (In scientific literature this "language mix" is often labelled pejoratively as *Deutschlisch* or *Denglisch*, see Ammon 1991). What are possible approaches to deal with English scientific terms? Is it possible to try to "germanify" or translate the term, or to stick to the English variant or to use either of the variants depending on the situation? Which strategies do Austrian university teachers use?

Table 3 summarises strategies used by university teachers of history, mathematics and sociology. The absolute numbers are followed by percentages in brackets.

Table 3: Strategies used by university teachers of history, mathematics, sociology (multiple answers possible)

	H	M	S
I translate the term	1(2)	4(9)	3(6)
I use the English term	7(15)	10(28)	3(6)
I use the term in a context-sensitive way	15(33)	7(15)	1(2)

The least used strategy of teachers is creating an independent terminology; surprisingly enough, interest in this strategy in a discipline defined as one oriented towards the national language (history) is rather low (with only 1 historian mentioning this option). At least in our sample, historians do not show great efforts to maintain and promote the "national language". Only a third uses the English term. Two of all interviewees opt for another "mixed" strategy: "I use the English term and explain it (its meaning) in German", "It depends on the term and its (common) usage".

The kind of technical terminology used is closely connected with the teaching aids that are available and used. Therefore one part of the questionnaire focused on them. By "teaching aids" those parts of technical literature are to be understood that were developed and produced for teaching. First the language used in written lecture material would show the anglophone profile of teaching most clearly because in most cases teachers would either author this material themselves or in other cases proof-read and approve manuscripts compiled by students. Moreover the lecture material addresses only a restricted group of readers / learners since they contain the contents of one lecture only.

Furthermore, standard text books, reference books and bibliographies are counted as teaching aids because they are indispensable for all students, no matter what science. In the context of considerations to use English as a language of instruction, teaching materials conceived for beginners, e. g. introductory works or exercise books, are of particular importance. Consequences of the introduction (or non-introduction) of an additional language of instruction would mainly affect first-year students and university teachers. As working in and with a second language is "a cognitive burden and a time problem" (Ammon 1998: 276) and as problems in understanding are to be expected, especially with beginners, it could become more problematic for students to cope with course contents. Appropriate teaching aids would hardly eliminate these problems completely but at least reduce them considerably. Electronic media are also included in the term "teaching aids". The generic term "electronic media" comprises media suited for teaching, i. e. overhead transparencies, slides and CD-ROMs. Scientific journals are also classified as teaching materials because the language used in a scientific journal shows the loss of status of a scientific language already at an early stage. Regarding the German language, Ammon 1998 stated that the percentage of articles published in German is marginal.

The respondents could choose between three answers (true – not true – I do not know) to answer the question: Are teaching aids in English available in your discipline? The answers are shown in table 4.

Table 4: Teaching aids in English language meet the demand for teaching materials in my discipline (figures in absolute numbers)

	H	**M**	**S**
True	1	4	2
Not true	22	12	4
I do not know	0	0	1

It is interesting that the statement "All teaching aids in my discipline are in English" is confirmed to some degree in all disciplines. University teachers could state additional languages in which relevant teaching aids were available. Only the historians listed further languages: Italian, Spanish, Portuguese, Slovenian and Hungarian. In all three disciplines teachers hold the view that relevant materials are available in German as well. In mathematics French and Russian seem to be of some importance (Russian has more scores in this discipline than in the departments of history and sociology!). As a matter of fact, in history relevant books, journals, etc. may be found in almost all languages.

If one attempts to make a general assessment, it could be said that a change towards bilingualism has partly taken place in the field of teaching aids. In every discipline at least some English teaching aids are used.

5.4. Language skills of students

In the penultimate section of the questionnaire we asked the teachers to state the languages they considered necessary for successful studies in their disciplines. Is the knowledge of German an absolute requirement even if English is the dominant research language? In a second step, teachers were asked to evaluate the students' language skills. Are teachers content with them as they are? Are enough technical-language classes offered? Are such classes necessary at all?

The answers given to the following statement "In my discipline other languages than English do no longer play an important role" support our theory of a tripartite classification of sciences as regards the use of English as a language of instruction. 5 mathematicians and 2 sociologists affirm that English plays an absolutely dominant role, an opinion not shared by historians. Above all historians did not agree with the following statement: "In my discipline other languages than English are no longer important", while more than half of all responding mathematicians and sociologists agreed with this statement. If English is not the only important language for a discipline, it can be assumed that teachers regard a good command of other languages as indispensable for students. We would like to briefly summarise which languages are considered necessary for studies.

All history teachers regard perfect reading, writing, speaking and understanding skills of German as necessary, all in all 21 out of 23 require perfect / good language skills. As for French and Russian, at least two thirds demand intermediate reading skills of history students. Students of mathematics have to have relatively advanced German language skills, 14 out of 16 teachers require a perfect/good command. Reading skills in French are recommended by 8 persons and 7 consider reading skills of Russian necessary. On the whole, within the three sciences discussed in this paper, Russian seems to play the most significant role in mathematics. As far as students of sociology are concerned, we expected a significant demand for foreign languages considering the curricula. The overall picture is rather diverse: German language skills are as important as in the other two disciplines. The high demand for French (6 out of 7 recommended very good / good reading skills) is striking. Russian, by contrast, does not play any role in this science.

Generally we can see that highly advanced skills of German are demanded by teachers of all disciplines. The anglophone profile of a science, i. e. the dominance of English in research, seems to have no great influence on the status of German as language of instruction. According to the teachers, students cannot study history, mathematics and sociology without at least a good command of German. This attitude mirrors the ambivalent view of politics and the media. The use of non-German teaching materials and the language use of teachers themselves (in lectures, classes at home and abroad) shows that "German only" is no longer a realistic option.

Reasons given for the necessity of German vary considerably, though. Mathematicians argue from a cognitive-didactic stance: "Studies would be too difficult if they were not taught in the mother tongue" – while sociologists and historians rather argue in a political-ideological manner: "In Austria German is spoken and science wants / has to communicate in this language publicly."

To conclude this subchapter, we want to comment briefly on how teachers judge their students' language skills and on teachers' attitudes towards supply and demand for classes in English for specific purposes. More than two thirds of all teachers interviewed were content with their students' command of English. In sociology high satisfaction on the teachers' part correlates with an adequate supply of language classes. In this discipline regulations laid down in the curricula seem to ensure that an adequate number of language courses is available to students. Moreover, this language policy seems to have created some problem awareness on behalf of the teachers: specific language classes are considered a desideratum or necessity.

5.5. Chances and risks of German / English / bilingual teaching

The last section of our questionnaire pointed out chances and risks associated with a possible change of the currently monolingual approach towards teaching. Teachers were asked whether they considered it necessary, desirable or not to continue teaching in German, to introduce bilingual (German, English) instruction or to teach exclusively in English. Not all respondents answered these questions. The following reasons were given: "too global", "depends on the subject (for some French is just as relevant !)." 38% of the teachers that did answer considered teaching in German language necessary and 28% desirable. This corresponds to almost 75% (17 out of 23) who prefer to stick to teaching in German.

Summing up, one could say that teachers of all disciplines advocate the introduction of English-language instruction (monolingual or bilingual [together with German]). Teachers argue in favour of bilingual teaching because it would enable students to meet the requirements of the international scientific community. Further arguments in favour of bilingual teaching refer to globalisation, internationalisation and, closely related to it, higher mobility, e. g. "For mathematicians an early training in the international scientific language (= English) seems to be sensible", "Studying scientific literature for an MA or PhD degree would be much easier for students [of mathematics]", "Phases of instruction in English to get used to [mathematical] technical terms", "A lot of [sociological] publications are written in English – English literature simply covers more", "One is no longer afraid of making harmless grammatical mistakes – early training for international contacts", "In connection with the general globalisation of science [history] and the internationalisation of universities I consider bilingual teaching an absolute desideratum", "internationalisation / globalisation also in the domain of science [history]". One person argued decidedly in favour of introducing bilingual teaching to facilitate "access to the world of Anglo-American science".

Which arguments are used against introducing bilingual instruction and the exclusive use of English? In history reasons are "discipline-specific", in mathematics there are cognitive-didactic reasons put forward against it: "As opposed to some other disciplines, historical publications are published in the important national languages [...] Especially for Central Europe Anglo-Saxon research presents a fringe [...] so teaching in English does not seem to be very conducive", "Other important languages exist for historians, e. g. Spanish", "No need [for it] in Austria, except if history is considered a testing ground for learning English", "Everything important can be said in German – learning of the [mathematical] content is most important – language is of secondary importance".

Which arguments are given in favour of the scientific language German? Interestingly enough, in mathematics more than 50% consider teaching in German still necessary. The necessity of classes in German is explained by the fact that instruction in a foreign language would make especially the first semesters much harder: "The difficulties for students should not be doubled by the use of a foreign language", "Otherwise start of studies too difficult", "Mathematics can probably only be studied in one's mother tongue including dialects", "At least in beginners' classes [instruction in German] because otherwise students would be overburdened", "Better understanding". Furthermore some persons refer to the tradition of German as a scientific language in mathematics and the advantage of being able to talk in one's mother tongue about mathematical subjects.

In history we find cognitive-didactic arguments: "Students can think more differentiated and can articulate themselves better in the mother tongue", "Science tries to delineate subtle and difficult facts in a differentiated way, which has to / should be done sensibly and almost necessarily in German by teachers and learners." Political-ideological arguments were as follows: "In German-speaking countries the knowledge of German is still an absolute principle", "History serves to assure people of their cultural identity – this is linked to a national culture", "In terms of a pluralist society and pluralist cultures on our globe", "In Great Britain and the USA it [history] is taught in the native language as well – we should be open for other cultures without giving up our own". The fact why such arguments are found in history most frequently may be explained as follows: History as a science is always (also) practised to legitimate the history of a state / nation / people.[24] The following arguments may be termed discipline-specific: "The subject 'benefits' after all from regional restraints and its reference to the roots", "There is no necessity inherent in the discipline", "Most historical sources of my research area are in German", "Because sources and literature relating to Austrian history are in German". A more general reason given is that "one-sidedness is always bad". In sociology cognitive-didactic reasons prevail: "I can express myself more precisely in German", "Students are hardly ever bilingual" "[...] all scientists should be able to express themselves precisely and clearly in their mother tongues".

In all three disciplines the (few) persons advocating monolingual (English) teaching consider the command of English and the students' ability to communicate in English at a very high level ("Graduates of our discipline must know English for their specific purposes") indispensable and therefore there is no need for teaching in German.

6. Outlook

Our small-scale empirical survey of English as a language of instruction at Austrian universities gives the following picture: The respective legislation provides fairly precise information on the mandatory knowledge of the "state language" German as well as Latin and Greek (required in some disciplines) but there are hardly any regulations concerning the knowledge of English as an entrance requirement for studies. The supply of language classes and specific language classes is relatively incidental and hardly sufficient if one considers the actual need for this language, the significance of English as a research language and as the language of scientific publications. English as a language of instruction is used to a certain extent particularly in the disciplines of natural science, most intensively at technology universities and least for humanities and cultural studies. According to our survey, some of the teachers would be prepared to accept other languages of instruction. The situation at Austrian universities reflects Austrian language policies in general: There is no targeted language policy planning. A systematic data collection on the (foreign) language needs in various study courses and disciplines would thus be urgently required. Furthermore, it seems necessary to discuss language policies in detail and in terms of a diversified European language policy.

7. Notes

1. APA homepage: http://www.univie.ac.at/apa/.
2. "The official language of the Republic is the German language, notwithstanding the rights of the language minorities which are guaranteed by federal law" (de Cillia 1998: 29).
3. § 16. 1 School Organisation Act (*Schulunterrichtsgesetz* [SchUG]).
4. Amendment of Citizenship Act (*Staatsbürgerschaftsgesetznovelle)* of 8 July 1998.
5. Information on the laws still effective and their content is provided on the homepage of the Federal Ministry of Education, Science and Culture: http://www.bmbwk.gv.at/.
6. See for example § 27 of the Rules of Procedure of the Federal Council (Upper House of the Austrian Parliament) which read as follows: "The German language is the exclusive language of negotiation of the Federal Council and its committees."
7. In Germany well-known newspapers, e. g. *Die Zeit* or *Die Frankfurter Allgemeine,* took interest in this question. This in turn stirred public interest. (see Ammon 1999: 230).
8. They confine themselves to individual disciplines, e. g. medicine or psychology (see Navarro 1996 on the language of medicine in Austria).
9. However, it should be noted that Skudlik's types refer to research rather than teaching.
10. Additional questions were asked concerning the language choice for MA and PhD (number of MA and PhD theses written in English, number of theses written in

an other language than German). As writing an MA or PhD thesis belongs to research rather than teaching we will not go into detail here.

11. Two reasons seemed to justify this classification: On the one hand we adopted a historically developed classification into "general universities" and "technical universities". This division was reflected in legislation before the "all-unifying" UniStg. Moreover, the situation of Arts Academies differs insofar as they were awarded the status of universities only in 1998. They have been organised quite differently (e. g. admission without a high- school diploma) and still are organised differently today (e. g. entrance examination).
12. Also partial replies were interpreted as positive replies, e. g. in the case of Innsbruck: Only the department of English and the faculty of catholic divinity replied.
13. Negative answers were those stating reasons why the data could not be evaluated, e. g. message from University of Salzburg: "[...] Unfortunately I have to inform you that your questions can not be answered because no statistical data are available to us [...]" (e-mail from Norbert Langler, GW-Dean's Office, 16 Dec 1998)
14. All data concerning universities of type 2 and 3 were gathered by the respective Academic Offices (*Studien- und Prüfungsabteilung*) and kindly provided to us.
15. Some data for illustration: in the fall/winter semester 34% of all beginners in Austria were enrolled at Vienna University. At this time 92,280 students were enrolled at Vienna University. (Source: Academic Office of Vienna University)
16. All data were taken from the official University Calendar, which appears twice a year (at the beginning of each winter/fall and spring/summer semester).
17. We did not count a) general courses offered by the faculty and related disciplines, which also can be chosen by students of sociology and b) the English language courses offered for Social science students.
18. The number of courses offered per year exceeds one thousand, therefore the total number will be given approximately.
19. Homepage of Leoben University of Mining and Metallurgy: http://www.unileoben.ac.at/.
20. We would like to thank all heads of department who distributed, collected and sent back the questionnaires for their co-operation.
21. See e. g. Budin 1993: 4 "[...]. All these classifications should not be overestimated. Due to the dynamic development of science, individual disciplines develop and change rather quickly so that sciences interact and thus create new "hybrid" disciplines."
    Unfortunately we cannot discuss in this paper whether the methods of individual disciplines are subject to the influence of English and a U. S. understanding of science. Nevertheless this question should not be ignored.
22. Percentages as well absolute figures are stated.
23. The last question was answered only vaguely by two persons for the following reasons: "I only give lectures at my home university", "I live in Austria". 3 more persons crossed out "tick where appropriate" and one person added a question mark.
24. We thank S. Manhartseder for this idea.

8. References

Ammon, Ulrich (ed.)

1989 *Status and Function of Languages and Language Varieties.* Berlin/ New York: de Gruyter.

1991 *Die internationale Stellung der deutschen Sprache.* Berlin/New York: de Gruyter.

1998 *Ist Deutsch noch internationale Wissenschaftssprache? Englisch auch für die Lehre an deutschsprachigen Hochschulen.* Berlin/New York: de Gruyter.

Atteslander, Peter

1995 *Methoden der empirischen Sozialforschung.* Berlin/New York: de Gruyter.

Bailey, Richard W./Görlach, Manfred (eds.)

1982 *English as a World Language.* Ann Arbor: The University of Michigan Press.

Budin, Gerhard

1993 *Wie (un)verständlich ist das Soziologendeutsch?: begriffliche und textuelle Strukturen in den Sozialwissenschaften.* Frankfurt a.M.: Lang.

Bundesministerium für Wissenschaft und Verkehr (eds.)

1999 *Hochschulbericht 1999.* Vol. 2. Wien.

–/Bundesministerium für Unterricht und kulturelle Angelegenheiten/Bundesgeschäftsstelle des Arbeitsmarktservice (eds.)

1999 *'99. Universitäten Hochschulen. Studium und Beruf.* Wien.

Bungarten, Theo (ed.)

1981 *Wissenschaftssprache. Beiträge zur Methodologie, theoretischen Fundierung und Deskription.* München: Fink.

Clyne, Michael

1995 *The German Language in a Changing Europe.* Cambridge: Cambridge University Press.

de Cillia, Rudolf

1996 Linguistic Policy Aspects of Austria's Accession to the European Union. *New Language Planning Newsletter* 10 (3): 1–3.

1997 "Alles bleibt, wie es ißt". Österreichs EU-Beitritt und die Frage des österreichischen Deutsch. *Jahrbuch Deutsch als Fremdsprache* 23: 239–258.

1998 *"Burenwurscht bleibt Burenwurscht". Sprachenpolitik und gesellschaftliche Mehrsprachigkeit in Österreich.* Klagenfurt/Celovec: Drava.

–/Anzengruber, Grete (eds.)

1993 *Fremdsprachenpolititk in Österreich, Mittel- und Osteuropa.* Schulheft 68.

–/Wodak, Ruth (eds.)

1995 *Sprachenpolitik in Mittel- und Osteuropa.* Wien: Passagen.

–/Menz, Florian/Dressler, Wolfgang U./Cech, Petra

1998 Linguistic Minorities in Austria. In Paulston, C. B./Peckham, D. (eds.) *Linguistic Minorities in Eastern and Central Europe.* Clevedon: Multilingual Matters: 18–36.

Drozd, Lubomir/Seibicke, Wilfried

1973 *Deutsche Fach- und Wissenschaftssprache.* Wiesbaden: Brandstetter.

Eichner, Heiner (ed.)

1997 *Sprachnormung und Sprachplanung. Festschrift für Otto Back.* 2nd ed. Wien: Edition Präsens.

Fishman, Joshua A./Cooper, Robert L./Conrad, Andrew W.
1977 *The Spread of English: The Sociology of English as an Additional Language.* Rowley: Newbury House.

Fluck, Hans-Rüdiger
1985 *Fachsprachen.* 3rd ed. Tübingen: Francke.

Landsiedler, Isabel/Kerschbaumer, Markus/Kettemann, Bernhard/Cossé, Markus
1998 L'Autriche. L'enseignement des langues étrangères. Faits, tendances et politiques. In *Bibliothèque des cahiers de l'institut de linguistique de Louvain* 92: 61–89.

Laponce, J.
1987 *Languages and Their Territories.* Toronto etc.: University of Toronto Press.

Muhr, Rudolf/Schrodt, Richard (eds.)
1997 *Österreichisches Deutsch und andere nationale Varietäten plurizentrischer Sprachen in Europa (empirische Analysen).* Wien: Hölder-Pichler-Tempsky.

Navarro, F. A.
1996 Die Sprache der Medizin in Österreich (1920–1995). *Wiener Klinische Wochenschrift* 108: 363–369.

Skudlik, Sabine
1990 *Sprachen in den Wissenschaften. Deutsch und Englisch in der internationalen Kommunikation.* Tübingen: Narr.

Universitätsdirektion der Universität Wien (ed.)
1996 *Vorlesungsverzeichnis der Universität Wien. Wintersemester 1996/97.* Wien: Holzhausen.
1997 a. *Vorlesungsverzeichnis der Universität Wien. Sommersemester 1997.* Wien: Holzhausen.
1997 b. *Vorlesungsverzeichnis der Universität Wien. Wintersemester 1997/98.* Wien: Holzhausen.
1998 *Vorlesungsverzeichnis der Universität Wien. Sommersemester 1998.* Wien: Holzhausen.

Wodak, Ruth/de Cillia, Rudolf/Reisigl, Martin/Liebhart, Karin
1999 *The Discursive Construction of National Identities.* Edinburgh: Edinburgh University Press.

The Presence of English at Swiss Universities

Urs Dürmüller

1. Introduction

The English language, it has been noted widely and repeatedly, has been spreading all over the world ever since the end of World War II. In Switzerland, a country with four national languages, it has not only become the most popular foreign language and the language of choice for international communication, but also a *lingua franca* available to the Swiss for intranational communication (Dürmüller 1992; 1997). The increasing use of English inside multilingual Switzerland runs contrary to the efforts in language policy meant to encourage "polyglot dialogue" among the Swiss. When striking up a conversation with a Swiss from a different language background, instead of using the second national languages taught at school, the Swiss tend to use English. This has led to a change of the status of English in Switzerland, which is due not only to spread factors like material benefits and cultural affiliation, but also to the fact that English has established itself as the language of choice in quite a number of different domains. It is the use of Englishes for Special Purposes (ESP) in so many areas that strengthens the position of English within the national language repertoire. The many ESP varieties appear as the branches of a well developed additional language. It is this English as an additional language (EAL) which is seen as also taking over *lingua franca* functions inside multilingual Switzerland.

Science has long been singled out as a field within which English has attained a commanding presence world-wide. That is why, in Switzerland, too, the language of science tends to be English. Academia is where one would expect scientific language to be used foremost. It is in the universities and their research groups that one would try to find evidence of the language of computer and medical science, of biochemistry and engineering, etc.

Switzerland has four language areas, but has universities only in three of them. In German-speaking Switzerland there are the universities of Basel, Bern, St. Gall and Zürich and the Swiss Polytechnic (Zürich). In French-speaking Switzerland there are the universities of Geneva, Neuchâtel and Lausanne and the Swiss Polytechnic (Lausanne). The University of Fribourg/Freiburg is French- and German-speaking, being located in a bilingual city and canton. Italian-speaking Switzerland has the Università della Svizzera Italiana. It is only the tiny minority group of Rheto-Romance speakers (6% of the population) that do not have their own university.

All these state universities would be expected to do their business – administration, teaching, research – in the local territorial language, thus following the tacit maxim of Swiss language policy which has made possible the peaceful co-existence of diverse language groups within one nation.

Officially, there is no place for English in Switzerland. Nevertheless, the language is present, with more L1 speakers than Rheto-Romance, and with more L2/3 speakers than any of the national languages. As there are no regulations whatever that would ban English from the Swiss universities and as the universities themselves do not have any clearly defined language policy, English can be assumed to have made its forays into academia, too.

To what extent then is English used in Swiss universities? Do the universities keep to the national languages or have they switched to English? If English is used, is this as an auxiliary language for clearly defined purposes? If there are ESP varieties in the universities, how far have they spread? Are they limited to talks among experts? To publications? Are they also to be found in academic teaching, replacing the local national languages as the languages of instruction?

These are some of the questions I want to examine in this paper. I am basing myself mainly on data made public by the universities themselves (course catalogues, webpages, research proposals, publications). Not all of the universities offer the whole range of academic subjects, the Università della Svizzera Italiana, e. g., is a school of architecture only. I will therefore not establish a compilation of data according to language territory, but use the German-speaking University of Bern as my main example and refer to other – especially French-speaking universities – as needed. Bern university representatives have also answered questions regarding the information transmitted by the data.

2. English in course listings and on webpages

In order to learn which languages are used for instruction, the titles of the courses listed by the universities would appear to be reliable sources of information. According to these course catalogues, all the Swiss universities list most of their courses in their local territorial language.

Table 1: Courses taught at the University of Bern in the academic year 1997/98 according to language of instruction in official course listings

	Languages used in teaching					
Language	**Total**	G	F	E	I	**Other**
Teacher training	996	560	306	81	49	
Natural sciences	990	967	13	10		
Humanities	1094	887	62	94	31	20
Veterinary	159	158		1		
Medecine	742	742				
Law/economics	275	260	5	6	4	
Theology	187	186	1			
Interdisciplinary/ continuing education	227	209	8	6	4	
Total	4670	3969	395	198	88	20

(G = German, F = French, E = English, I = Italian)

Out of a total of 4670, there are 198 courses, which, according to their titles, appear to be held in English. This is only 4% of all the courses taught. The number becomes even less important if one takes into consideration the fact that most of these courses are offered in teacher training or the humanities where they are clearly linked to English language subjects. By comparison, French appears to be doubly as strong as English, with a percentage of 8. But again, most courses are offered in teacher training and the humanities and are linked to French language subjects.

If one leaves teacher training and the humanities aside, this first set of data seems to leave no doubt about English being of only marginal importance as a language of instruction at the University of Berne.

A cursory inspection of the course catalogues of other Swiss universities – in German-, French- and Italian-speaking Switzerland – confirms this impression. The Swiss universities advertise their courses in the language of the territory in which they are situated. Other languages, Swiss and foreign, appear above all in the course listings of the humanities where their occurrence is motivated by the use of those languages as subjects of instruction.

The University of Fribourg might appear as an exception, but isn't. Fribourg, the canton and the city, are bilingual, and so is the university. The territorial languages for Fribourg are French and German, and accordingly, the courses are offered in these languages predominantly.

A different picture emerges if, instead of the printed guides to the academic programs of the universities, one turns to the homepages of the universities and their institutes/departments as published on the world wide web.

In all instances – except for the Università della Svizzera Italiana, which is Italian-only – the universities do not only have a homepage in the language of their territory – German or French – but also in other languages. Typically, one finds the three main Swiss languages, German, French and Italian, while Rheto-Romance (RR), the minority Swiss language, is never included. Instead it is the foreign language English which is always present and often appears as the language of the opening page with links to the pages in the other languages.

The fact that the selectors are frequently set to English rather than to the page in the territorial language – which is especially so in German-speaking Switzerland – indicates that the universities are aware of the function of English to transport information world-wide.

Although the reason for having English homepages is international communication, the fact that these pages exist and often come up first when one chooses a university webpage, may also strengthen the position of English as a language for intranational communication in multilingual Switzerland. The probability is high that Swiss readers looking for information about a university will not bother with moving to the webpage in their own or in another Swiss language if they are confronted with English first.

If one moves from the frontpage of the whole university to the level of the faculties and the departments/institutes, the English option becomes

Table 2: Languages used in university homepages

University	Swiss national languages				English
	G	F	I	RR	
Basel	x	x	x	0	x
Bern	x	x	x	0	x
Fribourg	x	x	x	0	x
Neuchâtel	x	x	x	0	x
Geneva	x	x	x	0	x
Lausanne	x	x	x	0	x
Zürich	x	x	x	0	x
St. Gall	x	x	x	0	x
Swiss Polytechnic Lausanne	x	x	x	0	x
Swiss Polytechnic Zürich	x	x	x	0	x
Svizzera Italiana	0	0	x	0	0

(G = German, F = French, E = English, I = Italian, RR = Rheto-Romance)

more of an English obligation. The farther down one moves hierarchically, the more probable it is that the webpage in question will be in English only.

While the universities are aware of Switzerland being a multilingual nation, offering their pages not only in English, but also in three of the four national languages, the faculties, departments and institutes, as a rule, do not bother with Swiss multilingualism. Their webpages are either just in the territorial language or in English. Links to a page with the identical information in another language are rare. The example of the departments/institutes of the University of Bern (www.unibe.ch) show a preference for German or English only, those of the University of Lausanne (www.unil.ch) for French or English only.

Which of these two languages, the Swiss territorial language or English, is given preference varies from department to department. There is a clear correlation between choice of language and academic discipline. The local territorial language (German or French) is favored by the humanities, by theology, and law, English by medicine, veterinary medicine, and the natural sciences.

However, the faculties and departments are not homogeneously either local language-oriented or English-oriented. Table 3 correlates academic disciplines and the dominant language only. The local territorial language or English may be used exclusively in all the pages or it may give way here and there to other languages. A closer look at the University of

Bern webpages in the humanities, e. g., shows that three quarters are indeed in German only, but one quarter in both German and English. English webpages are offered by psychology and some of its branches. On the other hand, the Research Center for Multilingualism – ironically – appears as monolingually German only. Law and economics open in English, but have all of their further pages in German only.

Table 3: Correlations between academic discipline and language of department/institution webpage

Academic discipline	Language of preference on webpage			
	Swiss territorial language		English	
	G: Bern	F: Lausanne	Bern	Lausanne
Humanities	X	X		
Theology	X	X		
Law	X	X		
Economics	X	X		
Medicine			X	X
Medical services	X	X		
Veterinary medicine			X	X
Natural sciences		X	X	

(G = German, F = French)

Medicine, by contrast, is more clearly dominated by English. Links to alternative pages in German are not found. However, German pages open for the public services in psychiatry and neurology, for medicine history, and medical continuing education.

The natural sciences are the most clearly English dominated ones in Bern, though not in Lausanne. In Bern there are very few links to alternative pages in German. Computer science, there, is the only branch that shows exemplary Swiss links: the page opens in English, but then one may be taken to German, French or Italian. The natural sciences at the University of Lausanne open consistently in French, and only a third have alternative pages in English; none of them show German or Italian options.

A cursory check of the webpages of the Universities of Basel (www.unibas.ch) and Geneva (www.unige.ch) confirms the following general observations:

1) The Swiss national languages (German, French, Italian) are found at the level of the universities only. Rheto-Romance is never included. English always is.

2) Faculties, departments and institutes are monolingual or bilingual, offering webpages in either the local territorial language or in English or in both.
3) In Italian-speaking Switzerland (Università della Svizzera Italiana: www.unisi.ch), only Italian is used. In French-speaking Switzerland, French is frequently the only language offered: it is dominant even in the Natural Sciences. In German-speaking Switzerland, German and English have about an equal share.
4) The choice of the language of the webpage is left to the departments and institutes.

3. English for research proposals and publications

In the humanities, publications are mainly in the local national languages (except for departments of English that would publish in English as a rule). Publications by staff/affiliates of the Department of History, University of Basel, for 1994, counted a total of 79, of which 70 were in German, and only three in English. A similar ratio applies to the Bern Department of History publications. In 1997 there were 100 "recent' publications, of which 83 were in German, seven in English.

Other parts of the humanities may have more English language publications, e. g. ethnology (Bern), where a quarter of the publications are in English. Similarly, in linguistics, a quarter of the publications are in English, including two dissertations.

Table 4: University of Bern linguistics publications (1970–1997)

Publications	Language			
	German	English	French	Other
1970–1987	8	1	6	0
1988–1997	14	7	1	0
Total 37	22	8	7	0

What is interesting here is that since 1988 the French-English ratio has changed: while from 1970 to 1987 there were six items in French and one in English, from 1988 to 1997 there were seven English and just one French items.

In economics, both in German- (Bern) and French-speaking Switzerland (Lausanne), most publications are in English. And in medicine and the natural sciences English appears as the language of choice for publications: At the University of Bern, in neurology, of the publications for 1996, 45 are in English, 17 in German, 1 in French, 1 in Italian. In visceral surgery, there are 110 publications in English, 27 in German. In internal medicine, there are 24 publications in English, 5 in German. Applied physics, laser department, lists 243 publications in English, 33 in German. Microwave physics lists 149 publications in English, 9 in German.

These statistics are in tune with the results reported by Murray and Dingwall (1997), who did a survey involving 110 junior and senior researchers at five Swiss universities in both French- and German-speaking Switzerland: There is a much higher number of English publications in the natural sciences and medicine than in the humanities and social sciences. "The average percentage of papers and talks in English reported by researchers in the humanities and social sciences was 32%, and in the natural sciences, 57%. Swiss biomedical researchers reported the highest proportional use of English, with an average of 79%" (55).

Murray and Dingwall also followed up the languages used in project proposals selected for financial support by the Swiss National Science Foundation in 1975, 1985, and 1995. They found a dramatic increase in the use of English over those twenty years. However, it was above all the life sciences that had switched almost completely to English (from 14% to 91%); the natural sciences went up from 2% to 58%; and the humanities, which were at 0% in 1975, moved up to 6% (1997:55).

That, overall, the local national languages, are still slightly in the lead is confirmed by the University of Bern data on the financing of research projects for 1997. Out of 529 proposals 287 were titled and described in German, 238 in English. It was only in medicine that English clearly outdid German, while in the natural sciences (which, in Bern includes earth sciences) German and English had roughly the same share.

Increasingly, research proposals and reports of research results in papers and other publications tend to be written in English rather than in the local languages. As most of these items come from the natural sciences and from medicine, one might get the impression that in the field of research Academia is dominated by English. Indeed, also psychology and economics have largely switched to English. However, in language and literature studies, in history and the social sciences, in art and music, in theology and law the local territorial languages do not appear to be seriously challenged.

Table 5: Financing of research projects at the University of Bern as listed for 1997: Titles/descriptions in language

Research proposals	Language			
Field	Total	G	E	F
Theology	3	3		
Law/economics	36	33	3	
Medicine	235	86	148	1
Veterinary	42	25	17	
Humanities	69	62	6	1
Natural sciences	132	67	63	2
Teacher training	1		1	
Interdisciplinary	11	11		
TOTAL	529	287	238	4

(G = German, F = French, E = English)

4. English in graduate and post-graduate teaching

Generally, teaching is in the local national languages, i. e. in German (at the Universities of Basel, Bern, St. Gall, Zürich in German-speaking Switzerland), French (at the Universities of Geneva, Lausanne, Neuchâtel in French-speaking Switzerland), German and French at the University of Fribourg/Freiburg in the bilingual canton of Fribourg/Freiburg, and Italian at the Università della Svizzera Italiana). In the humanities, the various language and literature departments will use their subject languages also as mediums of instruction, not only for the Swiss national languages in areas where they are not L1s, but also in the case of Spanish and English. Further languages and literatures may be taught, partially or entirely, in the local national languages.

Officially, then, English is not a medium of instruction at Swiss universities. Where it is, the subject-matter – courses in English language and literature – dictates its use. Nevertheless, a few English course names appear in the official listings published by the universities in fields not related to the teaching of English language and literature. This is the case in law and economics and in the natural sciences. These courses appear as exceptions to the rule.

However, if one turns to the detailed programs made available by the departments themselves, many more course titles and course descriptions appear, now above all in the natural sciences and in medicine. What the course catalogues of the universities do not say is that many courses in graduate and postgraduate education are held in English. Again, this is typical for the natural sciences and medicine.

At Swiss universities, undergraduate students are instructed in the local territorial languages in all disciplines (except, of course, in the language and literature courses mentioned), i. e. also in the natural sciences and in medicine. English is used in undergraduate teaching only if exceptional circumstances apply: new faculty or visiting faculty from abroad, short courses instructed by visitors, or if the subject itself is English. Additionally, English may be used in reading assignments everywhere, not only in the natural sciences and in medicine, but also in the humanities and in the social sciences. E.g., the ratio of English to German in the reading lists posted for proseminars, seminars and lectures in politology (Bern) is 5 to 3.

On the other hand, many advanced courses and seminars are taught in English, one reason being that this is often the only common language considering the relatively large number of graduate students, post docs, and visiting scientists from abroad, another reason being that English is indeed the only generally accepted common language of the scientific community and the language in which research is discussed. In many fields the most significant international science journals publish in English and just about all international meetings use the English language almost exclusively.

A third reason for the choice of English in graduate courses is that English can also serve as a *lingua franca* not only within the international scientific community, but also for Swiss students and scientists. Close collaboration in graduate and postgraduate teaching between the Universities of Bern (German-speaking), Neuchâtel (French-speaking) and Fribourg/Freiburg (both German- and French-speaking) as well as between all the French-speaking universities and the University of Bern favor the use of English as a language of instruction again. For such courses, not only the students at Bern, where there is no detectable reluctance to make use of English as a language of instruction, but also the students at the French-speaking universities – where the general use of English is not always welcomed – are addressed in English, and in English only (e. g. Coordination Committee of the Suisse Romande for doctoral students in Biological Sciences) and they are clearly warned that "This course will be taught in English" (e. g. geochemical modeling of natural and contaminated groundwaters, BeNeFri 7350, 1997/98). There is also collaboration between Swiss universities and neighboring French universities. Although, on the one side, French is a Swiss national language spoken either as L1 or L2 by the students involved in these programs and, on the other side, French language policy actively discourages the use of English instead of French, English is the language used in the lecture cycles for French and Swiss doctoral students (cross-border University Convention of the Rhône-Alpes Region of 1990).

English also makes participation by foreign students from other countries possible. Their contributions would be in English, and very frequently the Swiss students and instructors would then also switch to English. (Aegidius Pluess, computer science, personal communicaton, 2/98).

Apart from oral presentations and discussions also reporting of research in dissertations may be in English. In mathematics, e. g., the majority of doctoral dissertations are written in English (Joel Adler, mathematics, personal communication, 2/98,), while in computer science this is not the case (Aegidius Pluess, computer science, personal communication, 2/98). Outside medicine and the natural sciences, where English is the rule for dissertations, a dissertation in English is still an exception.

Murray and Dingwall (1997: 56) noticed that the number of dissertations written in English at Swiss universities had more than tripled between 1975 and 1991, from 135 to 422, or from 8% to 20% of all dissertations. Their findings confirm that this trend is particularly strong in the natural sciences – in 1996 61% of all natural science dissertations written at the University of Zürich were in English. The comparison with the University of Lausanne for the same year and the same field – 39% of the natural science dissertations written in English – once more points to the smaller popularity of English in French-speaking academic Switzerland.

A closer look at some of the departments shows that the use of English as the language of reporting and discussing research has affected the choice of the languages of instruction at higher levels. As graduate and especially post-graduate students are involved in ongoing research, English quite often takes the place of the local Swiss language as the language in which courses are taught. At the Institute of Geology, Mineralogy and Petrology, University of Bern, in 1998, e. g., there were three advanced seminars that were instructed exclusively in English. In three other seminars the speakers could choose their language of preference (German, English, or French). Several graduate courses were also instructed in English. Contributions to graduate and postgraduate seminars and colloquia were posted in English. The clearly Swiss names of the presenters were paired with English titles. Talks would be given in English, and the discussion would be in English, even if everyone present was a speaker of English as a second language only.

According to professors responsible for such seminars and colloquia, there does not seem to be a problem with the use of English as a second language, neither for the instructors, nor for the students. Students are reported to participate without any problems, choosing English over German or French also where using one of the Swiss languages might be possible (Aegidius Pluess, computer science, personal communication, 2/98). The level of proficiency in English is said to be satisfactory in all cases, and quite good in most cases. French-speaking students, however, or visitors from France, are reported to often lack adequate knowledge and usage of the English language (Urs Maeder, earth science, personal communication, 2/98).

Apparently, in medicine and the natural sciences, where the reporting of data is centrally supported by visual means (graphs, tables, slides, etc.), the verbal fluency of the presenters is less important than in the humanities and the social sciences. When, in 1985 at the University of Bern, students of English taped and transcribed a postgraduate colloquy in immunology and a lecture with question-and-answer period in nuclear physics (both held in English exclusively) so that the performance of the speakers could be evaluated, lots of shortcomings were found. However, these were mostly due to interferences from the speakers' L1s affecting their pronunciation of English words. As the choice of vocabulary was correct and sentence patterns simple, the flow of information was not affected. International Scientific English jargon, it seems, will tolerate quite a number of imperfections – which exactly, and up to what point, this remains to be investigated separately.

5. Conclusions

The spread of English has not stopped at the doors of the Swiss universities. As this spread is largely due to the increasing use of ESP varieties in a large number of fields, it is not surprising that English also appears as the language of choice in certain academic disciplines. There, English has definitely become the language of research. As research is an integral part of graduate and especially postgraduate education, English quite naturally also is the language used for presentations, for discussions, and for written reports. Which means that, quite often, English is also used for degree-required work, such as the writing of theses and dissertations, and, more rarely, in oral examinations, and it means that where a course is devoted to the discussion of research, particularly ongoing research, the language chosen for the course is English as well.

The academic disciplines, within which English is used in this way, are first of all the medical and natural sciences, then, disciplines like psychology, parts of the social sciences and economics; theology, law, and most disciplines in the humanities can still do without English most of the time.

The use of English at the Swiss universities on the graduate and postgraduate levels differs from discipline to discipline. It depends on whether English is already established as the language of that discipline outside the Swiss universities. Where it is and where there is a need to participate in the international scientific discussion, English will be used without reservations. Where English is only one among several languages serving a discipline, as, e. g., in history, the pressure to use it is much smaller. Here, the decision seems to be up to the researchers/professors and their eagerness to reach an international audience.

Although, in Switzerland, the universities are state institutions, they do not appear to be under any obligation to follow a particular state policy regarding the use of languages. All of them have accepted the language of the territory within which they are situated as their prime language. However, they do not control whether the local territorial languages are used by all the departments and institutes without exception. That is why the languages of the webpages depend on whoever takes the responsibility in these departments and institutes. It appears that the webmasters are autonomous with regard to the languages they use on their sites.

There are, however, certain attitudinal differences with regard to making use of English. In German-speaking Switzerland, outside the humanities and law, where the use of German is traditional, there are no reservations with regard to using English, whether on homepages, in teaching upper-level courses or writing papers. Pragmatism rules, what counts is easy exchange of information and the productive dialogue across language boundaries. English is accepted as a lingua franca, and as the language of the club. If you can express yourself in Scientific English, you belong. In this way, English has the same function as did Latin in the Middle Ages, giving access to the community of scholars and maintaining partisanship within it.

In French-speaking Switzerland, on the other hand, and especially, in Italian-speaking Switzerland, there is greater hesitation to write websites and to teach courses in English, or, to put it differently, the bond between the universities and the territorial languages is stronger. Thus, one is tempted to establish a correlation between the relative openness towards English and the minority/majority status of the territorial language. The larger the language territory is and the more speakers of the national Swiss language it has, the more open it is to allow English in. And the smaller the territory is, the less likely it is that English will be used by the university. For publications and research proposals, however, this correlation is not valid. Nor is it for teaching graduate and postgraduate students in the medical and natural sciences. Where the goal is to reach the international community of scholars, also the representatives of French-speaking Switzerland will use English. And where students and scholars meet to discuss research written up in English, also the French speakers will switch to English.

The spread of English in Switzerland has made Grin (1997) look at the languages taught in the Swiss universities and the economical and political benefits of knowing languages in general and English and the Swiss national languages in particular. He acknowledges the material benefits that come along with English language competence in Switzerland. He is able to quantify the financial benefits of a quasi native-speaker competence: If you are proficient in English, you can earn up to a fifth more than a person who isn't. Some financial benefits are also attributed to proficiency in German and French, though they are clearly much smaller. For Italian, Rheto-Romance and further, non-Swiss, languages these benefits are negative.

Grin forgets to mention that the motivation for Swiss citizens to acquire a good, even excellent competence in English is not simply material. There are cultural benefits also in the case of English, not only in that of knowing additional Swiss languages, as he suggests. And what seems to be especially relevant for Academia, there is a motivation for mastering English which is linked to the ego: Swiss academics who publish their research results in English, who address an international audience in English get recognition, are quoted, and can become members of the international academic community.

Researchers interviewed by Murray and Dingwall indicated that "not using English would lead to exclusion from international scientific forums, to a lowering of regard for their research, as well as to a loss of recognition for Swiss science in general." English, Murray and Dingwall conclude, is thus viewed "not so much as a matter of preference but as a communicative necessity in the academic world" (1997:56). This is recognized by those institutes/departments that consider themselves part of the international community of scientists: "I think it is important for students to be very familiar with scientific English. For graduate students it is a must. If possible, also undergraduate students should get exposed to the English technical terms." (Beda Stadler, biology, personal communication, 2/98)

6. References

Dürmüller, Urs

1992 The Changing Status of English in Switzerland. In Ammon, U./Hellinger, M. (eds.). *Status Change of Languages*. Berlin/New York: de Gruyter, 335–370.

1997 *Changing Patterns of Multilingualism. From Quadrilingual to Multilingual Switzerland.* Zürich: Pro Helvetia.

Grin, François

1997 Enseignement des languages à l'université: critères politiques ou économiques? *Babylonia* 4: 60–64.

Murray, Heather/Dingwall, Silvia

1997 English for Scientific Communication at Swiss Universities: "God helps those who help themselves". *Babylonia* 4: 54–59.

English as a Language of Science in Russia

Tatjana Kryuchkova

1. Introduction

The English language has been increasingly permeating through various communication spheres in Russia. This process tends to be similar to the expansion of the English language in European non-English speaking countries. However, there are some specific features in Russia caused by the fact that expansion of the English language in this country started somewhat later, and the society has neither determined the positive and negative aspects of this phenomenon, nor has it elaborated any reasonable language policy in this sphere.

The linguistic situation in any country is dependent on the totality of idioms functioning within the country's territory. Therefore, functional expansion of one of these (the English language, in particular) would bring about restructuring in the functional spheres of all other idioms. Let us consider the situation in the Russian Federation from this perspective. Over a hundred local languages are known to be functioning in its territory, greatly differing in the scope of social functions they service. Along with these, a number of "imported" idioms are also present, such as English, German, French and some other foreign languages. Among local idioms, only the Russian language is on a level with English in terms of functional development, whereas among "imported" idioms, this would be true of German and French. Accordingly, it would be expedient to analyze the connection between these languages.

Russian, like English, is a fully developed, multifunctional language, enjoying a written tradition of long-standing and servicing virtually every sphere of communication, including higher education and research in technology and natural sciences (as is known, quite a few of the world languages are functional in the above spheres). The Russian language enjoys a variety of thoroughly developed terminology systems in all branches of science and technology, as well as a wide assortment of functional styles.

During the Soviet age, functioning of the Russian language in general, as well as in science, in particular, was conditional on a number of social, economic, and ideological factors.

2. The predominance of Russian as the language of science in the Soviet Union

The sphere of sciences was considered by the government to be one of its top priorities. Research and development studies used to be adequately funded, which was particularly the case with technology and natural sciences, as related to the development of the military-industrial complex. A great number of research institutes were operating in the country within the framework of the Academy of Sciences of the USSR, representative of the major branches of the technical sciences, natural sciences, and the Humanities. In addition, there were institutes operating under the academies of particular branches of science, such as institutes of the Academy of Medical Science of the USSR, Academy of the Agricultural Science of the USSR, as well as a broad network of research institutes involved with applied sciences. Every Soviet Republic had its own academy of sciences with a more or less developed (depending on the republic's economic potential and size) network of research institutions. Furthermore, research and development work was conducted in the country's numerous universities and other institutions of higher learning. To cite an example, in 1970 (according to Tsunoda (1983), this year witnessed a record number of Russian language publications on natural and technical sciences), the USSR had 4985 research institutions, which employed 927,700 research workers, including 611,833 Russians (*The USSR National Economy* in 1970: 656, 658). Books on all branches of sciences were published in the USSR in a large number of copies, together with numerous scientific journals.

Throughout the country's entire territory in the sphere of science and higher learning, the Russian language undoubtedly occupied that place, which the English language currently holds or is striving to hold in a number of non-English speaking countries. Although science literature, including studies in natural and technical sciences, was produced in basically all languages of the title nations of the Soviet Republics, and sometimes in impressive quantities, the Russian language tended to be predominant in this sphere. Thus, in 1970 some 5518 publications were produced on mathematics and natural sciences in Russian, with the number of copies totaling 134,328,000. For the sake of comparison, it needs to be mentioned that in all other languages of the USSR, 1466 books on mathematics and natural sciences were published, with the number of copies totaling 41,725,000. The bulk of these publications was accounted for by textbooks and manuals. On the whole, two thirds of the books and other publications produced in Russia in 1970 were in Russian (*Printing Business in the USSR in 1970*).

Similar to that, was the role of the Russian language as a means of oral communication in science. According to Yu. D. Desheriyev (1976), communication on problems of science in the republican research institutions was mostly bilingual, since it was maintained in both the language of the title nation and Russian.

According to our observations, in the case that there was a single language prevailing in a particular sphere, this tended to be Russian. The factors contributing to that were as follows: a diversity of nations employed at the institutes, the fact that the Russian language was most developed as a language of science, and some specific features of professional training described below.

An overwhelming majority of scientific events, starting from sessions of the Academic Council to conferences and symposia at any level, both local and all-Union, were held in Russian. The text of a doctorate or a candidate's thesis as well as the procedure of defending a thesis could be both in Russian and in the title language of a republic. However, if a research paper was done in a national language, it needed to be translated into Russian so that it would be accepted for evaluation by the Higher Attestation Board. It would only seem natural that the majority of the applicants preferred to write their research papers in Russian only.

Successful graduation from any institution of higher learning, with the exception of agricultural institutes and some of the humanities departments in a number of teacher training institutes and universities, was also conditional on Russian language proficiency. In the majority of the Soviet Republics professional training was done in their title languages. However, as the students started majoring in a particular area of knowledge in their third year of studies, they faced the need to read scientific literature in Russian. This was impossible without sufficient language training, which was the focus of attention in their first and second years (Desheriyev 1976: 12).

Two circumstances accounted for this. Firstly, it was not cost-efficient to translate all the books and research papers into 14 languages in which teaching was done in the institutes. Secondly, there was no demand on the job market for professionals trained exclusively in their native language and lacking an adequate command of Russian. Their job opportunities were limited by their native republic, whereas the Russian language proficiency allowed possibilities for a professional career in any part of the Soviet Union.

Post-graduate training was almost exclusively held in Russian (with the exception of training in the national languages and national literature). Also, it needs to be taken into account that most promising students came to Moscow, Leningrad, and Russia's other major research centers to complete their post-graduate training. Owing to these factors, many scientists spoke Russian in the sphere of their professional communication far better than their native language.

3. Promoting Russian as an international language

At the ideological level, an idea was being consistently supported in the Soviet Union about the Russian language being the same type of international language as English, capable of servicing international communication needs. Therefore, it was not necessary for Russian language speakers to use foreign languages, such as English. It needs to be noted that low standards of foreign language training at schools and institutions of higher learning would render the use of foreign languages impossible anyway. An exception would be so-called "schools with a profound foreign language program" as well as foreign language departments at universities and teacher training colleges.

A theory describing Russian as "the second native language" for all the nations of the country had been fairly popular within the territory of the USSR for a long time. The Russian language was promulgated in every possible way in the countries of the socialist community, where it was a compulsory subject at schools, and Russian language teachers were trained in colleges and universities in great numbers. Russian was a working language at all sorts of events, including scientific conferences, held in the countries of the socialist camp. During this period, international co-operation in science was mostly limited for Soviet scientists by the same area, where, as noted before, the Russian language held a dominant position and, thus, could be used as an intermediary language in international research teams. Many foreign participants of such teams had also graduated from institutions of higher learning in Russia.

The Third World countries constituted yet another area where the Russian language was used. Numerous students from these countries were studying in Soviet institutes, following preliminary Russian-language training. In various countries of this region, Soviet cultural centers had been operating, where both children and grown-ups could study Russian free of charge. These countries were willing to hire Soviet experts to work in their industries and educational establishments. All this contributed to creating favorable conditions facilitating the functioning of the Russian language there as a language of international communication.

4. The Soviet Union's opening up to English

Before 1985, the Soviet Union was a closed society. Soviet scientists had fairly limited possibilities of direct communication and even correspondence with their colleagues from capitalist countries. Only few researchers were occasionally able to participate in conferences held outside the socialist camp countries. Instances of publication of articles or monographs by Soviet researchers abroad were extremely rare. In such cases, a research paper was written in Russian and then translated by professional translators either in the USSR or in the country where it was to be published. In essence, a foreign language publication of a research work was the result of editorial work rather than the author's efforts. Thus, the need to be fluent in a foreign language was in fact minimal.

Apparently, to keep abreast of the up-to-date achievements in science, researchers needed to read literature in foreign languages. However, to do so, it was sufficient to use a foreign language passively. Moreover, this involved the use of only one of the language subsystems rather than the foreign language in its entirety. Besides that, major research centers of the country had the so-called 'Translation Bureaus', where highly qualified specialists with a good command of a foreign language (mostly English) were regularly reviewing books delivered from other countries. At the request of the research workers, an article or part of a monograph could be translated into Russian.

Two specialized institutions were operating within the framework of the Academy of Sciences of the USSR: the All-Union Institute of Scientific and Technical Information and the Institute of Information on Social Sciences. A large body of staff at these institutions were responsible for reviewing both Soviet and foreign literature on virtually all branches of science, summarizing new publications on particular subjects, and drawing up lists of publications.

It needs to be noted that despite the fact that Russian was absolutely predominant in the sphere of science, the expansion of English into this sphere gradually started in the 70s. Despite an intensive ideological campaign in support of the Russian language as a means of international communication, scientists (even those who provided a theoretical foundation of this ideological campaign) were aware that outside the countries of the socialist camp and some of the developing countries, the Russian language was only understood within a limited academic circle of linguists specializing in Slavonic languages.

Therefore, reports to be delivered at international conferences hosting a wide range of participants were sometimes prepared in English. For instance, in preparation for the 8th International Congress of Sociologists held in Canada in 1974, the Soviet Association of Sociologists published all contributions of the Soviet participants in English. Moreover, in some journals, especially those published in the national languages of the Baltic or Transcaucasian republics, the contributions were accompanied by Russian, sometimes English, and occasionally German or French summaries. Some scientific journals duplicated their titles in English. Sometimes monographs were supplied with an English summary. However, these were never on a mass scale.

It should be pointed out that during the Soviet era, there was yet another area where the English language was functional as a language of science. The Main Publishing House for Foreign Countries was operating under the USSR Academy of Sciences, handling editing and preparation for publishing of a wide range of works on social sciences in English, French, Spanish, Portuguese, and Arabic. The bulk of the publications comprised topical collections of articles by Soviet leading experts on philosophy, history, economy, and other social sciences. These collections of articles were published in the USSR and then distributed in foreign countries. This activity, however, bore a propagandist and ideological nature and was aimed at dissemination of Soviet ideological conceptions abroad, mostly in the countries of the Third world. Besides that, *The Social Sciences Journal (Social Sciences and the Contemporary World* in the Russian version) was published in English, French, Portuguese and Arabic.

Another journal of a wider profile, *Science in the USSR* was published in Russian, English, French and German. This was a popular journal which carried articles on various aspects of natural and technical sciences to promote achievements of Soviet science. The languages in which the journal was published indicate that it was intended for somewhat different audiences and mostly distributed in Western countries.

In addition to the above mentioned academic agencies that handled publication of books in foreign languages, there were two major state publishing houses in the USSR, 'Progress' and 'Mir'. The former was responsible for translation both from foreign languages into Russian and from Russian into foreign languages, as well as for publication of books on social and political sciences, whereas the latter handled translation and publication of books on technical and natural sciences.

Thus, during the Soviet age, the English language in the sphere of science was functioning at two levels: unofficial and official. On the one hand, starting from the 1970s, scientists began using English not only for reading scientific literature but also for stating their own ideas. The process was rather slow, since the socially determined demand for the English language was restricted by the closed nature of Soviet society. Furthermore, most active at that period was the generation of scientists educated in the early 1930s to the early 1940s, when German used to be the main foreign language taught at all levels of education. Nevertheless, some books and articles were provided with English summaries. Reports and summaries of reports to be delivered by Soviet scientists at international conferences were translated into English. This occurred against the backdrop of another global process: English started to gain ground in secondary and higher education as a foreign language, and gradually ousted German and French from this sphere, which was probably in tune with the increasing Anglicization of the outer world. Therefore, the need to use English more actively started to be felt and accordingly articulated, along with the increase in the number of people who more or less knew English. It also needs to be noted that the demographic base of the English language was expanding to include the shrinking demographic base of other foreign languages.

At the state level, on the other hand, there existed a mechanism for selection of scientific literature to be translated into foreign languages, mostly in the field of the humanities, but also partly in the field of natural and technical sciences. The former was mostly intended for the Third World countries and therefore translated into the literary languages functioning there, English being one of these. Books on both natural sciences (mostly physics and mathematics) and technical sciences were translated into the three foreign languages most actively used in the area of scientific communication: English, German and French. As a matter of fact, the authors had nothing to do with publication of their works in foreign languages. The All-Union Agency for Copyright selected the books for publication, professional translators did the translation, and The International Book Agency distributed them abroad.

It needs to be acknowledged that both forms of the English language functioning in science at that period were directed outwards with the aim of presenting the achievements of Soviet science in the international arena, which had virtually no impact on the functional distribution of languages in the sphere of science inside the country. Here the Russian language was undoubtedly prevailing. The influence of Russian in this sphere was expanding into the countries of Eastern Europe, as well as into numerous countries of the Third World, since a considerable number of scientists from these countries obtained their degrees or went through post-graduate training in the Soviet Union in the Russian language and continued using Russian to read scientific literature and even communicate with their colleagues in the sphere of science.

An essential feature typical of this period was the fact that the English language was functioning almost exclusively in its written variety. This mostly resulted from lack of direct contact with western colleagues which would require a command of spoken English. On the other hand, language training in institutions of higher learning and post-graduate courses was basically directed at reading scientific literature. In point of fact, until now foreign language skills stipulated by a program for post-graduate students are limited to the reading of a scientific text and translating it into Russian with a dictionary, reading without a dictionary and producing a Russian summary of a smaller extract from a text based on the student's field of research, summing up in Russian the main points of a newspaper article, and giving answers in a foreign language to a number of elementary questions, mostly centering on the student's biography. As is known, developing speech habits in any language needs special training. In the case of the English language, with its typical gap between spelling and pronunciation models, the role of this aspect of didactic activity becomes critically important.

5. Post-Soviet Russia's shift towards English

The early 1990s witnessed the breakup of the Soviet Union, establishment of 15 independent states, and a radical change in their social and political structure, as well as transition of the socialist camp countries to the capitalist mode of development. These events, undoubtedly, influenced the language situation in the territory of the former USSR as well as that in the territory of entire Eastern Europe. This was reflected in the narrowing demographic and to a certain extent, functional power of the Russian language. First, it ceased functioning in the sphere of international relations, education and science in the countries of Eastern Europe, where it was replaced by English. In Latvia, Lithuania and Estonia, the Russian language was quickly losing its clout in said spheres, as well. In other republics of the former Soviet Union, a decrease in the scope of functioning of the Russian language went on at a somewhat slower pace. In that case, however, where the Russian language lost, English did not necessarily win. Sometimes Russian was replaced by the language of the title nation, and sporadically, by Turkish (in Turkic countries). One would suppose that this does not necessarily affect the status of the Russian language in Russia, which, however, is not true. Demographic and functional weakening of an idiom, the reduction of its area of usage always affects the idiom adversely, even in the territories and sphere where its position seems to be fairly stable. First of all, this affects subjective attitudes of the native language speakers, which could consequently result in further restriction of its use in certain spheres. To cite an example, at recent international conferences held in Russia, participants from the Baltic states and former socialist countries, known to have a good command of Russian, since they had obtained their degree or followed a post-graduate course in Moscow in the Russian language, delivered their reports in English, which was often not as good as their Russian. Apparently, for Russian-speaking audiences, this means a restricted access to scientific information and creates a motive to switch to the English language in the sphere of scientific communication.

However, this was not the key factor in facilitating an intensive expansion of English into the sphere of sciences in Russia. As compared to the Soviet Union, Russia is a far more open society. Even before Russia became an independent state, namely, since the start of perestroika in 1985, scholars got ample opportunity to communicate with their Western colleagues, travel abroad, participate in conferences, receive foreign colleagues in Russia and publish their own works abroad. It soon turned out that effective communication was impossible without a good command of English. Later, starting from the early 1990s, Russian scholars set their sights on the West, not only striving to broaden their scientific vision or make the results of their research available to an international scholarly community. To a considerable extent they were driven by purely economic interests. Science belongs to the spheres hit most hard by economic transformation in Russia. The payment for the work done by Russian scholars cannot be compared with the salaries of their foreign colleagues. Even in Russia, it would be difficult to find a job which is worse paid. Researchers in the field of natural or technical sciences found themselves in particularly dire straits, as they are badly in need of expensive equipment and materials which they cannot afford. Therefore, the only opportunity for Russian scholars to secure more or less adequate conditions of life and work is to find a job abroad, participate in joint projects together with Western colleagues, or obtain an international grant. To achieve any of these goals, knowledge of English is required. Typically, the knowledge of English allows the possibility to obtain an internship or a grant not only in universities and research centers of the United States and Britain, but also in Germany where a command of German or English is a mandatory requirement (as opposed to French universities which only require knowledge of French). As is known, however, Germany and the United States maintain the closest cooperation with Russia in sciences.

Some foundations make extension of a grant conditional on the applicant's English language proficiency. For example, guidelines on filling out grant applications under the Research Support Scheme of the Open Society Support Foundation are as follows: "The application and all related materials must be in English. Poor spelling and grammar do not give a good impression to evaluators. It is better to ask a native speaker to check the language." Since the native English language speakers willing to check a grant application or a translation of an article are few in Russia and the services of professional translators are too expensive for Russian scholars, they are faced with the need to improve their command of English.

To get an idea of what the level of their English language proficiency is, we distributed a questionnaire among 50 scholars (Doctors and Candidates of Science) of a chemical research institute of the Russian Academy of Sciences. Perhaps, in the opinion of sociologists, the number of respondents would not be statistically representative, and the results are only based on the respondents' own assessment. However, the data cited below are in indicative, in our view, of some trends in the development of the reference sphere.

Answers to the question: "How do you grade your English language skills" are summarized in Tables 1 and 2 (the data are represented as a percentage against the total number of respondents in every age category, and the number of respondents in the groups is different).

Table 1: Reading and translation skills in English

Age	Reading an English scientific text				Translation of a scientific text or letter to a colleague from Russian into English			
	Good	Satisfactory	Bad	Unable	Good	Satisfactory	Bad	Unable
27–35	75	25	–	–	50	25	–	25
36–45	45.5	36.3	18.2	–	18.2	27.3	9	45.5
46–55	28.6	57.9	7.1	7.1	7.1	21.4	35.7	35.7
56–65	35.3	47	17.6	–	5.8	35.3	11.8	47

Table 2: Skills in conversation and oral understanding in English

Age	Conversation on everyday and professional topics				Listening comprehension of a conversation on everyday and professional topics			
	Good	Satisfactory	Bad	Unable	Good	Satisfactory	Bad	Unable
27–35	50	37.5	12.5	–	50	37.5	12.5	–
36–45	–	36.3	45.5	18.2	–	36.3	36.3	27.3
46–55	7.1	21.4	42.8	28.6	7.1	14.3	50	28.6
56–65	5.8	29.4	29.4	35.3	5.8	17.6	29.4	47

As seen from the tables, the younger scholars top the list in all skills in all the varieties of the language, leaving the other age groups far behind, which is no wonder. These young people were educated and started on their professional career during the period when contacts with Western countries were expanding and knowledge of English became essential for professional growth. Typically, this age group, as distinct from older generations, is competent in active and passive skills in English that is they can speak, translate from Russian into English, and understand spoken English quite well. On the one hand, the growth in active skills was affected by motivation behind studying a foreign language, which has noticeably changed over the past fifteen years. Formerly, a foreign language was studied mostly in order to be able to read scientific literature, whereas the leading motivation factors to date are the need to participate in discussions with foreign colleagues during conferences, symposiums, or personal meetings, maintaining correspondence, preparation of one's own research works for publication abroad, etc. On the other hand, in response to the society's needs, methods of teaching English have changed drastically to be focussed mostly on the speaking skills, aural comprehension, and translation into a foreign language. It needs to be noted, however, that all these changes did not greatly affect the system of secondary and higher education other than that provided by specialized departments of foreign languages. Both education systems do not provide an adequate level of English language proficiency necessary in the modern world. Thus, many people strive to better their English language skills by joining various types of short-term schools. A relatively new form of education popular with the young people now is obtaining, alongside a diploma of higher education, another diploma as a "translator in the sphere of professional communication".

The second item on the questionnaire was: "Would you like to improve your knowledge of English?". Some 42 respondents (84%) answered positively. Those who answered negatively were requested to specify reasons for their unwillingness to improve English language skills. Four respondents (27–35 years category) said that they had been on a long-term internship in the United States and could speak English fluently. One respondent (46–55 years category) had worked in India for several years, and thus knew English well enough. One respondent (56–65 years category) was raised in a foreign country and had been speaking English fluently since he was a child. Two more respondents (56–65 years category) stated that it was too late to improve their English, since their professional career was nearing completion.

Thus, the need to have a good command of English was acknowledged by virtually all respondents who intend to continue their research work.

The third item on the questionnaire was about motivation behind studying English (respondents were encouraged to provide more than one answer). All, the respondents described approximately the same motives and no correlation was found between their sequence and the age of the respondents. The answers listed below follow the sequence in which they were given by the respondents, number 1, being the answer which the majority of the respondents put at the top of the list, etc.

1. Need to read foreign scientific literature.
2. Need to use English-language Web sites.
3. Wish to participate in discussions at international scientific conferences.
4. Wish to communicate with foreign colleagues.
5. Wish to get an internship abroad or secure a temporary/permanent job in a foreign country.
6. Need to translate their own research works into foreign languages in order to publish these abroad.
7. Need to fill out requests for grants and prepare supporting documents.

We believe that most interesting in the above answers is the fact that all needs listed are directed outward and linked very little, if at all, to the functioning of science inside the country, where, by all appearances, the Russian language continues to fully service sciences. The only exceptions are Response 1 citing the need that goes back to the Soviet era, when it used to be almost the only motive behind studying a foreign language, and Response 2, which is an "electronic analogue" of No.1.

Finally, the last query of the questionnaire was about knowledge of other foreign languages by respondents. The answers are summarized in Table 3.

Table 3: Knowledge of foreign languages other than English

Age Group	German	French	Other Languages
27–35	12,5%[1]	–	–
36–45	18,1%	9%	–
46–55	21,4%	14,3%	7,1%
56–65	35,2%	11,8%	5,8%

Given the fact that in Russia's secondary schools and institutions of higher learning a single foreign language has always been taught (two foreign languages were taught only in specialized language schools and at foreign languages departments of universities), the data listed above signal that during the last fifty years, English was gradually replacing other foreign languages, first of all in the sphere of education. This influenced indirectly the functioning of foreign languages in other social spheres.

We received interesting information at the Foreign Languages Chair of the Russian Academy of Sciences which is responsible for a one-year training course for post-graduate students prior to their compulsory exam in a foreign language. Some 80 to 85% of post-graduates studied English at a secondary school or institute and continue English studies at the post-graduate course. All post-graduates who used to study other foreign languages request an opportunity to start an English course as a second foreign language. Among post-graduates whose first foreign language was English, only few majoring in natural and technical sciences are willing to start on a second foreign language, whereas the number of post-graduates majoring in the Humanities and willing to master another foreign language is greater. It needs to be noted that among these latter, many join the post-graduate course with a command of two foreign languages, and depending on their proficiency level in each, they prefer to master either English or another foreign language.

The situation is similar with various short-term schools of foreign languages operating in Moscow. Many of these provide classes only in the English language. At schools where other foreign languages are taught (most often German, French, Spanish, and Italian), about 80% of students also study English, and only 20% of students account for all other foreign-language classes.

In the sphere of Humanities, the English language has not, thus far, replaced other foreign languages to the extent it replaced them in natural and technical sciences. Naturally, a scholar studying, for instance, G. Simmel's sociological conceptions needs, first and foremost, a command of the German language. However, in order to be able to deliver a report on this subject, say at the 13th International Congress of Sociologists (Bielefeld, Germany), this scholar would have spoken English, since English was the only working language of the Congress.

Our own experience of participating in various international conferences and workshops indicates that even when there are several working languages at a conference, chances are that the bulk of the reports will be delivered in English, since most participants would like their contribution to be appreciated by as many audiences as possible, and the English language is the most adequate means to reach this end. Talking to us, virtually all scholars involved with the Humanities agreed that the English language proficiency has by and large become a component of their professional competence, as important as the knowledge of the language of the country which is the subject of their research. Only Francophones could be described as an exception, since they claim that the French language is still very instrumental in facilitating their communication with foreign scholars involved with the problems of the French-speaking area. However, even French scholars acknowledge the need to know English in order to maintain international contacts, to read scientific literature on theoretical foundations of a particular branch of science, and to obtain information via the Internet. Thus, the English language is acquiring still greater importance in the field of Humanities, as well.

Computerization together with the spread of the Internet contribute significantly to the expansion of the English language, particularly, into the sphere of sciences. All PC's used in Russia 'communicate' with their users in English. The assortment of English commands is rather limited, of course, and it does not take any particularly high level of English language proficiency to cope with them. However, in our opinion, this fact weighs on the minds of users, since psychologically, English is associated with state-of-the-art technologies. This, in turn, boosts the prestige of English as the language of sciences.

The role of the Internet as the most up-to-date means of exchanging information has been steadily growing over recent years. It is particularly effective for exchanging information in the field of technical and natural sciences. As a rule, this information is presented in a structured and focussed manner, so that it does not take up much space and is easy to read or print out.[2]

The bulk of sites containing such information are in English. It is not by chance that all our respondents cited the need to make use of the Internet as one of the critical motives behind mastering English (this topped the list of answers by younger respondents, whereas the mention of the need to read research literature was ranked below). It needs to be noted that Russian scholars use English not only to obtain information via the Internet, but also to share their information through it. A lot of institutes under the Russian Academy of Sciences have their own sites on the Internet, where information is presented both in Russian and in English. Some scholars have similar personal sites, whereas others choose to present their information only in English.

The English language is fairly widely used now in Russian scientific periodicals. Virtually all of the scientific journals we examined, on the natural and technical sciences, as well as those on the Humanities, contained some information in the English language – summaries of the articles, contents, or at least the name of the journal and a list of the editorial board. The only exception were journals published in two versions, English and Russian that we will deal with somewhat later. Many journals, first of all those on natural and technical sciences, publish contributions by foreign scholars in English, although it is not always the case that these scholars are native English speakers. Among materials published in English, we noticed articles authored by scholars from the Baltic states or from East European countries, which supports our view voiced above with regard to the substitution of English for Russian in the sphere of scientific communication in these regions. Sometimes Russian scholars also publish their contributions in English, however this does not happen on a mass-scale. Most often these publications are part of a volume to mark the anniversary of a key figure in science, where Russian authors participate together with their foreign colleagues.

In books on sciences, the English language is used less often. Part of books on natural and technical sciences are published without any English summaries. The latter could be seen more often in books on the Humanities, which often include an English translation of the contents.

Usually, reports or summaries of reports by Russian scholars participating in major international conferences and symposiums held abroad are published in two languages, Russian and English. At international conferences and symposiums held in Russia, usually there are two working languages, Russian and English. Accordingly, programs and working materials are also prepared in both languages. However, in the course of discussions and spontaneous exchange of opinions, it is usually necessary to use the services of professional interpreters, since the number of bilinguals among the audiences, as a rule, is not large. On the other hand, we repeatedly witnessed the situations when two Russian scientists, involved in a discussion, were speaking English and caring very little if they could be understood by their fellow countrymen.

The English language is widely used in all kinds of reference and information publications. Reference materials informing about Russian publications usually duplicate in English titles of sections, books and articles, personal indexes, and tables of contents, in order to make the information available to foreign audiences. For instance, an abstract journal on physics published by the All-Russia Institute for Scientific and Technical Information provides English abstracts of most of the works by foreign scholars (these are probably abstracts of books or articles provided in the original English edition). Presumably, since the journal has a considerable press run, it is widely read by experts, who have nothing against such presentation of information.

In recent years, a new branch was formed on the basis of Russian science, functioning exclusively in English. In 1991, the Nauka (Science)

Publishing House together with a U. S. publishing firm set up the Nauka/Interperiodica International Academic Publishing House. The firm prepares English versions of 88 academic journals on various branches of natural and technical sciences.[3] These versions are absolute analogues of their Russian originals, which are translated into Russia by experts and edited by Americans. The journals are marketed mostly in the United States and some other developed countries. Besides that, the publishing house issues three original journals in English: *Laser Physics, Pattern Recognition and Image Analysis,* and *Russian Journal of Mathematical Physics,* also distributed abroad. Some of the journals independently seek partners abroad for arranging publication of an English version abroad. For instance the journal *Discrete Mathematics and Applications* has been fully translated into English since its sixth issue of 1999 to be published by the VSP Publishers (Netherlands). The editorial board of the journal suggests that the authors either submit their contributions both in Russian and in English or entitle the editorial board to arrange translation of the articles "by experienced translators under the control of the editorial board."

In Russia, several international scientific journals are being published both in a Russian and English edition. Besides that, the journals *Social Sciences* and *Science in Russia* (formerly, *Science in the USSR*) that were established in the Soviet times continue being published. Now they are only published in Russian and in English.

Projects to publish books in English turned out to be less successful. In 1998, the Nauka Publishers set up a section specializing in translation into English and publication of books on natural sciences. It used to publish some 12 to 15 titles yearly. However, in 1992, this section was closed due to financial difficulties. At present, the Nauka Publishing House produces 5–6 publications in English annually. These are mostly works by renowned Russian mathematicians, physicists, biologists, and other scholars specializing in natural sciences who have managed to obtain a grant for publication of their works. However, according to one of the publishing house workers, "books are now needed by aesthetes only", since electronic means of delivering information are acquiring increasing importance, whereas publication of books becomes increasingly expensive. This is true of books both in Russian and in English. Nevertheless, some private publishers occasionally undertake publication of books by Russian scientists in English translation.

The very fact of publication of journals and books in English in Russia can hardly be viewed as indication of the fact that English is functioning as a language of sciences in this country. All these books and journals could well be translated and published in English abroad. Probably, this is more cost-effective in Russia. However, these publications render indirect influence on the expansion of the English language in this sphere: some authors prefer to translate their articles themselves or check the adequacy of translations. A category of highly qualified publishing workers is being formed, capable of preparing scientific publications in English.

To sum up, a conclusion can be made that in Russian sciences, the English language is currently used far more intensively than before 1985. However, its main function in this sphere is to serve as a mediator language between Russian sciences and the international scientific community. In this function, it has basically replaced all other foreign languages in the sphere of natural and technical sciences, and is replacing them in the field of Humanities. This, of course, is a response to the prevailing position of English as a language of natural and technical sciences the world over, as well as to its increasingly prominent role as an international mediator language in numerous other areas of activity. Russian still remains the main language servicing all spheres of Russian science with no exception. Although following the break-up of the Soviet Union, the number of scholars actively using Russian in their professional activities has diminished, and the Russian language is now less often used as a means of international communication in sciences, its status as a language of sciences is still fairly stable. Russian remains a key means of scientific communication in 4110 Russian research institutions which employed 880,300 research workers (Russia in figures 1999). In 1994, 3195 titles of books were published in Russian on natural sciences only, with a total press run of 5,851,300 copies. In 1994, 540 scientific journals were published in Russia (*Printing Business in the Russian Federation 1994*).

However, there is every reason to believe that during the past 15 years, most scholars have come to consider the English language as an indispensable component of their professional competence and an essential condition of their career development. A younger generation of scholars have a better command of English than their older colleagues. Since the youth are the most socially progressive category of the population, which determines both the future of the country and that of science, in particular, it can be expected that within the next 10–15 years Russian – English bilingualism in Russian science will be increasingly expanding.

6. Notes

1. Since this age category included 8 respondents, 12,5% accounts for only one respondent.
2. Scholars involved with the Humanities tend to create voluminous monographs which are fairly difficult to read on a PC screen. Therefore, in this sphere books and publications will probably long remain a major source of information. However, the Internet is a source of information for scholars in the field of the Humanities, as well. This concerns particluarly the news on scientific conferences, various foundations. Besides, it makes electronic versions of scientific journals available, provides all sorts of data bases, etc.
3. Also, an English version of the economic journal *Studies of Russian Economic Development* is published.

7. References

Desheriyev, Yunus
1976 *Laws Governing the Development of the USSR Literary Languages in the Soviet Age. Development of the Literary Languages' Social Functions.* Moscow: Nauka.

Printing Business in the USSR in 1970 Statistical Data
1970 Moscow: Kniga.

Printing Business in the Russian Federation in 1994.
1995 Moscow: Rossijskaya knizhnaya palata.

Russia in Figures 1999.
1999 Moscow: Goskomstat.

Tsunoda, Minoru
1983 Les langues internationales dans les publications scientifiques et techniques. *Sophia Linguistica*, 144–155.

The USSR National Economy in 1970.
1971 Moscow: Goskomstat.

English in Brazil with an Outlook on its Function as a Language of Science

Vera Lúcia Menezes de Oliveira e Paiva and Adriana Silvina Pagano

1. English in Brazil

One might argue, as Pennycook (1994) does, that the spread of the English language and Anglo-American cultures throughout the world is the result of an effort of British and American governmental agencies, universities, religious organizations and publishing companies. Historically, the dissemination of English is linked to the political and economic policies of Britain and the United States through diverse projects of colonial and postcolonial expansion in search of new sources of raw materials and trade. Gradually, however, English has come to be adopted as the *lingua franca* of information flow and exchange in most of the western world, a role that has been substantially growing within the present context of globalization and increasing international links between western and non-western cultures and peoples.

Greatly propelled by the revolution in mass media communications that began in the 1950s with television broadcasting and advertising and has resulted in a new era of dissemination through the Internet, English is now an international language and has become the language of scientific knowledge production and technological progress. Being a tool for intellectual, cultural and commercial contacts, English has literally invaded every country on our planet. Fairclough (1989:21) reminds us that "a language has been jokingly defined as 'a dialect with an army and a navy.'" Exploring this war metaphor further, we can certainly add that each sign, be it a sound, a word, an utterance or a long text, can be seen as playing the role of a soldier in this linguistic invasive movement, a movement that operates in multiple directions and reveals a complex rationale. Word borrowing or loans is one of the most visible processes in sites of language contacts and interactions.

English vocabulary borrowing is nowadays a world phenomenon. In Brazil, English words can be found literally in every nook and cranny: advertisements, billboards, brands and labels, shop names, restaurant menus, foreign and national songs, T-shirts, buttons (in Brazil, *button* is spelled as *bottom* and pronounced likewise), packages, movies, soap-operas, bumper stickers, cartoons, comic books, graffiti, TV, radio, newspapers and magazines, and also in our oral and written interaction. In this bilingual environment, knowledge of Portuguese, the official language in Brazil, is no longer enough to understand our everyday context. The most famous Brazilian dictionary of Portuguese, popularly known as *Aurélio*, only registers 373 words borrowed from English or 'anglicisms', but many more terms can be found on the streets. In São Paulo, the biggest Brazilian metropolis, 15% of the names of bars and shops are in English and signs such as *sale, 30% off, open, fast food* and *delivery* can be easily found on shop windows.

Learning English in private courses is a customary practice of the middle and upper classes in Brazil. In São Paulo, there are 3000 English courses and in Rio de Janeiro, a school named Cardinal New York recently built a four-store building where English will be the official language in its premises, from the parking lot to the administration offices. The school hires Americans and also Brazilians who have lived in the United States in order to create an environment aimed at promoting monolingual interaction in the foreign language and reportedly speeding up learning. Further evidence of the power of English worldwide and particularly in Brazil is interaction on the Internet. According to Pedrosa (1999:39), 80% of its content is written in English. Most Brazilian homepages have a version in English, which is definitely the language of the information society. To talk about computers and to surf on the Internet, one needs to master a wide range of English words. As we pointed out earlier, English is today recognized as the *lingua franca* of the 20th century and is likely to continue to play this role into the next century, as it has become the means of communication of innumerous cultures. Consequently, the penetration of English terms and vocabulary together with the foreign cultural values and notions they carry can be predicted as an increasing phenomenon in our Brazilian society.

2. The Brazilian public opinion

Occasionally, the Brazilian press points out the influence of the English language in our linguistic habits and shows a deep concern for linguistic and cultural identities. These articles argue that there is an improper use of borrowed words either in oral or in written language. They acknowledge loan words as part of a necessary process for language development, but point out the replacement of our own words by imported ones as a highly undesirable practice. They blame those who, in the name of a misleading globalization, adopt English words instead of using their natural counterparts in Portuguese. This condemnatory tone is reflected in the picture below taken from one of those articles.

Figure 1: Portuguese under threat (Moreira 1999).

In a similar tone, *VEJA*, the most widely read Brazilian magazine, published an interesting piece of news in October 1999, its first sentence reading "Be careful with anglicisms." The text is addressed to those who are looking for jobs and advises them not to overuse words borrowed from English or anglicisms because, they argue, what was in fashion at the beginning of the decade is now considered a ridiculous practice. The text advises the reader, for instance, not to say *vou deletar uma idéia* (literally, I am going to delete an idea), *deletar* being a new Portuguese term coined after the English computer term 'delete'. However, despite these and other admonitions, the phenomenon of English word infiltration seems to be on the increase in everyday life in Brazil.

Intellectuals form two opposing groups when it comes to this issue. The first group views word borrowing as a consequence of natural language development caused by our interaction with other nations. They argue that anglicisms enrich a language and do not represent a threat to its structure or lexis. They base their arguments on the history of the English language itself, which enlarged its vocabulary with Latin and Greek-rooted words in the past and continues to do so through borrowings from modern languages such as Spanish, Hindi, and many others. While reaffirming the naturalness of this process, supporters of this view, however, are against the overuse of English words.

The second group wants to protect the Portuguese language and control the invasion of foreign words. A member of the Brazilian Federal Parliament, Aldo Rebelo is the author of a law project aimed at protecting the Portuguese language from foreign words. He contends that Portuguese must be used in public documents and events and also in mass media, product handling and shipping. If this law is implemented, all unnecessary or misleading use of foreign expressions will be treated as violations of the law as their effect on the Brazilian cultural patrimony is deemed harmful and undesirable. Violators will be subject to fines amounting to 6,500 dollars, besides other civil and penal sanctions for the crime of disrupting the language. Rebelo states that his aim is not only to get the law passed but also to organize a national movement in defense of the Portuguese language. His project, he claims, must not be seen as xenophobia, but as an initiative to protect Portuguese against the corrosive bilingualism that is defacing it and making Brazilians believe that it is ugly, limited and vague. Portuguese, he contends, has sufficient linguistic resources to describe the scientific discoveries, inventions and changes that transform the world. Needless to say, Rebelo's controversial project has divided public opinion and almost every week one can find an article on this issue in the press.

Control and punishment are certainly undesirable and negative practices when we deal with complex cultural phenomena, as is the case of linguistic contacts and exchanges in our modern societies. 'Tyrannical English', as Swales (1998: 125) puts it, no doubt "proves a serious obstacle to efforts to develop and sustain scholarly varieties in smaller scientific communities like Arabic, Hebrew, Malay, Philipino and Swahili" while also threatening scientific communication in major western languages other than English. However, we can argue that English is sometimes the only means of communication among researchers from distant nations and cultures. Similarly, the constant process of word borrowing from English can no doubt bring about considerable negative side effects for a language, the lesser of which is lexical impoverishment. In Brazil, for example, the practice of naming new concepts or even existing ones with English words (e. g., the adjective *diet*, borrowed in spite of the existence of the Portuguese term *dietético*) not only hinders natural processes of word coinage in Portuguese but also leads established word formation processes in the language into oblivion. On the other hand, it might be argued that word borrowing leads to the enrichment of a language and the strengthening of its capacity to interact with other cultures. Historically, this has been one of the main outcomes of travel enterprises and contacts. English, for one, can be said to have shown a reasonable degree of receptivity to other languages (see Swiderski 1996), borrowing and incorporating words from the many tongues that it has come into contact with throughout its history. Moreover, this incorporation of foreign words is usually followed by its transmission to other languages and tongues. In this sense, Swiderski (1996: 11) explains, "English appears to be playing on a much wider scale the role Portuguese once did with respect to English and other languages", that is, incorporating words and disseminating them in other cultures and societies. Swiderski is clearly referring to the expansion of the Portuguese language through Portugal's colonial enterprise, a comparable case to English and other European languages' dissemination. In this sense, it is useful to remember a further aspect of colonial Portuguese, which is its affiliation to a violent program of imperialist conquest. Making use of the war metaphor too, Houaiss (1983: 50) characterizes Portuguese as a language that triumphed in Brazil "by eating other languages and killing other cultures." This is certainly an issue that needs to be brought to the fore in the debates about national identity, with a view to problematizing the idea of the Portuguese language as being the sole and legitimate means of expression of the Brazilian nation.

3. Signs and wonders

Most of the research traditionally done on linguistic borrowing deals with the word as the operational unit of analysis. In this section, we would like to approach the subject from a different point of view and include as linguistic borrowing any realization of a foreign language sign in a discourse event, be it a morpheme, a phoneme, a syntactic feature or a text.

To illustrate our approach, we will take English language signs commonly used in Brazil, including those that users produce or receive under the belief that they are English signs in a process that makes them feel they are affiliated with English even though they cannot understand those signs. As a theoretical support, we are going to draw on Peircean Semiotics.

We can begin by classifying the English language signs found in Brazil according to Peirce's first trichotomy of the sign in itself. They are:

(1) Loan-signs in the form of qualisigns, that is, a quality, which is a sign and appears to us as mere sensation. As examples we have Portuguese words pronounced as if they were English, for example, *americano* [æmɚricænu] and *Manuel* [manwel], both found in a pop song by the Brazilian composer and singer Ed Mota. Another example can be found in a play named *Amante Inglesa* (English Lover), a play completely spoken in a "fake English", (i. e., the actors utter a sequence of sounds which resemble English but have no meaning in themselves). Only the context, facial expressions, intonation and other paralinguistic features convey meaning to the text, as the uttered signs do not belong to any natural language.

(2) Loan-signs in the form of sinsigns, that is, an actual object or event which is a sign). As examples we have names of shops, restaurants, pubs (such as *Gang, Mr. Blue, Grill, Pier*); names of buildings and shopping centers (such as *Golden Center, The First Place, Central Park, Santana's Contemporary Free Home, Elite Free & Flex*); names of several kinds of products: food (such as *Cream Cracker, Diet Coke, Brownie*); cleaning products (such as *Thunder, Comfort, Mr. Magic*), beauty products (such as *Clearskin, Lip Perfector, Moon Drops*). We can also find messages written in English on bumper stickers (such as *Fasten seat belts* and, believe it or not, *Fuck you!*), on buttons as (*I am married, not dead*), in souvenirs (such as *I love Rio*); on T-shirts (such as *Marvelous Moments or Green, I want you alive* [the latter worn by former Brazilian president, Fernando Collor de Melo, in a public event]); in advertising (such as *Complete the*

day with Black & White); in names of magazines (such as *Playboy, After Eight* and *Movie*); expressions in oral interaction (such as *yeeees, okay, good bye),* in graffiti (such as *KILL AND CRIME* painted during a revolt on the walls of a juvenile hall building for delinquent teenagers and published by the newspaper *Correio Braziliense* on October 27th, 1999: 12), and also in a series of articles written in English and published by several Brazilian academic journals. Other examples can be found in a list of English words, syntagms and sentences spread over a great number of lyrics in Portuguese. Some examples are: "Nosso amor é do tipo *one way"* [our love is a one way type]; "Quero passar um *weekend* com você" [I want to spend a weekend with you]; "Vivo num *clip* sem nexo" [I live in a meaningless Video-clip]; "Vê se me entende, olha o meu sapato novo/ Minha calça colorida, meu novo *way of life"* [Try to understand me, look at my new shoes/My colorful pants, my new way of life]; "*I don't want to stay here / I want to go back to Bahia* /Eu tenho andado tão triste"; "Quero ser feliz, *bye, bye* tristeza" [I want to be happy, bye, bye sadness], "Nem sempre é *so easy* se viver" [To live is not always so easy].

Figure 2: Graffiti on the walls of a juvenile hall building for delinquent teenagers in Brazil

(3) Loan-signs in the form of legisigns (signs which are laws, usually established by society). Three kinds of legisigns can usually be found everywhere: (3.1) the genitive case, as in *Delfim's boys* (in a newspaper report), *Kit's Cozinhas Planejadas* (in the telephone directory) or *Juca's bar* (a bar in Belo Horizonte); (3.2) the anteposition of the adjective as in *Tropical Bar* (instead of the Portuguese syntax Bar Tropical); (3.3) word-formation rules which eventually produce words which resemble English, as the meaningless examples *DIRD* and *SEIL* found on T-shirts.

The use of discursive practices such as those described above are widely spread among the different social groups in the Brazilian society. The purpose for which they are put to use and the way in which this is done, however, can be roughly mapped to vary according to the aims or desires underlying the constitution of each group or class.

Members of the Brazilian economic and intellectual elite have traditionally used English words as a strategy to differentiate themselves from the so-called middle and popular classes. By using in their language English words and concepts that they acquire thanks to the advantages of their privileged social and economic status (travels and study abroad, foreign language education in private institutions), members of the elite signal the boundaries of their social space and display their symbolic strength. For the great majority of the Brazilian population, however, the English language signs are just manifestations of "firstness", that is, mere phonological and morpho-syntactic qualities. The interpretants for those signs are rhemes, that is, open possibilities of signification. Thus, the English language signs that surround the Brazilian people who cannot speak English are perceived as rhematic iconic qualisigns. Those signs can only yield the sensation of being in contact with the English language as far as no relation among the signs and the objects they represent is established. A good example illustrating this was found in a Brazilian soap-opera. A female character joins a group protesting against a neighbor. When she produces a sign that reads *ELZA, GO HOME*, she is asked what it means, to which she replies that she does not know. She says in Portuguese: "I don't know, but I saw it once on TV and I liked it very much". A similar phenomenon takes place when songs in English are played. Although most of the audience does not know what the lyrics mean, they enjoy the melody of the words wrapped with the sound of the musical instruments. A Brazilian composer, João Bosco, for example, once said in an interview that when he was young he used to go to the movies and enjoy the sounds in American songs, although he did not understand them. The sounds remained in his mind and he tried to reproduce them in some of his songs. As a result some of the lyrics he composed are mingled with meaningless paralinguistic sounds that resemble English.

This apparently clear-cut division of the Brazilian society into elite and popular classes is actually an overly simplified way to approach the incorporation of English signs in the Brazilian society. Like most modern Latin American societies, the Brazilian can also be characterized, using Canclini's argumentation (1992), as a hybrid culture, a complex network of social groups that continually negotiate their identity and space. Group affiliation sometimes produces crossings and intersections of what are traditionally considered different social classes, as is the case, for example, of groups based on gender affiliations. These different groups can also be considered in terms of 'discourse communities', a term we borrow from Swales (1990), that is, groups who share aims, rules of interaction, and rhetorical principles for communication. In this sense, it is interesting to notice the role English plays in different communities in Brazil for which English words are part of their jargon of communication, such as homosexual groups, academic circles, computer users, Internet chatters, etc. Used in diversified contexts by different types of users and varying in their form (qualisigns, sinsigns and legisigns), English signs can be seen as fulfilling a role that transcends the mere adoption of foreign words in an effort to share in the process of consumer modernity disseminated by mass media. This is particularly true of the role of English signs in Brazilian music, a discussion we propose to undertake briefly before focusing on the role of English in science and education.

4. Brazilian popular music

Rap music, among other modern urban forms of musical expression, is an interesting example of the complexity of English vocabulary borrowing in the Brazilian society. In this type of music, English language signs are for most Brazilians only rhematic iconic qualisigns. As receivers of this music, teenagers imitate the sounds of the original lyrics of the rap songs without necessarily understanding the meaning of what they are singing. When these same songs are played in their instrumental versions, they create new refrains in Portuguese following the rhythm of the original song but replacing the lyrics with funny or nonsense Portuguese and pseudo-English, that is non-English words that sound like English. *You talk too much*, for instance, becomes *taca tomate*, which can be roughly translated into English as *Throw tomatoes*. The Brazilian rock band *Kid Abelha* also plays with analogous sounds in both languages (e. g., "too much/tomate") in their song called *Tomate* which says *Too much, um falso cometa* (…) [Too much, a fake comet]. This kind of pun produced by similar sounds from different languages is a resource also used in the production of Portuguese phrases that may be sung following the rhythm of some English lyrics. Thus, *Tell me once again* becomes *Telma, eu não sou gay* (literally "Telma, I am not gay") and *Yellow river* becomes *E ela é horrível* [And she is very ugly]. Playing with the words of a foreign language, however, is not the only function of English in this context. In fact, the use of English words and the choice of particular musical genres, as is the case for instance, of rap or heavy metal rock, have a further role for teenagers who seek to assert their identity by affiliating to the tone of protest and challenge of conventional society implicit in those musical movements.

But the presence of English in Brazilian songs dates back to previous decades, as was the case of samba, another musical genre. Some Brazilian samba lyrics, mainly the ones from the thirties, a decade influenced by the arrival of talking motion pictures, criticized the excessive use of English words in daily conversation. Some of the lyrics of that time reveal their authors' prejudice against the lower classes as far as the use of such words is concerned. As the lower classes were the target of criticism because of their failure to speak standard Portuguese, there was also a strong bias against their attempt to borrow English words. Examples can be found in verses such as the following ones: *Amor lá no morro é amor prá chuchu / As rimas do morro não são I love you* [Love in the slums is very intense / the slum rhymes are not I love you]; or *Good-bye, good-bye boy, deixa a mania do inglês / Fica tão feio prá você, moreno frajola / Que nunca freqüentou os bancos da escola* [Good-bye, good bye boy, get rid of these English words / they don't suit someone like you who has never been to school], both from the thirties; or one recorded in 1986 by João Nogueira, which says *Tem gente que qualquer dia / Fica mudo de uma vez / Não consegue falar gringo / Esqueceu do português* [Some people will eventually become dumb someday / They cannot speak English / and have forgotten Portuguese].

Why do they need to learn English if they can't even master their own mother tongue, Portuguese? was a common question posed by certain sectors of the Brazilian society in connection to the lower classes' aspirations to learning that foreign language that was increasingly permeating their everyday life. A fallacy traditionally used by those who wish to perpetuate their privileges and the boundaries of their social space, the question continues to be posed today not only in certain social sectors but also in educational policy and planning debates, revealing profound misunderstanding and an outdated view of the role of foreign languages in education, a problem to which we shall return later in this article, when discussing English within the context of science and education in Brazil.

If the point in learning English by the popular classes is still a controversial issue, for the middle classes in Brazil learning English is undoubtedly regarded as a basic need. Again this concern is reflected by Brazilian music, as in the Tropicalist song, *BABY*, by Caetano Veloso, which reads:

> "Baby, baby? How long / Você precisa aprender inglês / Precisa aprender o que eu sei / E o que eu não sei mais / Não sei, comigo vai tudo azul /Contigo vai tudo em paz / Vivemos na melhor cidade / da América do Sul? Você precisa / Você precisa / Não sei, leia na minha camisa / Baby, baby? I love you / Baby, baby? I love you" [Baby, baby / How long / You must learn English / You must learn what I know / And what I don't know any more / I don't know/ I'm okay, You are okay / We live in the best city / in South America / You need / You need / I don't know, read on my T-shirt / I love you / Baby, baby / I love you].

The lyrical *I* in the song addresses his girl friend as "baby" and emphasizes the importance of learning English in the new society. The verses show that learning English is the only way for the beloved woman to decode the proposition *I LOVE YOU* printed on his T-shirt. The lyrical I employs a new way to show love to someone, replacing oral discourse by printed text on a T-shirt, text written in English and not in the mother tongue, Portuguese.

Tropicalismo, a cultural movement that in the late sixties and early seventies proposed a critical reading of Brazilian culture and the recreation of Brazilian music through the interaction with traditional rhythms and new musical expressions such as rock and roll, made substantial use of English language signs, superimposing them onto the Portuguese code in such a way as to portray, on a symbolic level, urban Brazilian culture. Unlike the musical production of the thirties, the sambas that used English signs to satirize American cultural influence on the lower sectors of the population, *Tropicalismo* focuses on urban life in Brazil, the middle classes and mass media and incorporates English within its creative and critical reading of modern society. In *Batmacumba,*[1] by Gilberto Gil and Caetano Veloso, the super hero *Batman* is linked both to *macumba* (a Brazilian variant of voodooism) and to *iê-iê-iê* (a denomination of Brazilian rock in the sixties which is a transcription into Portuguese of the sounds *yeah, yeah, yeah*), producing the verse batmacumbaiêiêiê batmacumbaobá and its variations. The verse progressively loses a letter or syllable until it becomes *Ba*, an acronym for the state of Bahia, birthplace of the authors of the Tropicalist movement and the land of *macumba*, a syncretist religion that further signals the complex recreation of western tradition in Brazil. The verse begins to be repeated again, adding a letter or a syllable until it *becomes* again *batmacumbaiêiêiê batmacumbaobá*. The arrangement of the verses gives birth to a concretist[2] poem in the form of bat wings. The word *Bat,* also present in the lyrics, is a clear reference to the Brazilian anthropophagous movement[3] of the twenties. The sound *bat* yields a chain of interpretants as it can be analyzed as *bat* (the flying mammal) or *bate* (a form of the Portuguese verb *bate*r which means 'beat' – 'beat the drums' in that context). *Bat* is also an index of *Batman*, the American super hero. *Ba*, the morpheme which links the two wings of the poem is also a double index, of *Batman* (its initial syllable) and of macumba (its final syllable), and represents the synthesis of the two cultures. The melody is also made up of superimposed intercultural signs. Together, the Brazilian macumba rhythm and the foreign electric guitar produce a syncretic melody.

Torquato Neto also looks for this cultural synthesis when he joins in one word the Brazilian *bumba-meu-boi* (Brazilian folklore dance), a genuine sign of Brazilian culture, with the Brazilian version of rock named *iê-iê-iê* and creates "bumba-iê-iê-boi" in the refrain of his song *Geléia Geral* [General Jelly] which reads *é bumba-iê-iê-boi / ano que vem mês que foi / bumba-iê-iê-boi é a mesma dança*. Another famous composer and singer, Caetano Veloso, composed many songs in English and one of them *London, London*, is a good example of the use of a foreign language as a tool for political resistance during the years of dictatorship in Brazil. By describing his "wandering around" London, Caetano highlights the fear and lack of peace, felt in Brazil in the sixties and seventies, in contrast with the fearless and peaceful – though to him meaningless – life he led in London during his exile.

Together with *Tropicalismo* and other cultural movements that made use of English as a means of talking about hybridity in the Brazilian culture and as a means of expression within a context of censorship and repression, the seventies also witnessed another musical trend that involved the use of English. A group of Brazilian composers and singers adopted English pseudonyms because they had realized that some Brazilians disdained their own language and culture and overvalued the English language and the American culture. The pseudonyms worked as masks to hide their Brazilian identities and contributed to reinforce the myth that foreign products are always better than national ones. As those artists could see no possibility for success using their own image and their own language, they borrowed American names/identities and the English language, which was seen as a true passport for success. Among many artists who joined this movement, we can mention Morris Albert, composer of worldwide famous songs at that time such as *Feelings*, *Conversation* and *She's my girl.*

In more recent times, English has also played different and at times contradictory roles in Brazilian music. Some rock bands, as was the case of the internationally famous trash band *Sepultura,* compose their songs in English as they consider Portuguese inadequate for that purpose. On the other hand, other bands make use of English loan-signs in their lyrics in a very creative way. The song *Eu Sou Free* [I am free], by Patrícia Travassos and Ruban, for example, makes a pun with the signs *Eu sou free/sempre free/ eu sou free demais*, which can be understood in different ways: *Sou* is the first person singular of the verb *Ser* [to be], thus *Sou free* can mean 'I am free'. But the utterance *Sou free* also sounds like the past tense of the verb *SUFFER* in Portuguese and then can also be interpreted as *I suffered. Free* is also a popular brand of cigarettes in Brazil.

The ideology of any society can be read in the discourses it produces, which are multiple and complex. In the case of popular music, we have seen that samba, on the one hand, struggled to defend the Brazilian culture by protesting against the fashions brought from America, but, on the other hand, contributed towards legitimizing an ideology that justifies class divisions. The lyrics despise the lower class dialect and criticize the linguistic behavior of people from the slums. In the seventies, during the political repression of the military government, Morris Albert and his partners pretended to be Americans in order to achieve success. *Tropicalismo*, on the other hand, incorporated English signs, words and musical instruments, as a cultural project aiming at critically reading and recreating Brazilian culture. Popular lyrics are documents of a culture's negotiation of multiple identities, inviting, accepting, recreating or rejecting a foreign language depending on the issues at stake in the political and social agendas of the time.

5. *Wearing* English: T-shirts in Brazil

The incorporation of the English language into everyday life in Brazil can also be seen on the T-shirts worn by Brazilians, many of whom do not know the meaning of what is printed on them. T-shirts with messages written in Portuguese are hardly found on the streets in Brazil and messages in English are as common as the practice of wearing T-shirts. Needless to say, the very word "T-shirt" is a consolidated loan word in Brazilian Portuguese. A samba composer, João Nogueira, refers to that fact with the following verses:

> "Gerusa comprou uma blusa / Dessas made in USA / E fez a tradução / A frase que tinha no peito / Quando olhou direito era um palavrão" [Gerusa bought a T-shirt / one of those made in USA / and then she translated / the sentence she had on her breast / when she looked at it carefully / she saw it was a four-letter word].

According to Lima (1988: 45), Brazilian people's habit of wearing clothes with statements printed in English may signal a desire to become part of what they see as the most important part of the world, the American consumer society, which for them epitomizes the global consumer society.

Paiva (1991) interviewed printing firm representatives to verify their attitude towards the English language and found that their preference for it was supported by their prejudice in regarding Portuguese as an inferior language and certainly by the gross profits they envisaged from selling shirts with English on them. Interestingly, some of the representatives explained the use of English on T-shirts as a status symbol pursued by some groups who feel the need to differentiate themselves from the popular sectors. On the other hand, the use of Portuguese on T-shirts was also linked to the desires of certain groups to differentiate themselves from the large number of people who now wear English on their T-shirts.

In the T-shirt world, the English language is not seen as a linguistic code, but as an aesthetic sign, a rhematic sinsign, that is, an actual existent whose meaning is the mere possibility of meaning. The words are seen as signifiers only. Although some stylists say that the meaning of the words do not matter, the messages – whether talking about the products themselves, a metalinguistic function, or talking about what is happening around the world – are there to be read. Often, those messages do not belong to the stylists who produce the print but are taken from magazines and newspapers. Thus, wearing clothes with printed messages on them is also a way of broadcasting someone else's discourse. The T-shirt discourse is by nature deictic because it points to the person wearing the message. "A man's portrait with a man's name written under it is strictly a proposition, although its syntax is not that of speech," states Peirce (1974: 320). Likewise, signs printed on T-shirts – billboards people carry on their fronts or backs – are also propositions and those who wear the T-shirts are seen as their presupposed enunciators. On the streets of Brazil, one can see hundreds and hundreds of people 'advertising' all kind of slogans and sayings, English signs they post on their body as a signal for a desire to belong to a global world of consumerism and mass media. This desire for belonging to an English speaking virtual community of what is perceived as modernity is also present in people's attitude towards the language. Paiva (1991) analyzed written compositions by a group of university applicants who had been asked to write about the role of the English language in Brazil. Besides acknowledging the importance of English in the world, some of them stated that for them English is more beautiful and more sonorous than Portuguese.

6. The language of science: English in Brazilian universities and research labs

A long preamble such as the one hitherto provided before the discussion of the topic of this article is essential for understanding the role of English as the language of science in Brazil and the educational policies that are currently being discussed in connection to English language teaching in Brazilian schools. Besides its insertion in mass media and pop music, in Brazil, as in many other countries, English is also the language of science. It is the language of science textbooks and of technical manuals of imported machinery and an important job requirement in many fields. It is the main language chosen by students who apply for university entrance examinations and an obligatory language for candidates in many post-graduate programs. Most MA and PhD courses require only reading skills, but some demand oral and writing proficiency skills and students are expected to present papers in English and publish abroad as a requirement for their PhD title. The lack of English proficiency is sometimes a problem for some students, mainly for those who come from lower social classes and were not able to afford to study English in private courses. Some of them have no access to study-abroad programs and sometimes when they are awarded a scholarship to pursue their masters and doctorates abroad, they are not able to meet the TOFEL or IELTS requirements. Our universities interact with their foreign counterparts and receive visiting professors who teach courses in English. Naturally, students having poor language skills find it difficult to attend the lectures and discuss the English texts.

Scientific production is usually rated according to their place of publication and the number of citations it inspires. In order to be read, Brazilian scientists write in English, their reasons for this being analogous to the ones listed by the respondents in a study carried out by Jernudd and Baldauf Jr (1996: 11). Their survey revealed the following set of reasons:

(1) to reach a specialist audience, which only exists overseas as a community or network of scholars and which is composed of individuals who use other languages in their daily lives and who can only conveniently be reached through international languages, foremost among these languages, English, and through the relevant international journals;
(2) to reach the broadest (widest) possible readership; and
(3) to gain "prestige".

Brazilian scientists are evaluated by their number of citations and publications overseas and 'publish or perish' is the constant menace over researchers' heads. Publishing abroad, nevertheless, is not a simple matter. Discourse analysts such as Swales (1990) and Myers (1990) have done extensive research on the complex aspects involved in publishing within scientific communities, which range from linguistic and rhetorical issues to power negotiation. As Pennycook (1999: 330) rightly asks: "who has the right sort of linguistic and cultural capital (Bourdieu, 1991) to have an article published? Who knows the relevant codes of language and research – the 'secret language' of academic work? Into which discourses does one need to gain entry in order to become a member of this community?"

English is the official language of *Ciência e Cultura* [Science and Culture], the journal of the largest Brazilian scientific association, SBPC (Brazilian Society for the Advancement of Science). The decision to publish in English was taken in 1991 after a heated controversy among its members. The coordinator of the journal project argued that English was the language of science and that to publish in Portuguese would be a kind of provincialism. The president of SBPC at that time offered a conciliatory suggestion, that is, to allow the authors to choose the language, though he himself believed that all of them would choose English. In spite of all the protests and some of the members' contention that Portuguese is the official language in Brazil, the decision was taken on April 12, 1991. The option for English was motivated by the possibility to disseminate Brazilian scientific production abroad. (Vieira, 1999; *Folha de São Paulo*, April 19, 1991).

Besides *Ciência e Cultura*, there is a considerable number of Brazilian journals that publish only articles written in English. Among them, we can mention the following: *Journal of the Brazilian Chemical Society, Journal of the Brazilian Computer Society, Brazilian Journal of Chemical Engineering, The Brazilian Dental Journal, Brazilian Electronic Journal of Economics, Brazilian Journal of Genetics, Brazilian Journal of Medical and Biological Research, Brazilian Journal of Materials, Journal of Microwaves and Optoelectronics, Brazilian Journal of Physics, Brazilian Journal of Plant Physiology, Brazilian Journal of Probability and Statistics, Journal of Venomous Animals and Toxins, Brazilian Journal of Veterinary Research and Animal Science.*

7. A path to higher education and work opportunities: English in schools

The panorama sketched above clearly shows the major role English plays in Brazil as a gateway to science, culture, and technological advancement. Together with our mapping of everyday practices and popular music, the discussion helps compose a picture of the varied and complex insertion of English in the Brazilian society. English circulates in Brazil as a hard currency. Learning English is seen as a means of ascending the social pyramid since the language has been and is still seen as symbol of status, power, prestige, culture and technological development. On the other hand, English is also a means for constructing identities, establishing affiliations, critically reading the national culture and speaking out when the natural means of doing that, one's own mother tongue, has been silenced and repressed, as was the case during the military dictatorship in Brazil.

Within this context, the question posed by some of the sectors in the Brazilian culture, previously reproduced by us when discussing common prejudices as to who has the right to learn English, demands closer attention, especially at a time when misapprehensions of the public opinion and educational policy makers are currently biasing debates on the best ways to teach English in Brazilian schools. In a country where only those who can afford to pay for it are able to get some kind of English instruction, a policy focusing solely on the teaching of reading skills in public high schools, such as the one currently promoted by some education planners in Brazil, is certainly a cruel way of curtailing students' potential for developing their foreign language skills to their full extent.

Why do they need to learn English if they can't even master their own mother tongue, Portuguese? was a question raised in the thirties that has unfortunately thrived until today and subtly infiltrated present day discussions on the teaching of English in public schools. *Why do they need oral skills if they will never get to travel or use English in their everyday lives?* is perhaps an updated version of the old and pervasive question, a question we problematize and answer by pointing out, among other things, the variety and complexity of uses to which the language is put in Brazilian people's everyday interaction with English.

Some influential Brazilian English teachers have insisted on the teaching of reading based on an ESP/instrumental motivation. They argue that there are almost no opportunities for the average student at a public school in Brazil to interact with native speakers of English, a reason for which reading skills only are the sole basic need. Problematizing this view and unthreading some of its political and ideological underpinnings, Cox and Assis-Peterson (1999:437–8) state:

> "Instrumental reading is seen as a form of escape from the effects of assimilation and acculturation inherent in communicative language teaching and integrative motivation. If the expansion of English in the world is considered not the mere expansion of language but also the expansion of a set of discourses in which ideas of development, democracy, capitalism, neoliberalism, and modernization circulate (Pennycook 1994; 1995), the instrumental reading orientation of English teaching is only a Trojan horse. After all, nothing conforms more to these discourses than the pragmatism of learning to read in English in order to access information, technology, and so on."

One cannot but agree with the authors and highlight the potential harms such a limited educational policy might entail for future citizens and applicants in an increasingly competitive labor market. The choice for teaching reading skills not only underestimates the capability of Brazilian children and teenagers who attend public schools but also denies them the right to have adequate instruction and the language and cognitive skills that are essential to meet the requirements of the present and future workplace or site of higher or further education. In this sense, it must be remembered that learning a foreign language represents a cognitive process that promotes other cognitive and metacognitive operations at work in students' learning careers. Learning a foreign language leads, for instance, to reflecting upon one's own mother tongue as well as to developing cultural awareness of different cultures and peoples in our globalized world.

If borrowing words from English reflects a desire to establish an affiliation with a world and a language that markets and advertises the products of our global consumer society or a world and a language that probes into natural and cultural phenomena in order to provide a scientific explanation for them, limiting students' access to that language in all its potential appears to us an undesirable and pernicious policy, as is a law or an official decree that purports to stop and hinder cultural phenomena such as those inherent to linguistic contacts and exchanges.

Deep down, these policies seem to us to bring back the echoes of an old samba song of the thirties: *Good-bye, good-bye boy, get rid of these English words, they don't suit someone like you, someone who has never been to school.*

8. Notes

1. BATMACUMBA
   By Gilberto Gil

   Batmacumbaiêiêiê batmacumbaobá
   Batmacumbaiêiêiê batmacumbao
   Batmacumbaiêiêiê batmacumba
   Batmacumbaiêiêiê batmacum
   Batmacumbaiêiêiê batma
   Batmacumbaiêiêiê bat
   Batmacumbaiêiêiê
   Batmacumbaiê
   Batmacumba
   Batmacum
   Batma
   Bat
   Ba
   Bat
   Batma
   Batmacum
   Batmacumba
   Batmacumbaiê
   Batmacumbaiêiêiê
   Batmacumbaiêiêiê bat
   Batmacumbaiêiêiê batma
   Batmacumbaiêiêiê batmacum
   Batmacumbaiêiêiê batmacumba
   Batmacumbaiêiêiê batmacumbao
   Batmacumbaiêiêiê batmacumbaobá

2. *Concretismo* was a literary movement in the late fifties and sixties in Brazil, mostly associated with the names of Augusto de Campos, Haroldo de Campos and Décio Pignatari, who proposed recreating Brazilian poetry from the perspective of its rewriting tradition and seeking its insertion in modern international culture.

3. Anthropophagy was a metaphor used by Brazilian Modernism to talk about external influences on Brazilian culture. The movement does not want to reject foreign culture, but to absorb and transform it.

9. References

Canclini, Néstor García
1992 Cultural Reconversion in Yúdice, G., Franco, J./Flores, J. (eds.) *On Edge: The Crisis of Contemporary Latin American Culture.* Minneapolis/London: University of Minnesota Press, 29–43.

Cox, Maria Inês P./Assis-Peterson, Ana Antônia
1999 Critical Pedagogy in ELT: Images of Brazilian Teachers of English. *TESOL Quarterly* 33: 433–452.

Fairclough, Norman
1989 *Language and Power.* London/New York: Longman.

Houaiss, Antônio
1983 *A Crise de Nossa Língua de Cultura.* Rio de Janeiro: Tempo Brasileiro.

Jernudd, Björn H./Baldauf, Richard B., Jr.
1996 *Language Selection in Science Communication.* Paper presented at the conference on Knowledge & Discourse '96, held at the University of Hong Kong, Hong Kong.

Lima, Vitalina A.
1988 *A História da Camiseta.* Rio de Janeiro: Marprint, 43–49.

Moreira, Pedro
1999 O Português Ameaçado [Portuguese under Threat]. *Estado de Minas.* Belo Horizonte, 13 dec.: 9.

Myers, Greg
1990 *Writing Biology: Texts in the Social Construction of Scientific Knowledge.* Madison: University of Wisconsin Press.

Paiva, Vera Lúcia M. O.
1991 *A Língua Inglesa enquanto Signo na Cultura Brasileira* [The English Language as a Sign in the Brazilian Culture]. Ph.D dissertation, Faculdade de Letras, Universidade Federal do Rio de Janeiro.

Pedrosa, Mônica M.
1999 Do you Speak English? *Internet.br.* 38: 30–41.

Peirce, Charles S. [1931–1958]
1988 *Elements of Logic* [Collected Papers, vol. 2]. Cambridge, MA: Harvard University Press.

Pennycook, Alastair
1994 *The Cultural Politics of English as an International Language.* London: Longman.
1995 English in the World/The World in English. In Tollefson, J. W. (ed.) *Power and Inequality in Language Education.* Cambridge: Cambridge University Press, 34–58.
1999 Introduction: Critical Approaches to TESOL. *TESOL Quarterly* 33: 329–348.

Swales, John
1990 *Genre Analysis: English in Academic and Research Settings.* Cambridge: Cambridge University Press.
1998 *Other Floors, Other Voices: a Textography of a Small University Building.* Mahwah: Lawrence Erlbaum.
Swiderski, Richard
1996 *The Metamorphosis of English: Versions of Other Languages.* Westport: Bergin & Garvey.
Vieira, Cássio Leite
1990 SBPC Busca Agilidade com Novas Publicações [SPBC (Brazilian Society for the Advancement of Science) Looks for Agility with New Publications]. *Folha de São Paulo*. São Paulo, 19 jan.: 4.

English as a Language of Science in Japan. From Corpus Planning to Status Planning

Fumio Inoue

1. Introduction: theoretical background

1.1. General framework: Japan without colonial experience

In this paper the social function of the English language in the field of science will be investigated. Japan will be characterized here as a nation where the English language is not used for scientific instruction. Rather, scholars have continued to translate technical terms from English to Japanese for more than a century. Consequently, Japanese people could develop their own technology relying on their own mother tongue owing to these long-standing efforts of translating technical terms.

Among Asian nations Japan is exceptional along with Korea, China and Thailand in that scientific education in universities can be conducted completely in the students' native (official) language. This is related to the fact that these countries have not been colonized by western nations. Other Asian countries have experienced years or centuries of colonization under various western nations since the Ages of Great Voyages. Even after modernization and independence in the nineteenth and twentieth centuries, university education has been performed in English or other European languages.

This fact can be ascertained not only by examining educational systems themselves (McConnell et al. 1994) but also by examining textbooks, reference books, specialized technical books and encyclopedia in publication. Glossaries of technical terms are also a good clue to the sophistication of language in scientific fields as shown below. The educational system is thus reflected in the lexical system. A nation can do without English as a Language of Science only when the linguistic system (or lexicon) has been modernized to adapt to new technology. The Japanese language is thus exceptional in its autonomous status of native language as Language of Science. This phenomenon will be historically explained here, and the steady efforts of coining technical terms by the scholars in the past will be investigated.

It should be noted that the word "science" is interpreted here mainly as "natural science" in order to characterize the situation more impressively. However, "humanities" and "social sciences" are also taken into consideration to a certain extent in this paper.

1.2. Framework of language policy: corpus planning and status planning

First, the basic theoretical framework of this paper will be clarified in order to better describe the characteristics of Japanese technical terms. Relying on studies of language policy and language planning, the processes of introduction of foreign factors into a language will be dichotomized here (Daoust 1997).

(1) The first type, "corpus planning", is concerned with individual linguistic units which constitute a language system, or more concretely, loanwords. They signify the ordinary widespread phenomena of borrowing words by words.

(2) The second type, "status planning", is concerned with the use of English or any other foreign language as a whole, as a means of communication, or more concretely, the use of foreign languages in higher education.

The latter type may be actualized as personal bilingualism at first, and later as social diglossia. However, use of a foreign language (especially English) in modern scientific instruction relies often on limited linguistic ability and it rarely develops into coordinate or subordinate bilingualism. This means that use of English in higher education does not directly lead to widespread national bilingualism. It has been observed that the English language actually used in scientific fields is grammatically simpler than that in daily use, making machine translation easier and more reliable than literature for example. In this sense the English language in scientific fields has something in common with pidginization. It is one type of limited social diglossia.

Consequently, the Japanese situation is not typologically opposite to other Asian and African situations, but constitutes one portion of a continuum of global diglossia of English and all the other native languages. As is illustrated by Graddol (1997), languages of the world can be arranged hierarchically in social functions. This ranking is closely connected with social domains where individual languages can be used. The English language is always ranked as the "Highest" worldwide, when socially (and technically) higher domains are taken into consideration, air-control being a typical case. The language of science which will be discussed here also belongs to one of the highest domains (Jernudd et al. 1996). Japanese belongs to the upper portion of the pyramid of world languages, being utilized in many "High" social domains.

1.3. Status planning of the English language: market value of languages in Japan

The general situation of languages in Japan has been discussed in Inoue (1997, 1998). Language use in the media, in language schools and on the street were described among others, and the powerfulness of English was ascertained. Some scholars even advocate the term of "impersonal bilingualism" or "institutional bilingualism" to characterize the linguistic situation of present-day Japan, where the English language is often used as the "High" sociolinguistic variety, though real bilingual individuals are rarely found. For these scholars Japan is not a simple monolingual community.

In this paper another facet of language use in Japan will be analyzed in the domain of scientific research. However, we should not forget that the Japanese language has the economically high value of second or fifth in the world. The top again is English, if Gross National Product is recalculated according to official languages of the nations (Graddol 1997).

1.4. Status planning of the English language: low level of English and university education

Due to recent globalization of trade and communication, people's idea of language has changed in Japan, and the "market value" of language, especially that of English, seems to be unconsciously admitted (Inoue 1998).

Taking the standpoint of status planning, low English ability among Japanese students is famous or notorious. This is often explained by its geographical location as isolated islands on the periphery of the Eurasian Continent. Language contact seldom occurred on these islands, and very few intimate borrowings were observed during its history. Foreign languages were learned typically in order to introduce technology and ideas from outside of Japan. Thus English language skill was often interpreted as a symbol of cultured intellectuals and practical use for daily conversation was not highly valued in the past. Though Japan was occupied by the United States (or more precisely by Allied Forces) after the Second World War until 1952, few influences on the English ability amongst ordinary people were observed. Recent economical development and globalization have brought some changes in the people's attitude toward foreign languages and especially for English, and practical language skill has become emphasized to a greater extent. This is demonstrated by the recent prospering of the language industry, such as English language schools and companies which sell language learning instruments (again mostly English). Studying abroad to acquire practical English skills has also become fashionable among young women.

In this connection low achievement of Japanese examinees of TOEFL is often lamented and discussed. International comparisons of average levels by countries has shown that Japan belongs to the lowest group. In the late 1990's Japan was the lowest among Asian countries. This tendency can be explained by the linguistic dissimilarity of English and mother tongues of the examinees, as statistics show that countries of Germanic languages and then of Indo-European languages achieve higher points in TOEFL. The Japanese people are at a disadvantage because of great differences in grammatical typology and word order. Linguistic relative difficulty can be objectively measured as was discussed by Inoue (1998).

A non-linguistic factor, the absolute number of examinees should also be taken into consideration. In Japan, even those students who are not good at English take examinations just in order to apply for studying abroad or to have objective criterion of their own English skills. It can also be contended that Japanese students are not good at English because university education is conducted only in Japanese. Higher education is given exclusively in Japanese without recourse to English so that students do not have contact with sophisticated communication in English. However this cannot be the sole reason because there are also other Asian countries where university education is given in the native (official) language, as in Korea and China, where better TOEFL results are obtained.

1.5. Principles of corpus planning: etymological classification of words

As for the problem of corpus planning, long-standing Japanese efforts in the translating of western scientific technical terms will be investigated below. The historical background should be taken into consideration in order to understand the present situation of Japan. If Japan is considered globally among many countries and cultures in the world, it can be characterized as a sociolinguistically autonomous (or independent) country, though it has received strong influences from Chinese and English in the form of individual loanwords.

Another non-linguistic factor is often pointed out to explain the present situation, because the linguistic situation in Japan is said to be different from former Asian colonial countries like India, Philippines, Burma, and Vietnam. This factor of political independence certainly has given impetus for constructing an autonomous higher education system conducted in the Japanese language and for translating necessary technical terms into the Japanese language. Thus, non-use of English in scientific fields (a problem of status) is closely connected with coining technical terms (a problem of corpus).

In order to discuss the processes of translation of technical terms, another theoretical distinction is again necessary and convenient. For introducing new ideas, three main means of borrowing are pointed out in linguistics as follows.

(1) Translation, or complete integration of the words into the recipient language, sometimes making use of corresponding words which have

existed like "tetsu" (iron), and sometimes coining a new word like "koogaku" (engineering),

(2) Calque and expansion, or making use of form or meaning of the target language as with "bariki" (horse power), "mushi" (bug) etc.

(3) Simple borrowing in the form of loanwords like "ponpu" (pump), "aamu" (arm) etc.

Such means lead to etymological dichotomy on the part of the recipient language, that is dichotomy into native words (1 and 2) and loanwords (3) in outside linguistic forms.

However, the situation is not so simple in Japanese, because the Japanese language has accepted many words from Chinese in the form of Sino-Japanese words. The lexical items in Japanese are conventionally classified into four categories according to etymological principles.

(a) Pure "Japanese", written mainly in the hiragana syllabary (sometimes with Chinese characters), like "teko" (lever).

(b) "Sino-Japanese" or Chinese words borrowed and integrated into Japanese lexical system, written mainly in Chinese characters, like "teisuu" (constant), "mokutan" (charcoal).

(c) "Western-loans" or loanwords from European languages, written mainly in katakana syllabary, like "natto" (nut).

(d) Etymologically "mixed" words in which roots of the above etymology are combined into one word, like "tsugite boruto" (coupling bolt) and "doren kan"(waste pipe).

Most of the existing words can be classified into the four categories stated above although there are exceptional words if considered in more detail. However, as this fourfold classification is convenient and widespread among Japanese linguists, it will be utilized throughout this paper. (a) and (b) correspond to (1) or (2), and (c) corresponds to (3) of the means above.

Hereafter the history of the linguistic situation in Japan will be reviewed in two ways; that is, as corpus planning in sections 2 and 3, and as status planning in section 4. Topics of corpus planning will be related first.

2. Introduction of Chinese and Western ideas through translation: corpus planning

In this section long-term efforts of the Japanese to adopt foreign ideas through translation will be pointed out. It is generally understood that these efforts began systematically after the start of modernization at the time of the Meiji Restoration in 1868. However a similar phenomenon can be observed even before modernization, so we shall start the explanation from the first millennium when early historical documents are found.

2.1. Ancient and classical ages (from 0 AD to 1185)

In the first centuries AD, the spreading of Chinese cultural area reached the archipelago east of China where non-literate people were beginning to constitute primitive countries. The Chinese writing system was introduced and became the first means to record the Japanese language, perhaps first accepted by emigrants of Korean and Chinese professionals, and was later adopted by trained Japanese natives. In the second half of the first millennium, stronger waves of formation of East Asian cultural area arrived at the Japanese archipelago. Chinese loanwords (or Sino-Japanese words) began to be introduced steadily into the Japanese language. Several Sino-Japanese words were already in use in the first anthology of verse in 8th century as is shown in Figure 2 below. Chinese loanwords were mainly used for things which had been foreign to Japanese culture.

In the second millennium formal documents were recorded in (grammatically partly Japanized) Chinese sentences, and some nobles were so good at Chinese as to make verse in impeccable Chinese. This situation is similar to Latin in medieval Europe. On the other hand, the Japanese syllabary system, "kana", was invented and used among Buddhist monks and court ladies. Even lay people and farmers acquired the ability of reading and writing later.

2.2. Medieval ages (1185 to 1603)

The ability of reading Chinese philosophical and religious documents was widespread among monks and scholars throughout the medieval ages of war and turmoil. Chinese technical terms for humanities were introduced directly without translation into pure Japanese, in the form of Sino-Japanese loanwords written in Chinese characters and pronounced in adapted Japanized pronunciation of the Chinese characters. The meaning of the words was easy to understand by eye if the reader knew Chinese characters well, but hard to recognize by ear, because of many homophones. (The pronunciation and tone systems of Chinese were largely simplified in Japanese).

The last decades of this period saw the beginning of the westernization of Japan. In 1543 a Portuguese ship was stranded on Tanegashima Island at the southern edge of Kyushu Island. A Chinese member of crew and a Japanese monk on the island were said to be able to communicate by writing ideographical Chinese characters, and the technology for making guns (muskets) was immediately introduced to Japan. This technology soon triggered great social change, and a force of warriors succeeded in unifying the whole of Japan.

2.3. Edo Period (1603 to 1868)

A new warrior government of the Tokugawa shogunate was established in Edo in 1603 and its rule continued until 1868. The Edo Period can be characterized as the incipient age of westernization in Japan. Civilized terms with materialistic characteristics were introduced from the beginning of western contact and continued to be imported during the Edo Period. Many of them were translated into Japanese or Sino-Japanese but some were introduced in the form of western loanwords.

Christianity was first introduced by a (Basque) Spaniard Francisco de Xavier in 1549 and later by other Catholic priests, and was propagated nationwide for about a century. Religious and spiritual terms were introduced during these times, sometimes directly as loanwords and sometimes translated into Japanese or into Sino-Japanese forms. Christianity was then prohibited in 1639 by the Tokugawa shogunate due to fear of intrusion by western powers. National seclusion continued from 1639 to 1854, protecting Japan from becoming a colony.

Only the Dutch or Netherlanders (and the Chinese and the Koreans) were allowed to visit and stay in Japan on the condition that they would not perform any religious activities. Even in the atmosphere of national seclusion, many European things and concepts were introduced to Japan by way of trade. The effort to coin technical terms continued using Sino-Japanese words, but many western loanwords were also introduced, first from Portuguese and Spanish and later from Dutch. In the latter half of the Edo Period systematic efforts to translate technical terms were accelerated. Several books were translated from Dutch, and many technical terms were introduced into the Japanese language in the form of Sino-Japanese words coined in Japan.

In the last decades during the Edo Period the opening of the country was coerced and, due to the Kanagawa (trade) treaty with the United States in 1854, Japan opened several ports to foreign ships. These outside movements brought about the establishment of Bansho shirabe-sho (Institute for the Investigation of Barbarian Books) in 1856. Dutch and later English, French and German were studied and taught there for a limited number of applicants. Though there were some interpreters and several experts in reading European books, there was no effort to use any foreign language as the medium of instruction. Foreign languages were mostly considered as a means of introducing western technology.

3. Meiji Restoration (1868) as a turning point of western influence: corpus planning

In 1868, the Tokugawa shogunate in Edo (later Tokyo) collapsed and the modern Meiji government was established after a kind of revolution called the "Meiji Restoration". The new Meiji government aimed at modernization in many ways, and opened the country's doors worldwide. A formal education system was also considered to be important for modernization. The establishment of Tokyo University in 1877 with unification of the former Bansho shirabe-sho of the Tokugawa shogunate was symbolic because it was the first westernized institution of higher education. Use of western languages for instruction at the university will be discussed in the next section, as it is a phenomenon of status planning. In this section invention and enrichment of technical terms as a phenomenon of corpus planning will be investigated.

3.1. Coining technical terms from English

Creation of new technical terms was necessary to discuss scientific matters which developed in the western world. Dutch language interpreters were active during the Edo Period in translating western books into Japanese. However, in the last decades of the Edo Period, English language use surpassed Dutch partly for political reasons of diplomacy and economical reasons of trade. Young intellectuals during the first ages of modernization turned to English even after learning Dutch.

After the Meiji Restoration many new terms were translated systematically from English into Japanese. Many English-Japanese and Japanese-English dictionaries were compiled from this time on, but only a small number of dictionaries for other languages appeared. This demonstrates the powerfulness of the English language.

The basic principle of coining new terms on the basis of Sino-Japanese did not change much and actually many Sino-Japanese words appeared after the Meiji Restoration. Use of Chinese characters (or words of Chinese origin) was still predominant and only a few efforts were made to make use of pure Japanese linguistic resources. The background to use of Sino-Japanese was the widespread literacy of Chinese classics among intellectuals at the end of the Edo Period. Higher education or culture still involved skills in reading Chinese books on philosophy and religion.

3.2. Lexical characteristics of technical terms and grammar

Coining a scientific technical term is different from translating novels for example, because a strict condition applies to technical terms. One-to-one correspondence of words is obligatory in technical terms. Original English (or any other) technical terms should be retrievable from translated Japanese technical terms. In this sense the same obligation was felt even in the Edo Period in translating Chinese encyclopedia of medicine and agriculture. In this case strict one-to-one correspondence of Chinese and Japanese names of plants, animals, medicine and instruments was essential, and several versions of graphic encyclopedia were published, some of which later developed into the first dialect dictionary.

Japanese technical terms have another characteristic of being distinguishable from ordinary lexical items. Special word-form was given to technical terms in order not to be confused with ordinary words. This situation is quite different from German for example, in which everyday terms and technical terms have often the same form. There is an episode where a Japanese philosopher studying in Germany was astonished to find that even lay people spoke in philosophical terms in Germany. This characteristic of Japanese technical terms brought about an enormous number of lexical items in Japanese which are minutely and exactly defined, and have one-to-one correspondence to European (actually English) words.

Grammatical expressions also developed which correspond to English expressions. For example, usage of the passive voice spread in order to secure one-to-one correspondence with the English passive voice, and new expressions to translate relative pronouns were developed and used widely. By these means faithful translation from English sentences to Japanese were secured.

3.3. Basic significance of adoption of western concepts

When the means of introduction of western words into Japanese is discussed in the field of linguistic borrowings, classification into (1) translation, (2) calque and (3) loanwords, as discussed above, neglects an important viewpoint. Though there are several means of introduction, the important fact is that western concepts (words) have been fully integrated into the Japanese linguistic system. One-to-one correspondence of translations became possible, and anything could be fully expressed in Japanese. In other words, corresponding concepts of western terms (words) have been introduced into Japanese no matter what kinds of means are used (Japanese, Sino-Japanese, western loans or mixed; translation, calque or loanwords). The most important thing in connection with this is that internal content or meaning of the technical terms is completely western, because western concepts can be fully expressed in the Japanese language, even though the external linguistic forms of such technical terms may superficially and etymologically look Japanese, Chinese or English. This is regarded as an inevitable process of "modernization" of a language, which is still in progress in newly established independent nations.

Thus in the field of natural sciences one-to-one correspondences of terms and ideas have been established owing to the sustained effort of scholars. The situation in the humanities may be a little different especially when things related to religion and the spiritual world are concerned, because deep understanding is sometimes said to be impossible in the field of the human mind. However, even in the field of humanities, technical terms based on western concepts have been established.

3.4. Trends of increase in technical terms

Thus, the total number of technical terms can symbolize the degree of modernization of the language concerned. The historical process of increase in technical terms in Japanese is reflected in Figure 1, where official movement of standardization of technical terms is shown. Figure 1 shows the number of technical terms authorized by the Ministry of Education (based on raw data in Abe 1990).

Standardization and unification of technical terms are obligatory because various translations have been sometimes applied to one and the same English term by different scholars and in different fields of science. Sometimes the same Japanese translation has been applied to different English technical terms in different fields of science. These may not only cause misunderstanding but can also cause serious accidents. One-to-one correspondence of English and Japanese technical terms is ideal, but actually one to several relations is common as the study of the National

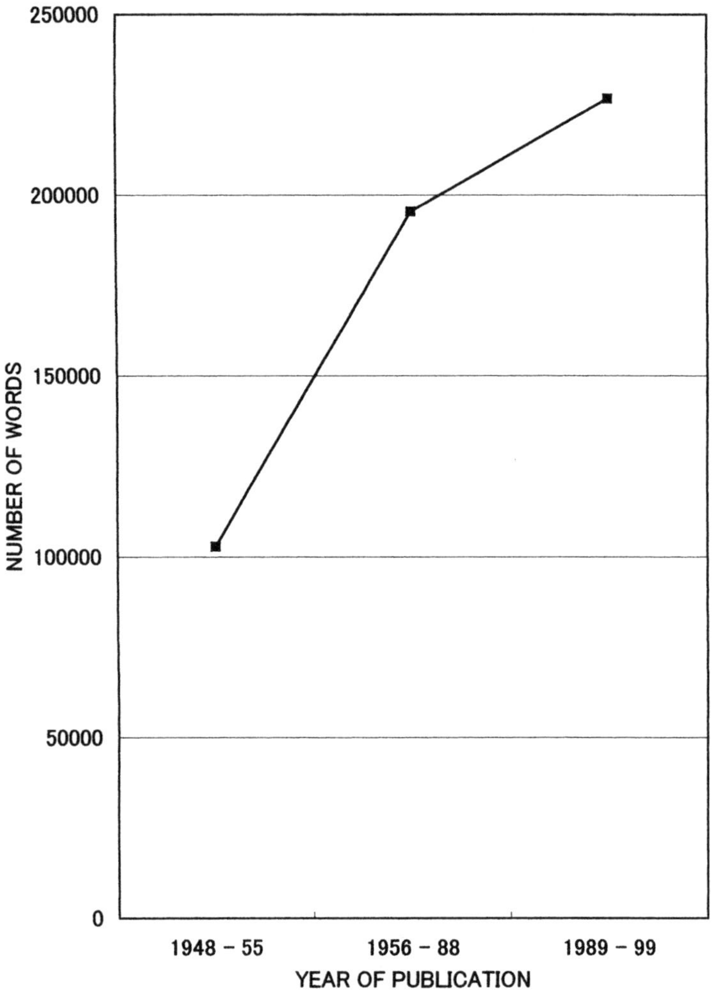

Figure 1: Number of authorized technical terms

Language Research Institute (1981) shows. The ministry of Education has published officially authorized glossaries of technical terms in order to solve these problems. As is shown by the data as of 1989, the technical terms which were officially authorized doubled in comparison with the period just after the War. Glossaries now cover 24 fields containing about 182,440 words in sciences, 11 fields containing about 444,250 words in humanities, and 35 fields containing about 226,690 words in total.

The movement of accumulating technical terms is also observed in commercial activities. A company accumulated and published a large dictionary of 350,000 technical terms of scientific technology. In the year 2000 the same company published a CD-ROM dictionary of 650,000 technical terms not only of scientific technology, but also of medicine and business (Alphabeta 2000). This large number of lexical items (as many words as in one of the largest dictionaries of Japanese) signifies that the Japanese language has developed terms fully adapted to modern technological society. The glossary is necessary not only for the Japanese but for foreigners for the purpose of efficient exchange of ideas and technology.

It is worthy of note here that the English language has been regarded as a basis for comparison of technical terms for Japanese. German, French and sometimes Russian terms were added only in reference to English.

Sophistication of lexical items to adapt to modern technology was made possible by incessant efforts of scholars and technologists who introduced and coined new technical terms. These efforts were necessary for practical industrial activity, and publication of a large glossary was commercially possible because Japan is actually producing high-tech products such as airplanes, artificial satellites, computers and so on through its own technology. If these actual demands of industry were lacking, coining technical terms only for instructional purposes would be insubstantial and commercially impractical. These circumstances perhaps lead to the situation of some Southeast Asian countries where science education in university is not conducted in native official language although there is a continued conscious effort to refine the official language.

3.5. Trends of increase in western loanwords

The increase of technical terms was treated as a whole above, because the contents or concepts expressed are important for modernization. Next, the word-forms will be analyzed from an etymological point of view, because the technical terms of etymologically English words seem to be used more frequently these days in Japan, reflecting prestige of the English language in the field of science in Japan. The tendency of turning to western civilization can be detected from the etymology of roots of technical terms.

As a typical example, the field of machine engineering will be investigated. Four glossaries of technical terms of the field of machine engineering were compared by the National Language Research Institute (1981).

Two hundred words were sampled from the words which were found in both the glossaries of 1886 and 1955. Corresponding terms on the basis of the English technical terms were entered from the other glossaries, and classified according to an etymology of roots into Japanese, Sino-Japanese, Western-loan and the mixed. Etymologically mixed words were later sub-classified into the three categories above, and divided numerical values were allotted to the three categories. (In place of one Japanese and Chinese mixed word, a numerical value 0.5 was allotted to Japanese and 0.5 to Chinese; in place of one Japanese, Chinese and western mixed word, a numerical value 0.3 was allotted to Japanese, 0.3 to Chinese, and 0.3 to western.) Inoue added more recent data of the revised glossary in 1985 on the same principle.

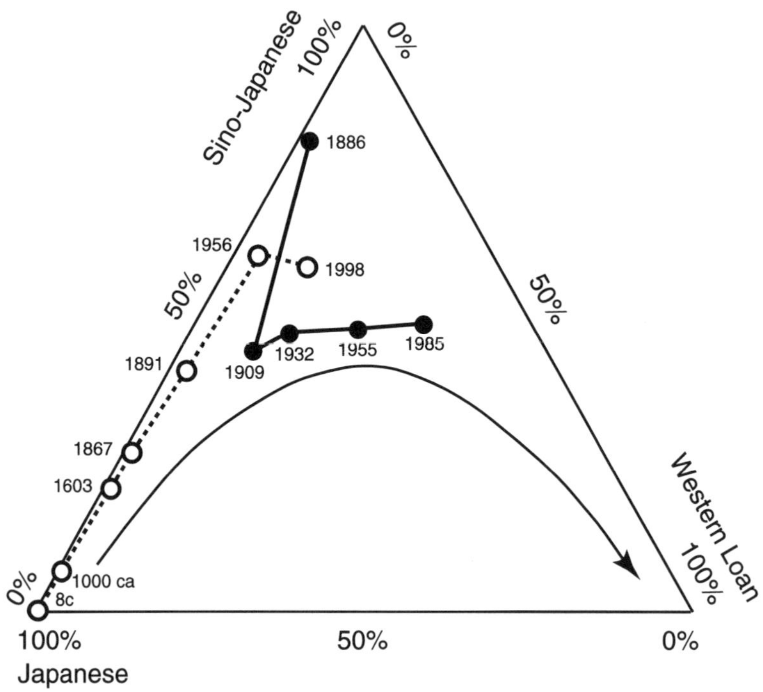

Figure 2: Change of etymology of technical terms of machine engineering with data of dictionaries

Figure 2 shows the result. The numerical values of the three categories, which become 100% in sum, can be conveniently represented in a "triangram" as is shown in Figure 2. Percentages of the three etymological categories of words from the five glossaries were entered by bold circles and linked by lines. The glossary commercially published in 1886 is plotted near the upper end of the triangram, showing that it was full of Sino-Japanese terms. The later glossaries are situated near the center of the triangram, suggesting that they contain similar ratios of Japanese, Sino-Japanese and western terms. However, if considered in more detail, the newer glossary is situated at the right-hand side, suggesting that it has more western loans than former versions. The glossaries published by the Society of Machine Engineering in 1909 and 1932 have less western loans than the glossaries published by the Ministry of Education in 1955 and 1985.

Figure 2 shows a typical tendency. At the beginning of modernization many Sino-Japanese words were used, and later more western loanwords are used. Through this trend, one-to-one correspondence of technical terms is easily established and original (English) terms become easily recognizable. The basic background mechanism is general diffusion of knowledge of English and also comparatively short time left for coining appropriate Japanese expressions.

In order to better understand the general trend, statistical data of type-words from historical documents and dictionaries are also indicated in Figure 2 for comparison (based on data of Hayashi 1982). One of the oldest documents of Japanese from the eighth century, which is composed almost totally of etymologically Japanese words, is situated lower-left. A novel from the eleventh century included a small ratio of Sino-Japanese words. The ratio of Sino-Japanese words continued to increase and the ratio of western-loans also began to increase in dictionaries published in 1603, 1867, 1891, 1956 and 1998, which are situated near the center of the triangram. Many similar lexicological studies consistently show this general trend of Japanese vocabulary, pure Japanese words succumbing to Sino-Japanese words, and recently Western words introduced overwhelmingly. This general trend is suggested graphically in this triangram in the form of an inserted curved arrow, starting from lower-left to upper-middle and then to lower right.

The actual data from the five glossaries seem to follow this general trend, forecasting that future glossaries will contain more western (actually English) loans. In comparison with the historical data of documents and dictionaries which include frequently used basic words too, the technical terms in machine engineering make use of significantly more western loans. This signifies that technical terms are vanguards of lexical change. If we extrapolate from this tendency, the Japanese language will include a higher ratio of western-loans (mostly from English) in the future, as is suggested by the curved arrow in Figure 2.

3.6. Recent tendency of adoption of English loanwords

In order to inhibit the English language as a whole from invading an industrialized community, it is necessary to introduce technical terms and scientific expressions into the native language instead. This is an evil dichotomy which a nation must select in order to become a member of the modern technological world. However, when trends of newer technical terms are rapid and powerful, efforts to make use of native lexical elements to coin words cannot catch up, and simple borrowing of forms as loans from English will prevail. When knowledge of the English language is widespread enough among people, this will not cause inconvenience, and invasion of English lexical items will spread steadily. Thus development of technology lead by English speaking nations will eventually lead to so-called English imperialism in non-English speaking nations in the field of science.

Recent trends in information and globalization give more impetus to this tendency. For example, terms for computers are full of loanwords from English. In this new field of technology, one-to-one correspondence with English technical terms is esteemed to be of ultimate necessity, so that direct introduction of loanwords seems to be more useful.

The tendency of reliance on English loanwords suggests that it is becoming more difficult to coin appropriate expressions in the Japanese or Sino-Japanese lexical stock owing to rapidly appearing new technical terms. It is sometimes argued that technical expressions in Japanese are so full of English loans, that actual colloquial sentences among scientists are lexically English and grammatically Japanese. As English and Japanese are grammatically so different (Japanese is agglutinative and English is inflectional in the classical linguistic typology, and word order is almost opposite), there is no possibility of mixed language or a pidgin being born. Japanese is not yet considered to be an endangered language generally, though Inoue (1998) foresees this possibility in the future.

4. Use of English as a language in science: status planning

4.1. Status planning for scientific communication

So far the use of English for individual technical terms has been under consideration as a problem of corpus planning. Hereafter matters connected with status planning of English in Japanese scientific fields will be discussed.

Japan has been treated as a typical monolingual country (Sibata 1998), with almost complete Trinity between language, nation and country (national borders), being the second purest Trinity country after Iceland, though reservations are necessary (because of the Ainu and Korean languages). In the field of education as a whole, and especially in the scientific field in higher education, the Japanese language has been mostly the sole medium of instruction in its long history. The only exception to this purebred tradition was the incipient ages of higher education just after the Meiji Restoration (1868), as has been pointed out in section 3 above. European scholars were invited as professors when the new Meiji government established Tokyo University in 1877.

4.2. Ephemeral use of English by government hired foreigners

At the beginning of modernization after the Meiji Restoration, a large number of foreigners were employed by the Meiji government, most of them being technologists and language teachers. They were hired in order to introduce western civilization and technology. When the first modernized institution of higher education was initiated in the form of Tokyo University in 1877, ephemeral use of government hired foreigners flourished in order to maintain levels of education at the university equal to the western standard. However, western technologists and professionals were more highly paid than ministers of the government. The high salary of the hired foreigners at Tokyo University soon became an obstacle to the new government. Then the Meiji government adopted a more economical system of sending excellent young scholars to western countries to let them take academic degrees there in a short period of time.

Japanese professors who returned from abroad after taking degrees took the place of the government hired foreigners very soon. This short period of foreign language usage for science can be seen from the numbers of government hired foreigners in Figure 3. The English language was especially necessary for those who wanted to learn from those government hired foreigners. When more detailed statistical data was inspected and foreign teachers and professors were classified according to nationality, English dominance became clear. Professors from the United States and England were by far the most common, though there are exceptions in the field of medicine, chemistry and so on where German was used.

English was often used even by Japanese professors returning from abroad as the following episode suggests. There is the legend of a profes-

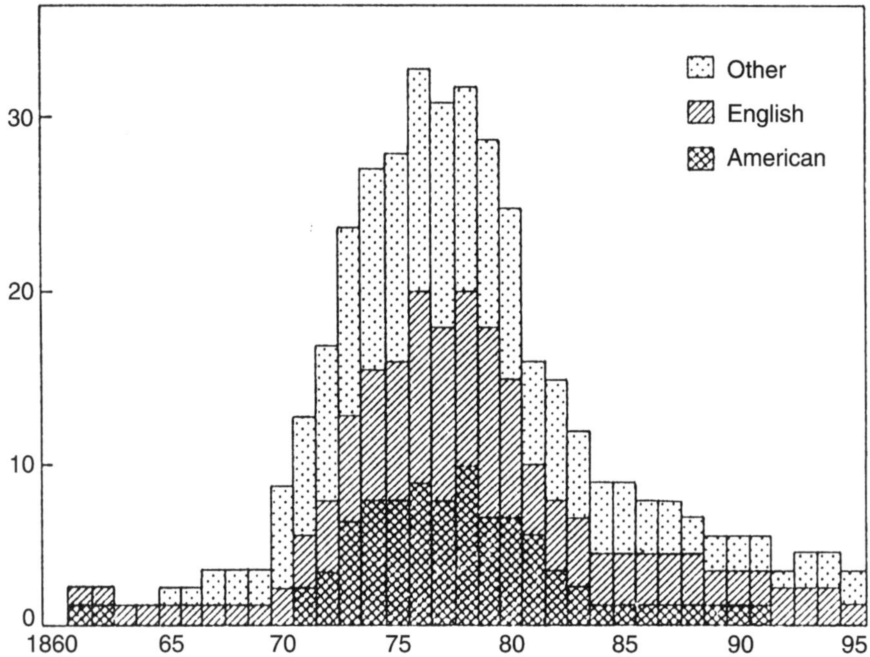

Figure 3: Numbers of government hired foreigners in science of university level

sor in mathematics who gave a lecture in French after coming back from Europe. Students complained to him that they did not understand French well. Of course they wanted to be taught in Japanese. However, the next week the professor began teaching in English, which was still hard for the students to understand. Later on, foreign professors were not accepted until the end of the twentieth century, as all government employees were stipulated to be Japanese by citizenship except language teachers.

This trend of dismissing hired foreigners and giving lectures in Japanese corresponds to the movement, discussed above, of coining new technical terms in Japanese.

4.3. *Usage of the English language in academic periodicals*

A short period of English dominance at the beginning of modernization in the nineteenth century has been reviewed so far. Changing the subject matter to academic periodicals, the writing of academic papers in English

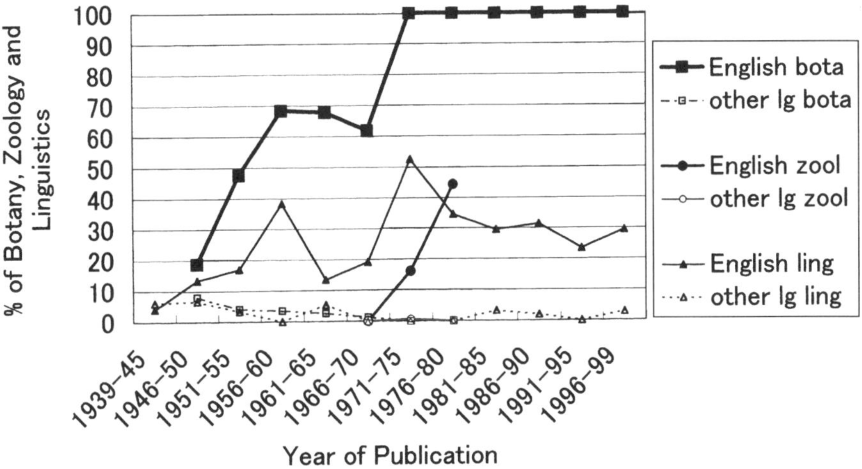

Figure 4: Ratio of foreign languages in academic periodicals

or by some other European language persisted. There is a tendency that good papers of international quality are written in foreign languages in order to be widely read internationally (Ammon 1994). A study of Japanese periodicals in medicine showed a consistent tendency of use of foreign languages and also a shift from German to English.

The tendency of writing academic papers and dissertations in foreign languages is demonstrated by the following study. Three Japanese scientific periodicals of long historical tradition were selected. Statistical data was prepared by the National Language Research Institute (1981); the author made the graph after adding recent data from linguistics. In this graph only papers in English and other foreign languages are shown, the rest being papers in Japanese. In the field of botany, papers written in English increased and in 1972 the title of the periodical itself changed and the content became completely English. Zoology seems to follow the same route in the graph. This tendency can also be detected in the humanities. Linguistics is situated between the two fields of natural science. Papers in Japanese still prevail at the end of the 20th century but other languages are not used as before. English seems to have become the most powerful foreign language medium for international academic communication.

In many facets of academic fields, English papers began to increase, the periodical "Kokugogaku (Studies in the Japanese Language)" being one of the exceptions. As the periodical is concerned with Japanese linguistics, all the readers are supposed to be good at Japanese, so that all the papers are written in Japanese. Some other Japanese academic periodicals still include only Japanese papers.

The tradition of language selection of academic periodicals in Japan should be considered in the broader context of global change of language use. The worldwide trend of language use for more than a century was investigated by Tsunoda (1983, 1993). The graph reflects the general historical tendency of scientific study. English papers were always at the top and steadily increasing. The language in second place reflects the ebb and flow of national powers. French lost weight in academic fields at the beginning of the twentieth century, and German prospered during the two World Wars. Russian was used after the Second World War. Papers written in Japanese were not so many at the beginning of the twentieth century, but increased for a time after the Second World War. The general trend suggests that most of the papers will be written in English in the future. More recent data can be found in Graddol (1997). The domestic trend for more papers to appear in English can be interpreted as one facet of the international inclination for English papers.

4.4. Use of English for foreign students in Japan

At the end of the twentieth century, the reappearance of the English language in Japanese university education is observed mainly as a special service for overseas students. The plan by the Ministry of Education to accept 100,000 foreign students by the year 2000 was not successful, partly because of an economical recession in the 1990's.

Newer tactics were proposed in order to invite more foreign students. One of them was accepting more students for a shorter period of one year or less in the form of the ISEP or "International Student Exchange Program" which began in 1995. As Figure 5 shows, the number of students is increasing every year because more universities (namely more professors) apply for this program. In the year 1999, 17 national universities offered ISEP courses, and 370 students were allowed. A different statistic shows that about 40% of ISEP students are from English speaking countries like the USA and Australia, and that about 35% are from Asian countries.

Another program UMAP (The University Mobility in Asia and the Pacific) will open courses for foreign students around 2000. By this program too students will be taught mostly in English. Some universities

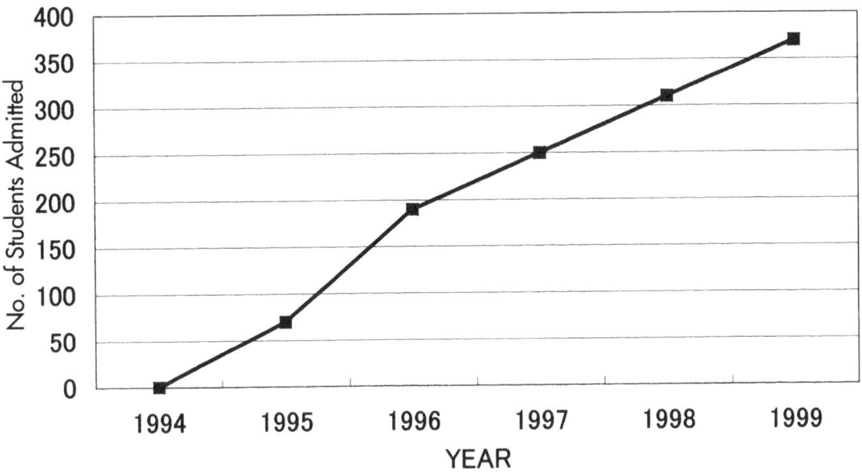

Figure 5: Increase of ISEP students

have begun still newer internaional programs of science in English for foreign (and Japanese) post graduate students.

Similar attempts have already been made in some private universities, which concentrate efforts on international student exchange and also on improving the English ability of Japanese students. Foreign students studying in Japan for a short period, with formal exchange of credits, are near 4,000 in total, which means that about 7% of foreign students in Japan take lectures in English. Until the implementation of this program, the increase of foreign students simply meant an increase in the students learning the Japanese language. However from now on, the English language will be utilized to teach students from many countries. Though the fields of lectures given in English are not limited to natural sciences (actually most of the ISEP courses are for humanities and social sciences), foreign students who want to learn from Japanese technology will receive credits from the lectures given in English. One of the greatest trade barriers of Japan, that is the language barrier, can be removed by such usage of English.

For this program, lectures in English were formally offered at universities in Japan. However, this program should not be interpreted as a simple repetition of giving lectures in foreign languages as in the time of the Meiji government. The greatest difference is that presently mostly Japanese teachers give lectures in English for foreign students. The nationalities of professors and students are just the opposite of government-hired foreigners at the time of Meiji Restoration. To add to this, the content of the lectures can also be transmitted in the Japanese language based on Japanese technical terms.

However, this signifies a new era of international exchange of academic ideas, because some of the lectures are open to Japanese students too. Advertisements on the street maintain that English is now a common world language or the global standard. In the fields of science in higher education in Japan, the English language has now definitely developed into a language for international communication for academic purposes, and has ceased to be the language of the former British Empire and the language of prosperous America.

5. Conclusion

The social status of English in Japan has been discussed in the field of science. In order to do without English, a systematic effort to prepare for many technical terms was necessary. And the process took a long historical time-span. However as higher education is becoming open to foreign students, sole use of the official domestic language may become another discriminative language barrier. The English language seems to have lost its nationalistic color of the past, and seems to have become a medium of global scientific communication.

6. References

Abe, Yoshimitsu
1990 *Jutsugo Senmongo* (Technical terms) *Koza Nihongo to Nihongo Kyoiku* 7 (Meiji Shoin).

Alphabeta
(2000) *130 Man Go Daijiten CD-ROM (Grand Dictionary of 1,300,000 words CD-ROM) (Alphabeta 2000).*

Ammon, Ulrich (ed.)
1994 *Die Deutsche Sprache in Japan.* München: Iudicium.

Burks, Ardath W. (ed.)
1990 *Kindaika no Suishinsha tachi – Ryuugakusei Oyatoi gaikokujin to Meiji.* Tokyo: Shibunkaku.

Adjust, Denise
1997 Language Planning and Language Reform. In Coulmas, F. (ed.) *The Handbook of Sociolinguistics.* London: Blackwell.

Graddol, David
1997 *The Future of English?* London: British Council.

Hayashi, Ooki
1982 *Zusetsu Nihongo* (Graphical Japanese). Tokyo: Kadokawa Shoten.

Inoue, Fumio
1997 Market Value of Languages in Japan. *Japanese Linguistics* (Nihongo Kagaku) 2: 40–61.
1998 Language Market and Its Basic Mechanisms. *Area and Culture Studies* 57: 83–103.

Jernudd, Björn/Baldauf, Richard B., Jr.
1996 Language Selection in Science Communication. *Knowledge & Discourse.* '96 Conference (Hong Kong University).

McConnell, Grant D./Roberge, Brigitte (eds.)
1994 *Atlas international de la diffusion de l'anglais et du francais: L'enseignement.* Québec/Paris: Presses de l'Université Laval/Klincksieck.

National Language Research Institute
1981 *Senmongo no Shomondai* (A Study of Specialized Terminology: The Problems with Technical Terms). Tokyo: Shuei Shuppan.

Sibata, Takesi
1998 *Sibata Takesi, Sociolinguistics in Japanese Contexts.* Ed. by Tetsuya Kunihiro et al. Berlin/New York: Mouton de Gruyter.

Tsunoda, Minoru
1983 Les langues internationales dans les publications scientifiques et techniques. *Sophia Linguistica* 13: 144–155.
1993 Les langues des publications scientifiques au 20e siecle: analyse et synthese quanitative. *Proceedings of the 15th International Congress of Linguists* (Quebec 1992). Vol. 4. Montreal.

Subject Index

Words which occurred through most parts of the text like *English, USA* or *Britain* were not incorporated into the index.